HOSPITALITY & TOURISM

Elbeyi PELİT / Hasan Hüseyin SOYBALI / Ali AVAN
(Eds.)

HOSPITALITY & TOURISM

MANAGERIAL PERSPECTIVES & PRACTICES

Bibliographic Information published by the Deutsche Nationalbibliothek
The Deutsche Nationalbibliothek lists this publication in the Deutsche Nationalbibliografie; detailed bibliographic data is available online at http://dnb.d-nb.de.

Library of Congress Cataloging-in-Publication Data
A CIP catalog record for this book has been applied for at the Library of Congress.

InGlobe Academy

ISBN 978-3-631-83490-9 (Print)
E-ISBN 978-3-631-84318-5 (E-PDF)
E-ISBN 978-3-631-84319-2 (EPUB)
E-ISBN 978-3-631-84320-8 (MOBI)
DOI 10.3726/b17900

© Peter Lang GmbH
Internationaler Verlag der Wissenschaften
Berlin 2020
All rights reserved.

Peter Lang – Berlin · Bern · Bruxelles · New York · Oxford · Warszawa · Wien

All parts of this publication are protected by copyright. Any utilisation outside the strict limits of the copyright law, without the permission of the publisher, is forbidden and liable to prosecution. This applies in particular to reproductions, translations, microfilming, and storage and processing in electronic retrieval systems.

This publication has been peer reviewed.

www.peterlang.com

Preface

As service enterprises produce more intangible products, it becomes harder gaining a good position in the minds and perpetuates it as against other business organizations. As a service-oriented industry and with its distinctive characteristics, tourism is based on experiences. In a limited period of time and a particular location, with agreement and participation of individuals, it is necessary to provide the highest level of service and to keep the quality of interaction at a certain level. In the tourism sector, differentiation and new perspectives are needed in order to increase the quality of experiences and to have a different position in the minds of individuals than competitors. There is a crucial role of efficient management of business processes (e.g., preproduction, during production, human resources and management structure) in order to ascertain this differentiation. Herein, especially in service/tourism industry, the pursuance of new tendencies will provide substantial benefits to the relevant enterprises at all of these processes. From this point of view, with this book, it is aimed to guide tourism organizations in terms of improving service encounter processes and quality of experiences by giving crucial tips about current issues and perspectives.

Dr. Elbeyi PELİT, Dr. H. Hüseyin SOYBALI, Dr. Ali AVAN

Contents

List of Contributors 11

Notes on Contributors 15

Kezban TALAK
The Role of Organizational Behavior in Tourism Enterprises 23

Bilal ÇANKIR and Nazlı EREN GÜN
Turning "Learning" to Innovation in Tourism: The Theory of the
"Tourism Learning Area" 43

V. Rüya EHTİYAR and Aslı ERSOY
Culture and Management Relations in the context of Intercultural
Differentiation in Tourism Industry: A View on Turkish Management
Culture 63

Ahmet BAYTOK, Özcan ZORLU and Ali AVAN
Service Culture: A Framework for Hotel Enterprises 75

Çağdaş ERTAŞ
Culture of Tourism Employees: A Research in Southeastern Anatolia,
Turkey 93

Tuğba GÜRÇAYLILAR-YENİDOĞAN, Hatice BARUT and Ayça ÇETİNKAYA
Linking Strategy to Business Model Renewal in the Hospitality
Industry: The Case of Land of Legends Theme Park in Antalya 115

Mustafa GÜLAYDIN and Gonca AYTAŞ
Investigation of Emotional Labor Behaviors in Hotel Enterprises 129

Esra ERENLER TEKMEN
Women in Management in the Tourism Industry and the Glass
Ceiling Problem (An Invisible Barrier in Workplace) 155

Gözdegül BAŞER
The Impact of Tourism on Turkish Economy 175

Gül YILMAZ
The Importance and Role of Gastronomy Tourism within the Context of Sustainable Destination Management 199

Oğuz TAŞPINAR
Branding and Image in Gastronomy Tourism 213

Suna Muğan ERTUĞRAL, H. Neyir TEKELİ and Sezgi GEDİK ARSLAN
Effect of Environment in Increasing Sustainable Competitiveness at Tourism Destination 235

Hidayet KIŞLALI and Mustafa KÖSE
Recreational Second Homes and Sustainable Tourism 251

Gonca MANAP DAVRAS
Website Usage in Event Tourism 269

Fatma Doğanay ERGEN
Information and Communication Technology for Event Tourism Management 287

Özlem ŞEN
The Role of ICTs for Effective Marketing Activities in Event Tourism 309

İbrahim ÇETİNTÜRK
A Study on Brand Loyalty in Accommodation Establishments 333

Ozan ÇATIR
Investigation of Brand Positioning in Hotel Managements by Text Mining: A Case of Izmir Province 349

Berna KIRAN BULĞURCU
Determining the Relative Importance of the Factors for the Selection of Winter Tourism Centers 369

Yasin KELEŞ, Seden DOĞAN and Mutlu KAYA
Graduation Metaphors of Undergraduate Students Studying Tourism: A Study at Ondokuz Mayıs University 409

Eda Rukiye DÖNBAK
Tour Guiding Profession Perceptions of the Students Receiving
Tourism Guidance Education 423

List of Figures 447

List of Tables 449

About Editors 453

List of Contributors

Editors
Prof. Dr. Elbeyi PELİT
https://orcid.
org/0000-0002-6418-801X
Afyon Kocatepe University
Faculty of Tourism
elbeyipelit@aku.edu.tr

Prof. Dr. Hasan Hüseyin SOYBALI
https://orcid.
org/0000-0002-5929-0933
Afyon Kocatepe University
Faculty of Tourism
hsoybali@aku.edu.tr

Asst. Prof. Ali AVAN
https://orcid.org/0000-0003-4510-39
Afyon Kocatepe University
Faculty of Tourism
aliavan@aku.edu.tr

Chapter 1
Asst. Prof. Kezban TALAK
https://orcid.
org/0000-0002-7837-5852
Yeditepe University
International Business Administration
ktalak@gmail.com

Chapter 2
Assoc. Prof. Dr. Bilal ÇANKIR
https://orcid.
org/0000-0001-5126-8769
İstanbul Medeniyet Universtiy
Department of Business
bilal.cankir@medeniyet.edu.tr

Res. Asst. Nazlı EREN GÜN
https://orcid.
org/0000-0001-5690-9617
İstanbul Medeniyet Universtiy
Postgraduate Education Institute
nazlieren.gun@medeniyet.edu.tr

Chapter 3
Prof. Dr. Rüya EHTİYAR
https://orcid.
org/0000-0003-2719-2156
Akdeniz University
Faculty of Tourism
ehtiyar@akdeniz.edu.tr

Dr. Aslı ERSOY
https://orcid.
org/0000-0001-6750-261X
Akdeniz University
Faculty of Tourism
asliersoy75@gmail.com

Chapter 4
Assoc. Prof. Dr. Ahmet BAYTOK
https://orcid.
org/0000-0002-5826-7694
Afyon Kocatepe University,
Faculty of Tourism
ahmetbaytok@aku.edu.tr

Assoc. Prof. Dr. Özcan ZORLU
https://orcid.
org/0000-0003-3533-1945
Afyon Kocatepe University
Faculty of Tourism

Asst. Prof. Ali AVAN
https://orcid.org/0000-0003-4510-39
Afyon Kocatepe University

Faculty of Tourism
aliavan@aku.edu.tr

Chapter 5
Asst. Prof. Çağdaş ERTAŞ
http://orcid.org/0000-0001-9641-8054
Şırnak University
School of Tourism and Hotel Management
ertascagdas@hotmail.com

Chapter 6
Prof. Dr. Tuğba GÜRÇAYLILAR-YENİDOĞAN
http://orcid.org/0000-0002-4836-6904
Akdeniz University
Department of Management and Organization
gurcaylilar@akdeniz.edu.tr
Hatice BARUT
http://orcid.org/0000-0002-8129-7576
Akdeniz University
Department of Business Administration
hatice_barut_3507@outlook.com
Ayça ÇETİNKAYA
http://orcid.org/0000-0001-8699-9895
Akdeniz University
Department of Business Administration
aycetinkaya.2@gmail.com

Chapter 7
MSc Mustafa GÜLAYDIN
https://orcid.org/0000-0003-0177-9354
Afyon Kocatepe University
Faculty of Tourism
mustafagulaydin@gmail.com

Assoc. Prof. Dr. Gonca AYTAŞ
https://orcid.org/0000-0002-8221-4808
Afyon Kocatepe University
Faculty of Tourism
kilicgonca@aku.edu.tr

Chapter 8
Asst. Prof. Esra ERENLER TEKMEN
https://orcid.org/0000-0002-2509-3149
Cankiri Karatekin University
Department of Business Administration
esraerenler@hotmail.com

Chapter 9
Asst. Prof. Gözdegül BAŞER
http://orcid.org/0000-0002-1450-191X
Antalya Bilim University
Department of Tourism and Management
gozdegul.baser@antalya.edu.tr

Chapter 10
Asst. Prof. Gül YILMAZ
http://orcid.org/0000-0002-1882-867X
İstanbul Arel University
Department of Tourism and Hospitality Management
gulyilmaz@arel.edu.tr

Chapter 11
Asst. Prof. Oğuz TAŞPINAR
https://orcid.org/0000-0001-8891-8788
Çanakkale Onsekiz Mart University

List of Contributors

Department of Gastronomy and Culinary Arts
oguztaspinar@comu.edu.tr

Chapter 12
Assoc. Prof. Dr. Suna Muğan ERTUĞRAL
https://orcid.org/0000-0001-9872-3941
İstanbul University
Department of Economy
sertugral@yahoo.com
Asst. Prof. H. Neyir TEKELİ
https://orcid.org/0000-0002-4627-2840
İstanbul Kültür University
Department of Management and Organization
n.tekeli@iku.edu.tr
Asst. Prof. Sezgi GEDİK ARSLAN
https://orcid.org/0000-0002-4081-2551
İstanbul University
Department of Tourism and Hotel Management
s.gedik82@gmail.com

Chapter 13
Asst. Prof. Hidayet KIŞLALI
https://orcid.org/0000-0002-0978-496X
Afyon Kocatepe University
Department of Tourism Guidance
hkislali@aku.edu.tr
Asst. Prof. Mustafa KIŞLALI
http://orcid.org/0000-0003-3206-2508
Afyon Kocatepe University
Department of Geography
mustafakose@aku.edu.tr

Chapter 14
Asst. Prof. Gonca MANAP DAVRAS
http://orcid.org/0000-0003-0014-9983
Isparta University of Applied Sciences
Department of Tourism Administration
goncadavras@isparta.edu.tr

Chapter 15
Asst. Prof. Fatma Doğanay ERGEN
https://orcid.org/0000-0002-2818-8944
Isparta Uygulamalı Bilimler University
Department of Tourism Guidance
fatmaergen@isparta.edu.tr

Chapter 16
Asst. Prof. Özlem ŞEN
http://orcid.org/0000-0001-6506-170X
Antalya Akev University
Department of Gastronomy and Culinary Arts
ozlem.sen@akev.edu.tr

Chapter 17
Asst. Prof. İbrahim ÇETİNTÜRK
https://orcid.org/0000-0002-9078-7606
Çankırı Karatekin University
Department of International Trade
icetinturk32@gmail.com

Chapter 18
Asst. Prof. Ozan ÇATIR
http://orcid.org/0000-0003-3168-7338

Usak University
Ulubey Vocational High School
ozan.catir@usak.edu.tr

Chapter 19
Asst. Prof. Berna KIRAN BULĞURCU
http://orcid.org/0000-0002-9695-2668
Çukurova University
Business Department
bkiran@cu.edu.tr

Chapter 20
Assoc. Prof. Dr. Yasin KELEŞ
http://orcid.org/0000-0002-6879-448X
Ondokuz Mayıs University
Faculty of Tourism
yasin.keles@omu.edu.tr
Asst. Prof. Seden DOĞAN
http://orcid.org/0000-0001-8547-7702
Ondokuz Mayıs University
Faculty of Tourism
seden.dogan@omu.edu.tr
Asst. Prof. Mutlu KAYA
http://orcid.org/0000-0001-9165-0110
Ondokuz Mayıs University
Faculty of Tourism
mutlu.kaya@omu.edu.tr

Chapter 21
Asst. Prof. Eda Rukiye DÖNBAK
https://orcid.org/0000-0002-1752-0390
Adıyaman University
Faculty of Tourism
edonbak@yahoo.com

Notes on Contributors

Elbeyi PELİT was born in Gümüşhane, Turkey in 1978. He completed his primary, secondary and high school education in Gümüşhane. He received his bachelor's, master's degree in tourism management from Sakarya University and doctoral degree in tourism management education from Gazi University. Working in the fields of Tourism, Tourism Education, Tourism Management, Human Resources Management in Tourism and Tourism policy, Professor Elbeyi PELİT has published many articles, papers, books and book chapters in these fields and is the editor of the *Journal of Tourist Guide*, *Journal of Contemporary Tourism Research* and *Afyon Kocatepe University Journal of Social Sciences* (Turkey). He is still working at the Afyon Kocatepe University Faculty of Tourism.

H. Hüseyin SOYBALI was born in Afyonkarahisar province of Turkey in 1967. He completed his primary, secondary and high school education in Afyonkarahisar. He received his bachelor's degree from Uludag University, Balikesir School of Tourism and Hotel Management. He received his MSc degree in Tourism Management from the University of Surrey and PhD degree from Bournemouth University in England, UK. His main research interest areas include tourism development, tourism management, tourism policy and planning, human resources management and tourism education. Professor SOYBALI has published a book, many book chapters, articles and papers. He is working at the Afyon Kocatepe University, Faculty of Tourism as a senior professor and vice dean.

Kezban TALAK was born in Germany and has completed primary, secondary and high school in Turkey and Germany. She has a master's and Ph.D. from Marmara University. She works as an assistant to Prof. Dr. in International Business Administration in German. She studies on international management and organization.

Bilal ÇANKIR has completed his undergraduate, graduate, and Ph.D. degrees in Business Administration and Tourism departments. He has studied organizational behavior and positive organizational behavior subjects during his education. He also did a Ph.D. study at UT Dallas about burnout and organizational citizenship behavior. He worked as a research assistant at Kirklareli University from 2010 to 2017. He is currently working as an associate professor at Istanbul Medeniyet University, Business Department, and his active research topics are Positive Organizational Behavior, Management, and Burnout. He also serves as Deputy Director of School of Civil Aviation at Istanbul Medeniyet University. He is editor of the *Journal of Behavior at Work*. Bilal ÇANKIR has numerous peer-reviewed papers and conference articles.

Nazlı EREN GÜN was born in 1984, Bursa. She completed primary and secondary education in Bursa and high school education in Edirne. She received an associate degree from Kocaeli University Derbent Vocational School, Tourism Guidance Department; bachelor's degree from Gazi University Faculty of Commerce and Tourism Education, Travel Management and Tourism Guidance Education Department and master's degree from Istanbul University Social Sciences Institute Tourism Management Department. She is still studying for Ph.D. at Istanbul Medeniyet University Postgraduate Education Institute, Tourism Management Department and also working as a research assistant at Istanbul Medeniyet University, Tourism Faculty. She studies in the fields of tourism social psychology, heritage tourism, memory tourism, history of war, battlefield tourism, peace tourism, tourism and language alienation, cultural studies and identity. He published academic studies on "The Effects of Tourism Concepts on Language Alienation," "Examining the Attitude and Perception of the Ottoman People in the Balkan Wars Process," "Battlefield Tourism in the Paradigm of 'Peace among War-Torn Communities.'"

V. Rüya EHTİYAR is a professor at Akdeniz University Tourism Faculty. She received a bachelor of Business Administration degree from Anadolu University, a master of Tourism degree from Akdeniz University, and a Ph.D. in Tourism from Gazi University. She works in the fields of tourism and hotel management, organizational behavior and organizational psychology.

Aslı ERSOY holds a Ph.D. in tourism. She received a bachelor of Tourism degree from Adnan Menderes University, a master of Tourism degree from Akdeniz University and a Ph.D. in Tourism from Akdeniz University. She works on issues such as culture, cultural intelligence and managerial resourcefulness.

Ahmet BAYTOK is the head of department at the Tourism Management department at Afyon Kocatepe University and has received his associate professorship in Tourism. He currently interests in the research of leadership in hospitality, organizational behavior, management and sustainability in tourism.

Özcan ZORLU is the co-head of Tourism Guidance department at faculty of Tourism at Afyon Kocatepe University. He received his bachelor's degree in hospitality and tourism management at Balıkesir University and his master's degree in tourism management from Balıkesir University. He has received his associate professorship in tourism and written widely on specifically organizational behavior, knowledge management and alternative tourism activities in tourism.

Ali AVAN is the vice dean of Faculty of Tourism at Afyon Kocatepe University. He received his bachelor's degree in hospitality and tourism management at Mersin University, his master's degree in tourism management from Afyon Kocatepe University and his Ph.D. in business administration from Afyon Kocatepe University. He is assistant professor in Tourism Management department now, and his areas of research include consumer behavior in tourism, services marketing, tourism marketing and sustainability in tourism.

Çağdaş ERTAŞ is the head of Tourism Management department at Şırnak University School of Tourism and Hotel Management. He received his bachelor's degree, master's degree and a Ph.D. degree in Tourism Management from Mersin University. Çağdaş Ertaş, who has carried out research on tourism management and tourism sociology, continues his studies on sociological issues in tourism.

Tuğba GÜRÇAYLILAR-YENİDOĞAN is a professor of management and strategy in the Department of Business Administration at Akdeniz University. Her primary research interests include network governance, corporate strategy, sustainability and innovation management. She acted as the guest editor of the special issue on "Governance of Inter-Firm Networks in the Automotive Industry: The Static-Dynamic Efficiency Paradox" in IJATM. She has published in the *Service Industries Journal, International Journal of Innovation Management, International Journal of Automotive Technology and Management*, Springer Book Series.

Hatice BARUT completed her bachelor's degree in 2019 from the Department of Business Administration at Akdeniz University. She has completed business law, sign language and personal development trainings. She was part of social responsibility projects and organizations during university years. Her academic interests include business management, accounting and marketing. She currently works as a business tracking specialist responsible for production and marketing in her own family business.

Mustafa GÜLAYDIN was born in 1991 in Izmir. He completed his primary, secondary and high school education in Izmir. He received the bachelor of hospitality management and master of tourism management from Afyon Kocatepe University. He worked as the Front Office Manager at the Bronze Hotel in the Bodrum. His academic interests include tourism management, economy in tourism businesses, management, organizational behavior, human resource management, congress tourism, food and beverage management and business management.

Gonca AYTAŞ completed her primary, secondary and high school education in Ankara. She studied at Abant Izzet Baysal University School of Tourism Management and Hospitality. She completed her master's degree in tourism management at Sakarya University. She completed her doctorate at Gazi University in the department of tourism management education. She received the title of associate professor in November 2020. Her academic interest includes congress-convention tourism, thermal tourism, tourism management and organizational behavior.

Esra ERENLER TEKMEN is the assistant professor of Management Department at the Cankiri Karatekin University in Turkey. She received her BA in Tourism and Hotel Management in 1996 from Uludag University in Faculty of Tourism. After her six years of experience in hospitality industry, she started her academic career as a research assistant at Akçakoca Tourism and Hotel Management School in 1999. She received a master's degree from Abant İzzet Baysal University Business Administration Department in 2001 and her Ph.D. in management and organization in 2010 from Hacettepe University, Turkey. She joined the Cankiri Karatekin University in 2012. She is interested in employee behavior. Many of her researches have focused on management, organizational behavior and human resource management.

Gözdegül BAŞER has been working as an assistant professor at the Tourism and Management Department at Antalya Bilim University. She received her bachelor's degree in Management from Middle East Technical University, her master's degree in tourism management from Akdeniz University and her Ph.D. in management from Akdeniz University. Dr. Başer continues on her studies on tourism economics, tourism management and tourism marketing.

Oğuz TAŞPINAR has born in 1982, Istanbul. He completed first, middle and high school education in Istanbul. After finishing his bachelor's and master's degrees in Çanakkale Onsekiz Mart University Tourism Administration, he got Ph.D. of the same department from Istanbul University. He started teaching as a lecturer in 2012. In 2016, he started academic career as assistant professor at Çanakkale Onsekiz Mart University. He teaches courses of bar and beverage technologies, mixology, turkish and world cuisines, gastronomy tourism and basic gastronomy.

Suna MUĞAN ERTUĞRAL works at the Department of Economics and Department of Economic Development and International Economics at Istanbul University. She completed her education period in Istanbul University Faculty of Economics with his undergraduate education titled "Economics," which started in 1982 in the Department of Economics, and his Ph.D. thesis titled "The Effect of Tourism Investments on Employment" in 1986. He has been working as a research assistant at Istanbul University Faculty of Economics since 1988 and associate professor in the Department of Economics at the same Faculty. His current academic studies are regional development, rural development, tourism economy, sustainable tourism, employment relations and macroeconomics. He has 3 books and 12 book chapters as well as many international and national articles and papers. He speaks English fluently and intermediate Italian.

Sezgi GEDİK ARSLAN is Dr. Research Assistant in the Department of Tourism and Hotel Management at İstanbul University – Cerrahpaşa, Turkey. He received his bachelor's degree in the Department of Tourism Management from Anadolu University, his master's degree in Tourism Management Department from İstanbul University and his Ph.D. in Tourism Management Department from İstanbul University. Her areas of research are tourism economy, sustainable tourism and environment.

Gonca MANAP DAVRAS received her bachelor's degree in tourism and hotel management from Akdeniz University, her master's degree, and her Ph.D. in the Department of Business Administration from Süleyman Demirel University. She started to work as a lecturer at Süleyman Demirel University Isparta Vocational High School in 2004 and also she worked as the head of the Tourism and Hotel Management program. She still works at Isparta University of Applied Sciences, Tourism Faculty. She works on hotel management and types of alternative tourism.

Fatma Doğanay ERGEN graduated from Tourism Management and Hotel Applied Sciences of Abant İzzet Baysal University in 2010. She completed her master's degree in tourism and hotel management at Afyon Kocatepe University in June 2013. In 2014, Ergen started to work as a lecturer and as the head of tourism and hotel management program at Nisantasi University. She completed her Ph.D. in tourism and hotel management at Balıkesir University. After completing her Ph.D.s, she was appointed as assistant professor to the

Department of Tourism Guidance at Nişantaşı University and continued working at Nişantaşı University. In 2019, she was appointed as a professor assistant and head of the Department of Tourism Guidance at Isparta University of Applied Sciences. She is currently working at Isparta University of Applied Sciences. She works on alternative tourism types, health tourism and tourism management.

Özlem ŞEN is an assistant professor of Gastronomy and Culinary Arts at Antalya AKEV University. Her work focuses specifically on tourism, gastronomy and marketing.

Dr. İbrahim ÇETİNTÜRK is working as an academician in the department of management and organization in international trade of the faculty of economics and administrative sciences, Çankırı Karatekin University. He received his bachelor's degree from the department of tourism and hotel management, Adnan Menderes University, and a master's degree and a PhD from the department of business administration, Suleyman Demirel University. He has published academic studies over innovation skill, sustainable competitive edge, cyberslacking, sustainable tourism policy, customer value and brand loyalty.

Ozan ÇATIR is the head of department at the Tourism and Service Department at Usak University. He received his bachelor's degree in tourism management education from Gazi University, his master's degree in tourism management education from Gazi University and his Ph.D. in tourism management from Gazi University. Asst. Prof. Ozan Çatır, who has carried out projects on tourism management and hotel management, continues his studies on organizational behavior and human resource management.

Berna KIRAN BULĞURCU has been working as an assistant professor at Çukurova University, the Faculty of Economics and Administrative Sciences since 2016. She received her bachelor's and master's degree in business administration from Çukurova University. During her graduate education years, she got education from Sweden Linköping University and Istanbul Technical University at different times. In 2014, she completed her Ph.D. in operations research at Çukurova University. Assistant Prof. Berna Kıran Bulğurcu, who has studied about fuzzy logic, neural network and multi-criteria decision-making techniques, attaches importance to interdisciplinary approach in her studies.

Yasin KELEŞ was born in Samsun in 1985. He completed his primary, secondary and high school education in Samsun. He received his bachelor's, master's and doctoral degrees in tourism management education from Gazi University. During his master's and doctorate education, he was entitled to receive domestic master and doctorate scholarships from TÜBİTAK. He received the title of associate professor in tourism science in 2019. Working in the fields of tourism education, tourism management and tourism policy, Associate Professor Yasin KELEŞ has published many articles, papers and book chapters in these fields and is the editor of the *Journal of Contemporary Tourism Research* (Turkey). He is still working at the Ondokuz Mayıs University Faculty of Tourism.

Seden DOĞAN was born in 1978 in Karabük. She completed her primary, secondary, and high school education in Karabük. She completed her undergraduate, master's, and Ph.D. degrees at Akdeniz University in Tourism and Hotel Management Department. She worked at different five-star hotels and travel agencies between 1997 and 2013, and worked as a lecturer at Adnan Menderes University between 2013 and 2017. She has been working as an assistant professor at Ondokuz Mayis University Faculty of Tourism since 2017. She published several articles, conference papers and book chapters about technology in the hospitality and tourism industry. She is serving as an assistant editor at the *Journal of Hospitality and Tourism Technology* (JHTT) and as General Secretary at the Turkey Chapter of International Federation for *Information Technology and Travel & Tourism* (IFITT). She is currently conducting her post-doctoral research at the University of South Florida in the United States.

Mutlu KAYA was born in 1982 in Trabzon. He completed his primary and secondary education in Terme (Samsun) and his high school education in Ünye (Ordu). He completed his undergraduate education at Dokuz Eylul University, Department of Geography Teaching, and started to work as a Geography Teacher in Sinop in 2004. He completed his master's and doctorate education at Ondokuz Mayıs University and started working at Ondokuz Mayıs University in 2017. Working in the fields of tourism geography, cultural geography, city geography and sustainable tourism, Assistant Professor Mutlu KAYA has published a book, many articles, papers and book chapters in these fields and still works at the Ondokuz Mayıs University Faculty of Tourism.

Eda Rukiye DÖNBAK received her bachelor's degree from Anadolu University, School of Tourism and Hotel Management in 2000. Her academic career started in 2011 as a lecturer at Adıyaman University, School of Tourism and Hotel Management, Tourism Guidance and Travel Management Department. In 2017, Donbak received her philosophy of doctorate from Atatürk University,

Tourism Management and Hotel Department. Since 2007, she is a licensed tourist guide. Donbak is founder and head of Tourism Guidance Department of Adıyaman University Tourism Faculty since 2018. Her research interests are tourist guiding, ecotourism, strategic management, strategic alliances, organizational ecology, organizational behavior, organizational competencies and competitiveness.

Kezban TALAK

The Role of Organizational Behavior in Tourism Enterprises

Meeting the needs of individuals and societies and solving their problems require interaction and cooperation with other individuals and societies. Social organizations have emerged in the context of this imperative. In fact, the organization, which is described as a social phenomenon, has continued from the earliest times of humanity to the present day. In this context, all social units are organizational (Güney, 2000: 186). As a matter of fact, we have to be in constant contact with organizations for different reasons at every moment of our daily life. For example, we are likely to be in a company or an economic, social, or religious organization. We are sometimes an employee, sometimes a customer or a student in these organizations. It is observed that with the increase of human needs in the modernizing and globalizing world order, interactions within or between organizations are constantly increasing. In this context, it is possible to call the modern world a world of organizations (Can, 2005: 3). In this world order and organizational structures, the quality of interaction and communication among individuals is gaining importance day by day.

Today, the phenomenon of competition in the globalizing world order has been felt extensively in all national and international sectors. Accordingly, businesses in the tourism sector also need to develop some strategies to outperform their competitors. One of the factors that businesses can gain advantage in the tourism sector is the organizational behavior factor. Businesses, which can apply the organizational behavior factor correctly in their businesses, have the chance to gain superiority over their competitors.

The development of businesses in the tourism sector depends on their creation of new products and appeal, unlike their competitors. The main reason for this is related to the fact that today's industry is being shaped by new global conditions. The tourism industry is directly affected by technological developments, experienced customers, and environmental limits. In order for tourism businesses shaped on the basis of these factors to be ahead of their competitors, they must adopt organizational rules such as innovation adaptation and environmental common sense. Current developments in the globalization and technology age that we are in have a direct impact on the personnel–customer, employee–leader interaction within the enterprise. This situation emphasized the importance of organizational behavior and organizational communication for businesses.

Along with organizational behavior in the tourism sector, it has been determined that the need to take care of human resources and the hotel industry managers should be given more ways to lead high-value employees by using humanitarian approaches and by constantly supporting the growth and development of each individual.

In the work environment, the essence of building organizational citizenship behavior is indistinguishable from the existing commitment among employees. The challenges and contribution of organizational citizenship behavior will be the driving force behind the effectiveness of the hospitality industry. When we make an assessment in general, organizational behavior is defined as the systematic examination and application of information about how individuals and groups behave in the organizations they work for.

Understanding and shaping the behaviors of individuals within an organization emerges as a result of the society in which it is involved and the organization in which it operates. The main determinant of employees' integration with the organization emerges as organizational behavior. Organizational behavior phenomenon can be included in the field of many disciplines such as psychology, sociology, and industry psychology.

The aim of this study is to examine current issues and trends in organizational behavior in the tourism industry. In this context, in this study, the existence forms of organizational behavior management in the tourism sector will be discussed. In the study, first of all, organizational and organizational behavior facts will be examined in order to provide a conceptual framework. In the continuation of the study, the effects of organizational behavior, organizational communication, organizational commitment, job satisfaction, organizational ethical climate, and organizational citizenship behavior factors, which are the dimensions of organizational behavior, will be examined in the tourism enterprises.

The study will be carried out with the literature review method, which is one of the qualitative research methods. In this context, by analyzing the documents in the current literature, the application of the organizational behavior phenomenon in the tourism industry will be discussed. It is hoped that this study, which examines current issues and trends in organizational behavior in tourism, will guide businesses in the tourism industry.

1 Introduction

Examining the impact of individual behaviors, groups and organizational structure tends to increase organizational goal and organizational efficiency.

This motivation includes interpersonal communication, creating a common organizational structure, management, power, learning, development, process change, conflict management, designing attractive jobs for each individual, stress management, etc. Tourism businesses that meet the needs of users should deal with their employees as the most important part of today's economy (Organ, 1988: 18).

Each organization has its own goals that can be achieved through successful management with people. All these and many other activities should be a part of strategic and operational planning by human resources management in any subject wishing to apply the modern methods and tools of hotel management (Kotler, 2002: 58).

Tourism companies serve and market through the active use of human resources and the environment. It has become an imperative for businesses in the tourism industry to contribute to the sustainability of the environment and at the same time ensure their sustainability and vitality in the market. Due to both social awareness and legal obligations, tourism enterprises can show awareness for the external environment; however, it is not enough to be aware and sensitive only about the external environment (Örnek, 2009: 165). At this point, it is very important for the employees employed in tourism businesses to show organizational behavior sensitively. This situation should also be taken into consideration by tourism companies in terms of gaining competitive advantage, expanding their market shares, and ensuring the sustainability of the environment and tourism industry (Allison, 2001: 279).

Organizational behavior demonstrated by human resources and increasing the value of tourism businesses can have a direct impact on environmental sustainability. On the other hand, the insufficient level of organizational behavior in the axis of employees in tourism businesses can have a negative impact on the institution itself. These behaviors can be called organizational communication, organizational commitment, organizational citizenship behavior, organizational ethical climate, environmental awareness, technological awareness, and economic awareness. In this context, the aim of this study is to determine the role of organizational behavior in tourism businesses.

2 Organization Phenomenon

When we look at the existing literature, it is seen that many different definitions are made about the phenomenon of organization, and it is understood that the researchers do not agree on a common definition. Alpagun has defined the concept of organization as "the structure in which people and physical tools are brought together and different relationships are created to achieve certain goals" (Alpagun, 1996: 305).

Dinçer and Fidan: It calls the "social groups formed to achieve certain goals" as an organization (Dinçer & Fidan, 1995: 161).

On the other hand, Güney defines the unity of action and power created by two or more people as an organization (Güney, 2000: 187). Young defines the concept of organization as the structural process in which individuals come together to achieve organizational goals. The manager is responsible for the operation of this process. However, events occurring within the organization affect all individuals within the body. In this direction, it is understood that organizational structures define the mutual and dependent relationships (Young, 2004: 100).

It is understood from these definitions that in order for an organizational structure to be formed, more than one individual should come together, and cooperate for a specific purpose and take a certain action. As it is understood from this definition, the basic condition of an organizational structure is based on the existence of a certain purpose, cooperation, and action. Organizations are the products of the ability to organize something in human nature. Organizing is to set up a systematic arrangement for a particular purpose. Organizing nevertheless creates more permanent, social structures that we call the institution today (Güney, 2000: 187).

Organization, on the other hand, is a tool used by the business leader to mobilize employees or members in order to achieve the objectives of the business he is the manager of (Akat et al., 2002: 217). It is necessary to complete the processes such as determining the objectives planned to be realized in the organization process, making the necessary plans in this way, evaluating the human resources, physical opportunities, taking the necessary responsibility, and sharing the work (Uygur, 2007: 104). With the planning of this process, the members of the organization should act with the awareness of the responsibility that falls on them.

3 Organizational Behavior

Organizational behavior is the study of how human behavior interacts with the organization and the organization itself in organizational settings. Also in connection with this view, the concept of organizational behavior can be defined as understanding, predicting, and managing human behavior both individually and within a group in an organization (Tutar, 2004: 44).

Başaran (2000: 20), on the other hand, explains the organizational behavior as the conscious activities of the members of an organization on the way to achieve the objectives of the organization. Robbins and Judge, on the other hand, explains the fact of organizational behavior as the systematic study of the actions and attitudes of individuals within the organization. Robbins and Judge argues that this systematic study will take place when circumstances are under control and the evidence is carefully examined (Robbins and Judge, 2013: 2).

According to another definition, organizational behavior is explained as follows: "Using disciplines such as psychology, sociology and cultural anthropology to determine the behavior, perceptions, values and learning capacity of the individual while working in the business; examining human behavior, attitudes and performance on an organizational basis; is a discipline that investigates the impact of the external environment on the organization and its human resource, goals, mission and strategy" (Zel, 2006).

Organizational behavior is shaped in the axis of individuals, groups, organizations, and environment, propose that it aims to examine the behavior of the people in a business, the management process, the scope of the organization or organization involved in the management process, and the working order or work done within the organization process, the whole created by the interaction of the environment and the organization outside the organization (Özkalp & Kırel, 2001: 2).

Davis (1984) explains the purpose of organizational behavior as a healing attitude and action that aims to improve human-organizational relations, to effectively satisfy the needs of organizational members and to carry out teamwork to achieve organizational goals (Davis, 1984: 8).

Başaran (2000), on the other hand, explains the purpose of organizational behavior as the planning and action process for producing products in order to achieve the objectives of the organization and to provide successful performance and efficiency in the work done (Başaran, 2000: 21).

When we make an assessment in general, organizational behavior is defined as the systematic examination and application of information about how individuals and groups behave in the organizations they work for.

a. Features of Organizational Behavior

One of the most basic features of organizational behavior is that it is mainly focused on people. In this context, technical, economic and structural elements of organizations and groups are evaluated within the context of the human factor (Koçel, 1993: 256).

Another feature of organizational behavior is that it aims to develop human-organization relations. However, organizational behavior is influenced by the fact of motivation.

The existence of an environment in which individuals within the organization will be motivated is one of the main factors affecting organizational behavior. Motivating managers within the organization affects organizational behavior. Harmonization and collaboration of individuals within the organization is related to the desire of individuals to work together for the same purposes. The aim of organizational behavior, which is affected by these factors, is to meet the needs of employees and to achieve organizational goals (Koçel, 1993: 256).

In general, the main purpose of analyzing organizational behavior is to correctly analyze the relationships between the individual, technology, and business and to play a guiding role in communicating with each other. In this respect, the fact of organizational behavior provides a significant benefit to businesses (Davis, 1984: 7).

4 The Role of Organizational Behavior in Tourism Enterprises

The tourism sector, which has a dynamic nature worldwide, is expected to continuously develop and progress in line with this quality. This dynamic structure of tourism companies requires being active in a competitive environment in the axis of countries and cities. In this competitive environment, a number of factors significantly affect the dynamism of businesses. One of these factors is the organizational behavior factor. When the current literature is examined, it is understood that organizational behavior has a significant effect on tourism businesses. In this part of the study, the effects of the organizational behavior factor in tourism enterprises are examined in the axis of organizational ethical climate, organizational citizenship behavior, and servant leadership factors; a conclusion will be made about the role of organizational behavior in tourism businesses, which constitutes the focus of the study.

In the next part of the study, the concepts of organizational communication, organizational citizenship behavior, organizational ethical climate, and organizational commitment, which are among the basic concepts of organizational behavior discipline, will be handled in the axis of tourism enterprises; thus, the role of organizational behavior in tourism enterprises will be understood.

a. The Effect of Organizational Citizenship Behavior in Tourism Enterprises

It is not enough for tourism companies to fulfill only legal obligations in their competitive environment. Organizational citizenship behavior has an important effect on the development of businesses (Organ, 1988: 18).

When literature is examined, it is the first time that the organizational citizenship phenomenon; it was understood that it was used by Dennis Organ et al. in 1983. Organ (1988: 20) based on the concept of organizational citizenship, named voluntary individual behaviors, which are not considered directly and explicitly by the organization's formal reward system, but which positively affect the functioning of the organization as a whole, as organizational citizenship behavior.

Greenberg and Baron (2000: 68), on the other hand, call the organizational citizenship behavior when the organization does more than expected except for the duties it has determined.

When the literature on organizational citizenship behavior is analyzed, different approaches are encountered regarding the dimensions of the concept. The most common and accepted classification in the literature is the classification made by Organ (1988).

Organ (1988) deals with organizational citizenship behavior in five dimensions: altruism, civil virtue, conscientiousness, courtesy, and gentlemen (sportmanship) (Organ, 1988: 24).

Altruism employees: The attitudes and behaviors they exhibit to help other employees in problem situations and tasks that occur in their organizations are explained as a whole (Karacaoğlu & Güney, 2010: 139).

Conscientiousness is expressed as the voluntary contribution of the members of the organization to the functioning of the organization, except for the duties and roles they assume (Allison, 2001: 284).

Courtesy dimension, on the other hand, expresses the integrity and attitudes and behaviors that an organization exhibits in interaction and communication with other members and that its members are constantly in contact with due to the work they are responsible for (İşbaşı, 2000).

Civil virtue, another dimension of organizational citizenship behavior: It is explained as the fact that members of the organization act in accordance with the interests and benefits of the organization and voluntarily participate in activities that benefit the organization (Allison, 2001: 282).

"Gentlemen" means that employees avoid negative behaviors that may cause tension within the organization (Organ, 1990).

It is emphasized that the most important factors affecting the organizational citizenship behavior are job satisfaction, motivation, organizational support, psychological empowerment, personality, and organizational justice (Poyraz et al., 2009; Sökmen & Boylu, 2011; Chiang & Hsieh, 2012).

Studies on organizational citizenship behavior generally show that the concept has positive organizational results. Researches conducted in both tourism and other sectors reveal that organizational citizenship behavior affects the performance and efficiency of employees, organizational performance, organizational efficiency, coordination within the organization, organizational compliance, environmental solidarity among employees, organizational compliance, organizational commitment, intention to quit, and employee turn over (Keleş & Pelit, 2009; Sökmen & Boylu, 2011; Chiang & Hsieh, 2012).

When the interaction between employees and customers is considered in the axis of organizational behavior, a new dimension called organizational citizenship comes to the fore. If we examine this phenomenon through an example, individuals working in tourism businesses have to work in communication with each other. However, the level of service quality in businesses is significantly affected by communication and compliance among employees. Another factor that affects the organizational citizenship behavior in tourism businesses is the relationship established between the customer and the employee. In tourism businesses, it is up to the employees to fulfill their demands. In this direction, it is understood that customers are actively involved in this service relationship (Ma & Qu, 2011: 680–688).

Another factor that organizational citizenship behavior interacts with is the leader–employee relationship. In the current literature, opinion has been expressed that the relationship established by the employees with their leaders affects both sides to exhibit organizational citizenship behavior. In a study that examines the interaction between leaders and employees in hospitality businesses in the USA, it is concluded that the organizational citizenship behavior is also positively affected by the relationship between leaders and employees. Again, in this study, it was concluded that the jealousy experienced among the employees reduced the level of solidarity among the employees. When a general evaluation is made, it is understood that jealous behavior experienced within the organization negatively affects organizational citizenship behavior (Kim et al., 2011: 530–537).

b. Organizational Communication

The existence of an organization emerges when individuals are believed to be accomplished with a collective effort, the goals that individuals cannot achieve alone. In this context, organizational factors are brought together and organized to achieve certain goals. This can only be achieved by establishing a healthy communication within the organization and between organizations (Tutar, 2004: 44).

When we look at the current literature, the organizational communication factor has started to be considered as an important factor in the enterprises since the second half of the 20th century. This situation has provided important developments in communication tools with the development of technology in recent years, and this situation has significantly affected the communication in organizations. However, the gradual growth of organizations in recent years caused a decrease in face-to-face communication within the communication age (Bingöl, 1997: 281).

Organizations need communication both within the organization and between organizations in order to produce services and products. Ensuring the continuity of the organization is provided by transferring the information around the organization to the organization. In this context, it is understood that organizational communication is the process that connects the systems of the organization and ensures harmony among themselves (Üçok, 1992: 146).

Lack of communication between employees and managers lies at the root of many problems in organizations. In this respect, organizational communication factor is described as one of the most important tools of organizational management. The "planning, coordination, decision-making, motivation, and supervision" processes run smoothly in the enterprises where efficient organizational communication is provided (Tutar, 2004: 117).

For an organization to work efficiently and effectively, it is possible with a healthy communication between the manager and the members. In this context, it is necessary to send the information to the manager to make a healthy decision. Too much information means expanding the communication network, and accordingly, expanding the control area. As the control area expands, the number of communication channels will increase (Üçok, 1992: 146).

In general, the objectives of organizational communication can be listed as the announcement of the organization's policies and decisions, functioning, goals, social opportunities, and wage system to the members of the organization correctly and at the appropriate time (Bozkurt, 2004: 15). The functions of organizational communication are to ensure the information flow within the organization; influenceng and persuasion can be explained as commanding and instructive communication, adaptation, and unification. (Tutar, 2004: 120).

c. Organizational Ethical Climate

In order to explain the organizational ethical climate (BSI)[1] phenomenon, which was welcomed and researched by many researchers in the 2000s, we first need to explain what the concepts of "ethical climate" and "ethical morality" mean. Ethical and moral concepts, which are frequently used interchangeably, actually contain different meanings.

While the concept of ethics is described as the reflection of values of the reaction developed by the individual toward a situation faced by the individual, the concept of morality is explained as one-to-one application of these reflections of values. The moral phenomenon determines what is right and wrong in the relation of individuals and groups, how the behavior should be, and the standards in this direction. The phenomenon of ethics can be described as more abstract than morality and consists of written moral standards (Aksoy et al., 2017: 134). When we make a general evaluation, ethics is a total of guiding values in determining the way of doing business to people in general and to society in general. The concept of "climate," originating in Greek, means "orientation" and "tendency" (Büte, 2011: 172). The ways in which organizations manage routine behavior and actions that are expected and supported by employees are described as ethical climate.

According to the explanations of Bulutlar and Öz, the relationships, behaviors, and organizational results formed in this relation constitute the ethical climate phenomenon. The determinant of the widespread ethical climate that exists within the organization depends on the individual's choice between wrong and right (Bulutlar & Öz, 2009: 273). As DeConnick said, the behavior of the individual creates the phenomenon of ethical climate (DeConnick, 2010: 384).

Research on ethical climate does not only include ethical variables, but also these studies include organizational results (Martin & Cullen, 2006). Although there are many business climates within the organization, the ethical climate of the organization is important because it shapes the ethical behavior of individuals (DeConnick, 2011: 618). In this case, the determining element of the

---

1 Organizational Ethical Climate (BSI): The common, sound and morally meaningful thoughts of employees about the ethical practices of their organizations (Şahin ve Dündar, 2011:130)

organizational ethical climate is the preference made by the members who are members between wrong and right.

A positive organizational climate promises a climate with a high degree of confidence, strong working voice, and clear discourse norms within the organization with which it interacts. Schneider et al. acknowledge that a positive organizational climate "affects more than one policy, implementation, procedure (and) reward at the same time" (Schneider, 1996: 12). Based on this view, it is understood that employees and managers have a positive relationship with organizational ethical climate policies and activities in tourism businesses.

The organizational ethical climate phenomenon affects the participation, autonomy, integration, supervision support, and organizational learning in the organizational behavior axis. However, human relations climate, interpersonal trust, and organizational change studies, which include attitudes toward employee welfare, are also considered among the factors shaped by organizational behavior in tourism businesses (Wilson, 1989: 18).

According to Denison (1996: 67), a research community based on organizational behavior and psychology conducted a study examining the relationship between organizational behavior and performance in the work environment. According to this literature, there is a strong link between organizational ethics and innovation and performance. Accordingly, it can be claimed that this is also the case for each profession group working in tourism enterprises. As a result, organizational ethical climate affects and motivates employees in tourism businesses in terms of innovation and performance.

d. Job Satisfaction

When the literature is examined, it is revealed that there is a direct relationship between organizational behavior and job satisfaction.

Job satisfaction is a collection of positive and negative judgments and evaluations about the aspects of a person's job or work environment with other colleagues, customers, and the general business itself (Nerkar, 1996: 169).

Şimşek and colleagues explain the concept of job satisfaction as positive emotional responses to the professional roles of the employees in the workplace. According to this definition, job satisfaction is material; it consists of the pleasure that employees enjoy working together and the happiness of the work they have created together with their colleagues (Şimşek et al., 2003: 150).

In another definition, job satisfaction is defined as the situation arising from the difference between the idea of what the employee deserves and what he/she gets from the work environment as a result of the work done by the employee.

Therefore, job dissatisfaction arises in cases where the employee cannot get what he believes he deserves. The positive attitude of the employee toward his/her job at work creates job satisfaction (Schneider, 1996: 12).

Based on psychological, sociological and economic theories, Nerkar et al. divided the job satisfaction and aspects into the following three dimensions:

1. Instrumental satisfaction
2. Social satisfaction
3. Self-centered satisfaction.

In this regard, it can be said that the opinions of Nerkar et al. about job satisfaction reflect the positive responses arising from the fulfillment of one's job responsibilities in a way to support organizational goals (Nerkar, 1996: 169).

According to Herzberg et al. (1959: 52-55), instrumental satisfaction, similar to motivational factors, depicts progress. In the context of this view, the organizational behavior and job satisfaction that an employee performs while performing his duty to prevent performance deficiencies is directly proportional.

When individuals feel incompatibility in organizational settings, they experience a negative attitude change, which is immediately affected by this situation, which reduces job satisfaction and increases tension among their colleagues. When satisfaction decreases as a result of incompatibility, individuals are increasingly motivated to reduce the presence of incompatibility by doing one of three things:

1. Changing conflicting cognitions.
2. Preventing the same action from being repeated in the future.
3. Developing additional cognitions that help minimize differences in thought processes (Schneider, 1996: 12).

When we make a general evaluation about the relationship between organizational behavior and job satisfaction, organizations come to the fore as a widely accepted fact that they will not provide a high level of efficiency without meeting the expectations and needs of their employees. In this regard, it appears that job satisfaction is an important concept for today's organizations. The main reasons why the job satisfaction that the employee has while performing his profession is vital for employee-oriented organizations can be explained as follows:

The unhappy and uneasy person who does not have job satisfaction while performing his profession can turn the employee into a psychologically, physiologically and sociologically incompatible individual. Consequently, it may cause negative symptoms in employees such as alienation to work, indifference to work, desire to leave work, complaining from the current business, future anxiety, and hopelessness (Nerkar, 1996: 169).

In the study conducted by Tsai and Huang in 2008, it was stated that when ethical climate policies and rules help individuals to minimize cognitive dissonance, the individual's satisfaction with his colleagues, supervisors, wages, and general studies tends to increase (Tsai & Huang, 2008: 570). In this context, it is understood that job satisfaction of employees, which is one of the main factors of the existence of tourism enterprises, has an important place for businesses. Job satisfaction, shaped in the axis of organizational behavior, significantly affects the performance and productivity of the organization or company to which employees are members. Employees with high job satisfaction also have a high level of performance. Ultimately, this has a positive effect on businesses.

e. Organizational Commitment in Tourism Enterprises

The phenomenon of organizational commitment, which is an important extension of organizational behavior, has been frequently used in academic literature since the 1970s. However, a definitive definition on the concept of organizational commitment could not be achieved. The main reason for this situation comes from the fact that the organizational commitment phenomenon is tried to be included in different specialties such as sociology, psychology and social psychology apart from the organizational behavior discipline (Samadov, 2006).

Allen and Meyer (1996: 255) argued that organizational commitment is a dimension of psychology, and organizational commitment can be seen in the axis of the employees' relationship with the organization and the sense of organizational feeling.

Randall and Cote (1991: 198) defined organizational commitment as the work of the organization to adopt the goals and want to continue its existence within that organization.

Organizational commitment according to the classification is widely accepted in the literature; there are three types of organizational commitment: "emotional commitment," "attendance commitment," and "normative commitment" (Allen &Meyer , 1993).

The emotional commitment of the individual to the organization he/she is in is explained as the continuance commitment to join the organization in order to benefit from the opportunities such as emotional commitment, wages in the workplace, and retirement rights. The status of the individual to feel compelled to stay in an organization in terms of social opportunities and work obligation is explained as normative commitment (Gautam et al., 2001: 240).

Based on the above definitions selected from the literature, it can be said that the concept of organizational commitment is a vital concept for both tourism businesses and employees. The strong sense of commitment of the employees toward the organization to which they are members increases their usefulness within the organization and the effort required for the organization's objectives (Mowday et al., 1982: 139).

The commitment of the employee to the organization to which he/she is a member affects the efficiency, quality level, loyalty of the organization, and intraorganizational communication in a positive way (Randall, 1987: 464).

The concept of organizational commitment is of great importance since it has a characteristic that the personnel working in tourism businesses have a one-to-one relationship with customers. The employee attached to the organization is more prone to striving for the success of the organization from a perspective that takes into account the objectives of the tourism business. Loyalty to the organization is a concept that is likely to affect business performance, customer satisfaction, service quality, employee satisfaction, and intention to quit. For this reason, tourism businesses need to take this into account and make efforts to create loyalty in their employees. There are many studies on the factors that affect organizational behavior in the tourism sector. These studies are based on wage, satisfaction with the policies of the organization, working conditions, job satisfaction, progress and career opportunities, human resource management practices, job characteristics, justice, leadership type and behavior, chefs, coaching, training, organizational support, group harmony, managers and communication with employees, and unmet expectations revealed that factors such as meaningless and routine work affect organizational commitment (Feinstein & Vondrasek, 2001; Lam et al., 2002; Simons & Roberson, 2003).

It has been determined that organizational commitment affects intention to quit, employee turnover, customer orientation, guest satisfaction, and discretionary service behavior in tourism businesses (Kuşluvan et al., 2010).

Conclusion

When a general evaluation is made, as a result of this study, which examines the role of organizational behavior in the tourism sector, it is concluded that organizational behavior has an important place for businesses. The survival and success of businesses have a direct positive relationship with organizational behavior. The main reason for this is that customer–employee and employee–leader interactions are closely linked with organizational behavior discipline in the tourism sector. The harmonious, polite, and helpful relationship established by the employees in the enterprises with their colleagues directly reflects on their organization and creates the organizational behavior. Likewise, the relationship established between the members of the organization and the leader of the organization directly affects the organizational behavior and this situation is reflected to the enterprises. However, it is thought that the organizational citizenship phenomenon contributes significantly to the business performance of employees who do not act criminally and reward-oriented in terms of courtesy, altruism, conscience, civil virtue, and gentlemen. In this context, tourism enterprises employing seasonal workers prefer to work with individuals who tend to exhibit positive organizational behavior for the purpose of the business. At the same time, employees who tend to exhibit organizational behavior can be used as an important tool in tourism businesses in terms of adapting to the work environment and adapting to the organization more quickly.

Bibliography

Akat, İ., Budak, G., and Budak, G. (2002), *İşletme Yönetimi* (4. Baskı), İzmir: Barış Yayınları Fakülteler Kitabevi.

Aksoy, S., Erdil, O., and Ertürk, A. (2017), "Etik İklim: Kavramsal Gelişimi, Bireysel ve Örgütsel Etkileri", *Doğuş Üniversitesi Dergisi*, 18 (2), pp. 133–151.

Allen, N. J. and Meyer, J. P. (1993), "Organizational Commitment: Evidence of Career Stage Effects?", *Journal of Business Research*, 26, pp. 49–61.

Allen, N. J. and Meyer, J. P. (1996), "Affective, Continuance, and Normative Commitment to the Organization: An Examination of Construct Validity", *Journal of Vocational Behavior*, 49 (3), pp. 252–276, https://doi.org/10.1006/jvbe.1996.0043

Allison, E. H. (2001), Big Laws, Small Catches: Global Ocean Governance and the Fisheries Crisis", *Journal of International Development*, 13, pp. 933–950.

Alpagun, O. (1996), *İşletme Bilimine Giriş*, Trabzon: Derya Kitabevi.

Başaran, İ. E. (2000), *Örgütsel Davranış İnsanın Üretim Gücü* (3. Baskı), Ankara: Feryal Matbaası.

Bingöl, D. (1997), *Personel Yönetim* (3. Baskı), İstanbul: Beta Basım Yayın.

Bozkurt, İ. (2004), *İletişim Odaklı Pazarlama* (1. Baskı), İstanbul: Mediacat Akademi.

Bulutlar, F. and Öz, E. Ü. (2009), "The Effects of Ethical Climates on Bullying Behaviour in the Workplace", *Journal of Business Ethics*, 86, pp. 273–295.

Büte, M. (2011), "Etik İklim, Örgütsel Güven ve Bireysel Performans Arasındaki İlişki", *Atatürk Üniversitesi İktisadi ve İdari Bilimler Dergisi*, 25 (1), pp. 171-192.

Can, H. (2005), *Organizasyon ve Yönetim*, Ankara: Siyasal Kitabevi.

Chiang, C. and Hsieh, T. (2012). "The Impacts of Perceived Organizational Support and Psychological Empowerment on Job Performance: The Mediating Effects of Organizational Citizenship Behavior", *International Journal of Hospitality Management*, 31, pp. 180-190.

Davis, K. (1984), *İşletmede İnsan Davranışı Örgütsel Davranış*. Çev. Kemal TOSUN, İstanbul: İ.Ü. Yayın No: 3028.

DeConnick, J. (2010), "The Influence of Ethical Climate on Marketing Employees' Job Attitudes and Behaviors", *Journal of Business Research*, 63, pp. 384-391.

DeConnick, J. (2011), "The Effects of Ethical Climate on Organizational Identification, Supervisory Trust, and Turnover among Salespeople", *Journal of Business Research*, 64, pp. 617-624.

Denison, D. R. (1996), "What is the Difference between Organizational Culture and Organizational Climate? A Native's Point of View on a Decade of Paradigm Wars", *Academy of Management Review*, 21, pp. 619-654.

Dinçer, Ö. and Fidan, Y. (1995), *İşletme Yönetimi*, İstanbul: Beta Basım Dağıtım A.Ş.

Feinstein, A. H. and Vondrasek, D. (2001), "A Study of Relationships between Job Satisfaction and Organizational Commitment among Restaurant Employees", *Journal of Hospitality, Tourism, and Leisure Science*, 1 (4), pp. 1-20.

Gautam, T., van Dick, R., and Wagner, U. (2001). "Organizational Commitment in Nepalese Settings", *Asian Journal of Social Psychology*, 4, pp. 239-248.

Greenberg, J. and Baron, R. A. (2000), *Behavior in Organizations* (7th Ed.), New Jersey: Prentice-Hall.

Güney, S. (2000), *Yönetim ve Organizasyon El Kitabı*, Ankara: Nobel Yayın Dağıtım.

Herzberg, F., Mausner, B., and Snyderman, B. (1959), *The Motivation at Work*, New York: Wiley.

İşbaşı, J. Ö. (2000). *Çalışanlarin Yöneticilerine Duydukları Güvenin ve Örgütsel Adalete İlişkin Algılamalarının Vatandaşlık Davranışının Oluşumundaki Rolü: Bir Turizm Örgütünde Uygulama* (Yayımlanmamış Yüksek Lisans Tezi), Akdeniz Üniversitesi SBE, Antalya.

Karacaoğlu, K. and Güney, Y. S. (2010). "Öğretmenlerin Örgütsel Bağlılıklarının, Örgütsel Vatandaşlık Davranışları Üzerindeki Etkisi: Nevşehir Ili Örneği", *Öneri Dergisi*, 9 (34), pp. 137-153.

Keleş, Y. and Pelit, E. (2009), "Otel İşletmesi İşgörenlerinin Örgütsel Vatandaşlık Davranışları: İstanbul'daki Beş Yıldızlı Otel İşletmelerinde Bir Araştırma", *Ekonomik ve Sosyal Araştırmalar Dergisi*, 5 (2), pp. 24-45.

Kırel, Ç. (2007), "Sanal Örgütlerde Örgütsel Davranışın Geleceği", *Anadolu Üniversitesi İktisadi ve İdari Bilimler Fakültesi Sosyal Bilimler Dergisi*, 1, pp. 93-110.

Koçel, T. (1993), *İşletmeYöneticiliği*, İstanbul: Beta Basım.

Kotler, P. (2002), *Pazarlama Yönetimi* (Çeviren: Nejat Muallimoğlu), Beta Yayınevi: İstanbul.

Kuşluvan, S., Kuşluvan, Z., Ilhan, İ., and Buyruk, L. (2010). "The Human Dimension: A Review of Human Resources Management Issues in the Tourism and Hospitality Industry", *Cornell Hospitality Quarterly*, 51 (2), pp. 171-214.

Lam, T., Lo, A., and Chan, J. (2002), "New Employees' Turnover Intentions and Organizational Commitment in the Hong Kong Hotel Industry", *Journal of Hospitality & Tourism Research*, 26 (3), pp. 217-234.

Ma, E. and Qu, H. (2011), 'SocialExchanges as Motivators of Hotel Employees' Organizational Citizenship Behavior: The Proposition and Application of a New three-Dimensional Framework", *International Journal of Hospitality Management*, 30 (3), pp. 680-688.

Martin, K. D. and Cullen, J. B. (2006), "Continuities and Extensions of Ethical Climate Theory: A Meta-Analytic Review", *Journal of Business Ethics*, 69, pp. 175-194.

Meyer, J. P., Allen, N. J., and Smith, C. A. (1993), "Commitment to Organizations and Occupations: Extension and Test of a Three-Component Conceptualization", *Journal of Applied Psychology*, 78, pp. 538-551.

Mowday, R. T., Porter, L. W., and Steers, R. M. (1982), *Employee-Organization Linkages: The Psychology of Commitment, Absenteeism, and Turnover*, New York: Academic Press.

Nerkar, A. A., McGrath, R. G. and MacMillan, I. C. (1996), "Three Facets of Satisfaction and Their Influence on the Performance of Innovation Teams", *Journal of Business Venturing*, 11 (3), pp. 167-188.

Organ, D. W. (1988), *Organizational Citizenship Behavior: The Good Soldier Syndrome*, Lexington: Lexington Books/D. C. Heath and Com.

Organ, D. W. (1990), "The Motivational Basis of Organizational Citizenship Behavior", *Research in Organizational Behavior*, 12, pp. 43-72.

Örnek, A. Ş. (2009), Turizm İşletmelerinde Stres Yönetimi. İçinde N. Hacıoğlu, ve Z. Sabuncuoğlu (Eds.), *Turizm İşletmelerinde Örgütsel Davranış* (pp. 163-188), Bursa: MKM Yayınları.

Özkalp, E. and Kırel, Ç. (2001), *Örgütsel Davranış*, Eskişehir: T.C. Anadolu Üniversitesi Eğitim, Sağlık ve Bilimsel Araştırma Çalışmaları Vakfı Yayın No: 149.

Poyraz, K., Kara, H., and ve Çetin, S. A. (2009), "Örgütsle Adalet Algılamalarının Örgütsel Vatandaşlık Davranışlarına Etkisine Yönelik Bir Araştırma", *Süleyman Demirel Üniversitesi, Sosyal Bilimler Enstitüsü Dergisi*, 1 (9), ss. 71-91.

Randall, D. M. and Cote, J. A. (1991), "Interrelationships of Work Commitment Constructs", *Work and Occupation*, 18, pp. 194-211.

Robbins, P. S. and Judge, A. T. (2013), *Organizational Behavior* (14. Baskı), (Çev. İnci Erdem), Ankara: Nobel Kitabevi.

Samadov, S. (2006), *İş Doyumu ve Örgütsel Bağlılık Özel Sektörde Bir Uygulama* (Yayınlanmamış Yüksek Lisans Tezi), Dokuz Eylül Üniversitesi Sosyal Bilimler Enstitüsü, İzmir.

Schneider, B., Arthur P. B., and Richard, A. G. (1996), "Creating Aclimate and Culture for Sustainable Organizational Change", *Organizational Dynamics*, 24, pp. 7-19.

Simons, T. and Roberson, Q. (2003), "Why Managers Should Care about Fairness: The Effects of Aggregate Justice Perceptions on Organizational Outcomes", *Journal of Applied Psychology*, 88, pp. 432-443.

Sökmen, A. and Boylu Y. (2011), "Örgütsel Vatandaşlık Davranışı Cinsiyete Göre Farklılık Gösterir Mi? Otel İşletmeleri Açısından Bir Değerlendirme", *Gaziantep Üniversitesi Sosyal Bilimler Dergisi*, 10 (1), pp. 147-163.

Şahin, B. and ve Dündar, T. (2011), "Sağlık Sektöründe Etik İklim ve Yıldırma (Mobbing) Davranışları Arasındaki İlişkinin İncelenmesi", *Ankara Üniversitesi SBF Dergisi*, 66 (1), ss. 129-159.

Şimşek, M. Ş., Akgemici, T., and Çelik, A. (2003), *Davranış Bilimlerine Giriş ve Örgütlerde Davranış*, Konya: Adım.

Tsai, M.-T. and Huang, C.-C. (2008), "The Relationship among Ethical Climate Types, Facets of Job Satisfaction, and the Three Components of Organizational Commitment: A Study of Nurses in Taiwan", *Journal of Business Ethics*, 80 (3), pp. 565-581.

Tutar, H. (2004), *İşyerinde Psikolojik Şiddet* (3. Baskı), Ankara: Platin.

Uygur, A. (2007), *Yönetim ve Organizasyon*, Ankara: Nobel Yayın Dağıtım.

Üçok, T. (1992), *Yönetim İlkeleri* (3. Baskı), Ankara: Gazi Kitabevi.

Wilson, J. Q. (1989), *Bureaucracy: What Government Agencies Do And Whythey Do It*, New York: Basic Books.

http://paribus.tr.googlepages.com/h_yilmaz6.doc.

Young, K. S. (2004), "Internet Addiction a New Clinical Phenomenon and its Consequences", *American Behavioral Scientist*, 48 (4), pp. 402-415.

Zel, U. (2006), 'Endüstri Psikolojisi, Örgütsel Davranış ve İnsan Kaynakları Yönetimi', http://www.insankaynaklari.com.

Bilal ÇANKIR and Nazlı EREN GÜN

Turning "Learning" to Innovation in Tourism: The Theory of the "Tourism Learning Area"

1 Introduction

The idea that learning is equivalent to the meaning of being human and that we recreate ourselves by learning constitutes the main character of this research. It maintains the idea that the power that enables us to be able to do something we could never do is only possible through learning. The research is based on the idea that we re-grasp the world and our relationship with it through learning. In this context, it is believed that our capacity to be a part of life's creation and production process can be expanded only through learning. Therefore, it is observed that everyone has a deep hunger for such learning. This study emphasizes the need to understand the concept of "metanoia" in order to grasp the deeper meanings of "learning" before moving on to the approach of the "tourism learning area." This is because it is thought that the concept of *metanoia*, in other words, "mentality change" underlies the approach of tourism learning areas. The Greek word *metanoia* means a fundamental shift or change, even transcendence (meta-above and beyond) in the mind (from the root *noia*, *nous*,). In fact, it is argued that learning is actually a fundamental change in mentality (Senge, 2018: 32).

It is observed that a "learning-oriented mentality change" is needed in order that tourism areas can be developed within the scope of strategic mind and included among the future tourism trends. In this context, learning forms the basis of this study as the main factor that establishes the connections between *innovation, competitiveness, and sustainability*, which are the current and future concepts in tourism. The research consists of a conceptual analysis study that examines the theory of the "tourism learning area" and its place in future tourism strategies. The aim is to set forth the importance and necessity of the concept of "learning area," which is defined as the adaptation of organizational and interactive learning to areas and institutions, in terms of innovation and transformation in tourism.

It is seen that the approach of the "tourism learning area" was first discussed as a tourism-based regional development instrument in the regions of the European

Union. The European Commission's handbook, which was first published in 2004 and later revised in 2006, details the conceptual and application-oriented stages of the tourism learning-area approach and recommends the approach of tourism learning area in its member countries and supports the applications. It is seen that the objectives of the learning area approach are the objectives of all actors mentioned in the sector. However, all these different actor groups and the learning-area approach, which was initiated by their motivation, resulted in a single common goal: *reducing social division at the local and regional level and recreating common interests*. In this sense, it is envisaged that the "community philosophy," which will affect tourism development in a positive and sustainable manner in every sense according to us, can be strengthened with the approach of the learning area (Stahl, 2003: 18). In the context of current tourism trends, innovative perspectives, and suggestions that are expected to steer the future, this study, which I prepared on the approach of the "tourism learning area" by realizing its importance in tourism strategic mind, is thought to be a section that can add value to the international book project titled Current Issues and Perspectives in Tourism Management.

2 Conceptual Analysis of "Learning Organizations"

This study, where we approach with the perspective of turning "learning" into "innovation" in tourism, considers the "approach of tourism learning areas" associated with the learning organizations evaluated under modern management strategies. This is because, just like the learning organizations, the concept of the learning area is expressed as the mobilization and utilization of the potentials of all regional actors in local development from bottom to top, self-organizing, and self-responsible (Stahl, 2003: 11).

It seems that the tourism sector has to be continuously learning as it needs to adapt to ever-changing and emerging conditions and to continuously renew itself. In this context, it is known that touristic actions as to constantly prepare itself for any situation that may affect the tourism sector, including political, social, and economic conditions. So tourism enterprises should constantly prepare themselves for customer expectations, market, demand and innovation in changing conditions. In this regard, it is stated that the efforts of tourism enterprises in the sector to develop and improve themselves can only be possible by creating learning organizations (Demirkol & Çetin, 2014: 241). Therefore, it is thought that when the number of learning organizations increases in an area, that place becomes a faster learning area. Furthermore, a change of mentality, metanoia, underlies learning organizations. As expressed by Senge, it

Tab. 2-1: Comparison of Traditional and Learning Organizations

Comparison Field	Traditional Organizations	Learning Organizations
Attitude and Manner	An obedient attitude is prevailing. It is a hesitation that employees have trouble expressing themselves.	A questioning and critical attitude is prevailing. It is a hesitation that employees are comfortable in expressing themselves.
Right to Learning	Owners, managers, and organization employees have the right to learn.	In addition to the owners, managers, and employees, outsiders including customers, locals, public institutions, etc. have also the right to learn.
Motivation	The satisfaction of basic needs (food, shelter, belonging) is a source of motivation.	Besides the basic needs, the satisfaction of the need for respect and success is also a source of motivation.
Thinking	Thinking and acting groups are involved in the thinking process.	Everyone is expected to take part in the thinking process.
Change	It pursues an external environment- and competitor-oriented reactive change.	It pursues an actional and progressive change. It requires being conscious about taking care of all internal and external factors.

Source: Coşkun, 2000: 113

is interpreted as "a mentality change from seeing ourselves separate from the world to seeing ourselves connected with the world, from seeing our problems as problems caused by someone else or something else to seeing actually how our actions create our own problems" (Senge, 2018: 32). In order to better understand the learning organization model, it is thought that it would be useful to show its differences with the traditional organization model. In this context, Tab. 2-1 shows a comparison between traditional organizations and learning organizations from five different perspectives.

It is seen that "the concept of learning organization used for the first time in the book *Fifth Discipline*, 1990, by Peter Senge, took a wide place in the human resources literature in a short time. Senge states that he wrote this book to knock down "the illusion that the world is made of separate and unrelated forces." He points out that we can establish "learning organizations" only if we stop this illusion. He emphasizes that the people in such organizations will continuously expand their capacity to create the results they really want and that new and enthusiastic ways of thinking can be nurtured there, and thus *people are constantly learning how they can learn together.* According to Senge, "while the world becomes more interconnected and the business world becomes more complex, the business itself has to become more 'learned.'" In fact, he points out that organizations that have solved the mystery of benefiting from the learning desires and capabilities of people at all stages of an organization can achieve real success in the future (Senge, 2018: 21–22). This study maintains that, as with everyone and every field, tourism should also become more "learned" on an individual, organizational, regional, sectoral, and, of course, national, and international levels, and this is the only way to achieve success.

Watkins and Marsick (1993) define learning organization as an organization that develops a learning culture for continuous learning to take place at the individual, team, organizational, and social levels. Therefore, the organization can timely transform itself. For this reason, they suggest corporations create a learning culture. Below are listed the five critical elements within the organizational process, which are effective in the learning organization's successful and sustainable progress:

- To create a clear sense of direction and purpose, and convey it to the organization,
- To strengthen employees at all levels,
- To promote in-house knowledge and sharing,
- To collect and integrate information from outside,
- To challenge the status quo and enable creativity (Shin, Picken & Dess, 2017: 47).

The learning organization is basically defined as "an organization that constantly expands its capacity to create its own future" (Senge, 2018: 33). In this context, the learning organization is described as a group of people who are constantly developing themselves, both individually and together, in order to reach the common future they desire. It is argued that the ability to exist in the face of the brutality of increasing rivalry depends on our ability to learn in an inevitably and rapidly changing world (Çalkavur, 2016: 36). Therefore, it is seen that the learning organization develops on the basis of a philosophy that aims to predict and manage change, complexity, and uncertainty. It is pointed out that, considering the change and complexity in the environment, each organization is responsible for *continuous change, continuous learning, managing change and learning* and should implement them in the best possible way in future strategies (Coşkun, 2000: 110).

According to Senge, the rules that make up the learning organization disciplines are listed as follows:

- The problems experienced today resulted from the solutions of yesterday.
- The more squeeze is put on the system, the more it retrogresses.
- Behaviors produce good results before bad results.
- Getting out of a problem easily causes that problem to return.
- Treatment can cause worse results than the disease.
- Everything that is faster is actually slower.
- In time and space, cause and effect are not closely related.
- Small changes can produce big results.
- "You have both the cake and you can eat it," but the two are not the same.
- "You cannot have two small elephants by dividing an elephant into two."
- There is no such thing as "arraignment" (Senge, 2018: 81–92).

2.1 Five Disciplines of the Learning Organization

The starting point in the development of tourist destinations is expressed as rapid adaptation to the changes in tourist motivation and needs and segmentation of the tourist demand accordingly. It is observed that learning organizations can adapt to these changes quickly and continuously through new technologies and modern leadership types. It is argued that the relationship between tourism products and traditional contents (experience) and facilities contribute to the easier implementation of learning organizations with the help of modern information technologies, thereby increasing the competitiveness of the tourist destination (Jerković, 2019: 234). In this context, it is seen that there are five basic disciplines that reveal the logic, scope, and application area of learning organizations as described below:

2.1.1. *Self-Mastery, Self-Competence Discipline:* Self-mastery discipline constitutes the spirit and spiritual basis of the learning organization. This is expressed as the discipline of "continuous clarification of personal visual horizon, deepening it, focusing energies, improving patience and seeing the reality objectively" (Senge, 2018: 26). Therefore, it is expressed as having a personal vision and continuously improving ourselves to achieve this (Çalkavur, 2016: 38).

2.1.2. *Intellectual, Mental Models Discipline:* Mental models refer to the discipline that affects our actions and understanding of the world as completely rooted assumptions, generalizations, pictures, and images in our mind (Senge, 2018: 26). In this context, it enables us to see the effects of our beliefs and views of life on our relationships, decisions, and actions, to talk about them and to understand the perspective of the others (Çalkavur, 2016: 38).

2.1.3. *Shared Vision Discipline:* Shared vision is expressed as a common vision, being clamped around a common goal (Çalkavur, 2016: 38). It is the discipline of sharing a common goal, value, and duty senses within the entire organization since people, who gain an actual vision as a result of this sharing, go beyond themselves and learn just because they want it, not because they are told so (Senge, 2018: 27–28).

2.1.4. *Team Learning Discipline:* It is the discipline that expresses the capability of the members of a team to suspend assumptions and engage in a real act of "thinking together" (Senge, 2018: 28). In this context, it means obtaining much more than individual results as a team by using the enriching and strengthening effect of individual differences (Çalkavur, 2016: 38). The discipline of team learning starts with "dialogue." In Greek, *dia-logos* means free flow of the meaning in a group, so that the group can gain insights that cannot be gained individually. In modern organizations, the learning unit is not individuals, but teams, so it is noted that organizations cannot learn unless teams learn (Senge, 2018: 28–29).

2.1.5. *System Idea Discipline – Fifth Discipline:* Like all human efforts, the business world is also a system since it is stated that they are also interconnected by the texture of invisible and interrelated actions as for people. In this context, the system idea is a discipline consisting of a conceptual framework, a set of information, and tools. It provides the opportunity to see the entire event cycle more clearly and helps us see how they can be effectively changed (Senge, 2018: 25). Seeing the big picture helps to establish a thinking system that will ensure focusing on root causes and actions with high lifting power instead of temporary solutions that have a lot of side effects (Çalkavur, 2016: 38).

It is stated that the development of five disciplines together is vital for learning organizations. Therefore, the *fifth discipline,* especially emphasized by Senge is "system idea" because he suggests that the fifth discipline is the one that welds disciplines together, combining them as a whole of consistent theory and practice. In this way, it prevents other disciplines from being "disconnected hypes" and also prevents the organization from remaining "an enthusiasm for change." The system idea strengthens each of the other disciplines to constantly reveal that the whole is more than the sum of its parts. Senge points out that system thinking constantly needs other disciplines in order to realize its potential. In this context, he states that creating a shared vision encourages engagement in the long term. He states that mental models provide the necessary clarity to reveal the deficiencies in our perspective on the world. He points out that the ability of the groups to see the big picture beyond their individual perspectives has improved thanks to team learning. Finally, he argues that self-mastery constantly encourages our learning motive about how our actions affect our world. In conclusion, he suggests that the system idea brings "individuals and their worlds to a new way of understanding," which makes the finest aspect of learning organizations understandable (Senge, 2018: 31–32).

3 "Conceptual Analysis of the "Tourism Learning Area" Approach

The new economic environment that has emerged since the age of knowledge-intensive capitalism has shifted the competitive trend toward ideas, the areas can create economic advantages thanks to their ability to mobilize and exploit knowledge and ideas. Therefore, this new age of capitalism requires a "new kind of area" approach. These areas are expected to adopt the principles of knowledge creation and continuous learning, thereby becoming learning areas. Thus, it is seen that the learning areas provide a series of infrastructure that can facilitate the flow of information, ideas, and learning (Florida, 1995: 532). In fact, it is thought that "learning areas" can become "earning areas" over time. This study argues that the tourism sector should be competitive, and it should adopt the approach of the "learning area" and always carry on its activities with knowledge and in cooperation in order to develop and implement plans and policies, accordingly.

3.1 Background of Learning Areas

Today, it is seen that the success of the enterprises depends on how successful they are in creating, using, and exploiting information. Considering that the people who produce and develop the information are people in those organizations, it can be said that the human element in the enterprises has gained a new dimension of importance and awareness. In this context, it is observed that the knowledge and people who create the knowledge have become the most important source of competition in today's business life, and therefore organizations and areas are transformed into information-based businesses. The fact that the learning organizations approach has gained importance is attributed to the result of this knowledge-based development. In the tourism learning area approach, initiated under the leadership of the European Union, "areas" are considered as focal points for knowledge creation and learning in this age of global knowledge-intensive capitalism. It is hereby stated that the learning areas serve as information and idea collectors and repositories and also provide a basic environment or infrastructure that facilitates this flow of information, idea, and learning. In fact, despite the continuing predictions that the "end of geography" has come, it is understood that the areas have become more important forms of social, economic, and technological organization on a global scale (Florida, 1995: 527).

In recent years, it is increasingly accepted especially by the United States and the European Union (EU) that the building block in the development of the national economy is the wealth and prosperity created by the metropolitan regional economies. It is argued that the growth of the US metropolitan areas as the locomotives of the global economy is a model for modern development strategies in other areas. In this approach, social and environmental conditions are taken into consideration by providing a holistic development culture where communities and their workforce are seen as the main components of such economic activities at the regional level. The tourism industry is at the center of this transformation in the new world, where the new technologies created in the 21st century have significantly changed the modes of living and working. Tourism is used as an effective tool in local and regional development strategies, with the use of new technologies in developing the global economy and offering multiple products and services with increasing travel opportunities. Therefore, it is seen that the continuous growth and sustainability of the sector depends on its adaptation to these new conditions (European Communities, 2006: 3). This is where the importance of turning "learning" into "innovation" emerges. Considering the global tourism trends, it is clearly seen that "learning" and "innovation" are necessary for the competitiveness and sustainability of the tourism sector.

3.2 Theory of Learning Area

It seems that the concepts of learning and area should be understood first in order to explain the theory of the learning area. In this context, "learning" is expressed as an improvement in the perspective, knowledge, understanding, behavior, attitude, skills, and values of the individual. In this context, the learning area coincides with the idea of "lifelong learning," which is beyond the classical/traditional education approach. Lifelong learning, on the other hand, is a process in which education never ends and is open to all age groups, and it is defined as the continuous "reskilling" of people in every segment of society for a more modern and flexible economy. In this sense, it is expressed as an effective "pedagogical methodology" that ensures sustainable production and consumption competitiveness. Thus, lifelong learning includes learning for employment-related purposes as well as for personal, civil, and social purposes. Finally, lifelong learning implies increased investments in people and knowledge (European Communities, 2006: 12).

The concept of "area" can be described in two ways: First, it is a certain place the position and boundaries of which are known. The term hereby is used to refer to a "terrestrial area" such as "an administrative zone, a geographical region, or a tourism destination." Second, the term area is used to refer to a conceptual expression in the sense of a thematic subject/issue or focal point. When viewed from this aspect, the concept of tourism learning area is rather considered as "a region that represents a thematic subject matter." However, the European Commission uses the concepts of "learning" and "area" together in the context of their meanings applicable to both "geographic region" and "thematic region" in its relevant handbook. "The tourism learning area of Algarve (Portugal)" or "tourism area that learns on sustainable production and consumption" can be shown as examples of this (European Communities, 2006: 13). We also adopted this approach in this study. When explaining the tourism learning area from this perspective, two main points should be focused on: the "learning area" and "learning society."

The concept of "learning society" is another important part of the learning area approach. The learning society is defined as a city, town, or region that creates an energetic/lively, participatory and culturally aware human environment without economic concerns by going beyond any legal status required to provide education and training to everyone who needs and by promoting learning opportunities, justifying and actively developing these opportunities to improve the potential of all of its citizens (European Communities, 2006: 14).

The learning area idea was first developed as the "learning area" or "learning societies" by the European Commission Directorate-General of Education and Culture through the European Center for the Development of Vocational Training (Cedefop) in studies conducted in synergy with research on the regional development of the US metropolitan area. It is stated that these regions were created within the scope of "regional lifelong learning" initiatives. It is pointed out that the commission group themed "Working together for the future of European tourism" developed by the European Commission's Directorate-General for Enterprise and Industry in 2001 focuses on the question of "how to improve education to increase the skills in the tourism industry." This is because it seems that some issues in stability and growth of tourism cause certain difficulties in the European tourism sector. These issues are listed as follows: a) attracting, retaining, and developing a skilled workforce; b) supporting micro-enterprises/entrepreneurs to improve their competitiveness; and c) providing the quality of region/destination in an emerging and global market (European Communities, 2006: 4)

Learning area theory is basically expressed as a problem-solving method. In this context, learning area theory is considered as a suitable tool for the development of the tourism sector since as mentioned before, it seems that it emerged with the expectation that it will provide solutions to certain difficulties faced especially by the tourism sector in Europe. It is stated that the problem of retaining skilled labor (high labor turnover rate) is the leading one among these problems, and even in some countries, there are intense labor problems. Therefore, it is noted that the tourism industry has a weak image due to issues such as low wages, seasonal, and unstable employers (especially in the accommodation and catering sectors) in many countries. In this context, it is emphasized that tourism experiences a "skill gap." At the same time, despite the improvements in the level of formal education, it is asserted that tourism involves a relatively lower education level compared to other sectors of the economy. Besides, lately, it is increasingly thought that only empirical information is no longer sufficient for enterprises to be competitive. Finally, it is stated that it is necessary to be more aware of the impacts of the sector on regional environmental resources and cultural heritage issues, especially in many areas of Europe. It is pointed out that this situation forces tourism stakeholders of both present and future generations to be more responsible for the protection of natural and cultural heritage (European Communities, 2006: 6).

The "Learning Pyramid-a" in Fig. 2-1 shows all levels of action that occur in a learning area. It also shows how each level is integrated into a regional networking process. The fact that all levels are included provides a significant added value to both the whole and the parts. That is, a better flow of information and a more advanced learning resource offer higher skills and innovative opportunities that help all stakeholders to improve their performance. The "three dynamic wheels" in this diagram consists of 1) *participation of all stakeholders in the touristic region;* 2) *using a holistic approach to achieve learning; and* 3) *developing all factors with learning and entrepreneurial innovation for competition.* Figure 2-2 shows an operation of this triple dynamic wheel in a single process in a tourism learning area to visualize the stage of "learning citizen, learning team, and learning organization" *forming the business level*; the stage of "learning organization, learning network, and learning society" *forming the*

Turning "Learning" to Innovation in Tourism 53

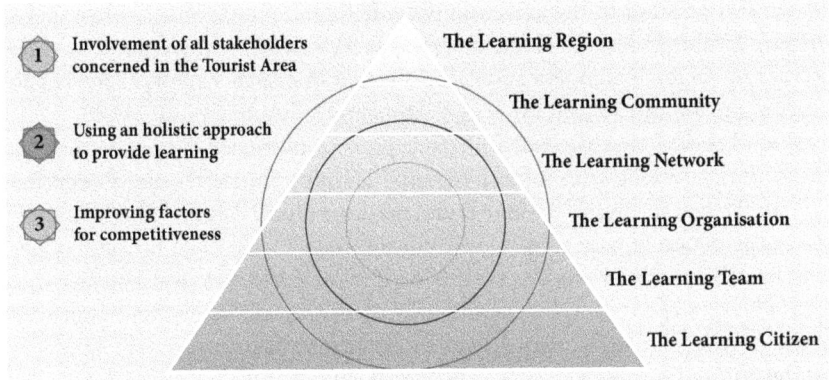

Fig. 2-1: Learning Pyramid-a
Source: European Communities, 2006: XIV

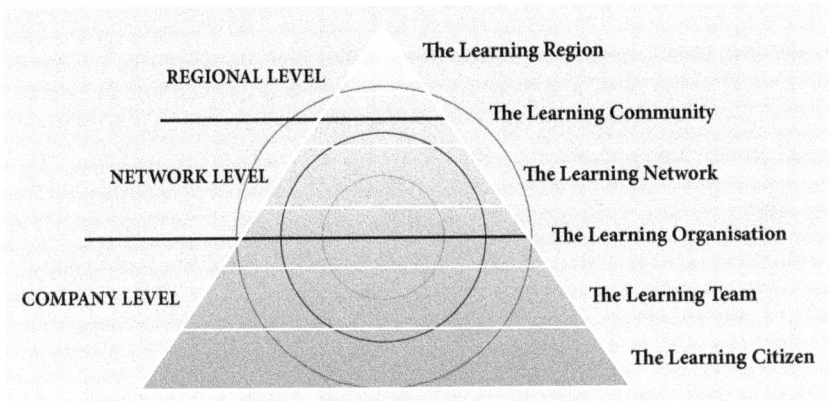

Fig. 2-2: Learning Pyramid
Source: European Communities, 2006: XIV

network level, and the stage of "learning society and learning area" *forming the regional level* from bottom to top. (European Communities, 2006: XIV).

As shown in the Learning Pyramid-b in Fig. 2-2, the learning area develops dynamic synergy among the individual, business, and area, improving the current status and profile of learning activities. This synergy is seen in the ongoing efforts of many areas and municipalities to improve local living conditions, attract skilled labor to the area, and provide them opportunities to live there, thereby creating a human resources base, as well as attracting enterprises and increasing regional competitiveness. It is stated that such synergy is the key to developing "innovation" in the area (European Communities, 2006: 10).

It is observed that the learning area process ultimately brings/will bring social changes along with it since regional learning initiatives aim to develop and strengthen the living standards and quality of local communities (through the participation of people from different interest groups) in an integrated way with an economic and social perspective. In this context, the "tourism learning area" emerges as a system that needs to be constantly developed and improved in the era of global influences, and as an approach that can address challenges on the basis of human resources and, most importantly, on a social basis, to offer and provide sustainable solutions (Moles, 2003: 129).

3.3 Theory of Tourism Learning Area

The Handbook titled "Innovation in Tourism – How to Create a Tourism Learning Area" which was prepared by the European Commission in 2004 and revised in 2006 (European Communities, 2006) first discusses the "theory of tourism learning area" and gives suggestions for its implementation in detail. The tourism learning area is described as a multi-stakeholder, cross-industry problem-solving approach that aims to increase human potential and performance to the target degree at the region/destination level in the tourism sector. To this end, it brings together all region-based stakeholders in broad terms in a top-down/ bottom-up process, creating a consistent network of information and collaboration. The tourism learning area is described as a way of dealing with and solving the current problems of the tourism sector by developing intraorganizational or interorganizational cooperation with a set of holistic learning opportunities (European Communities, 2006: XV).

It is stated that issues such as human capital, complex systems, innovation, entrepreneurship, and sustainability should be taken into consideration in order to understand the tourism learning area. In this context, first of all, the competitiveness and sustainability problems of the tourism sector are addressed through the multiple learning processes offered to private and public administrators, tourism stakeholders, and the tourism workforce. The necessity of a learning area approach arising from rapidly developing education and regional development areas is suggested. The purpose of the theory of the tourism learning area is described as ensuring the quantitative and qualitative development of stakeholder and employee performance in the tourism industry in cooperation with country/regional governors and locals (European Communities, 2006: XIII–XIV).

The theory of "tourism learning area" is defined as a region that constantly changes, keeps up with the globalized world, follows innovations, and adapts, and can organize the continuing learning efforts well. Besides, it is expressed as "a system that allows for being more advanced compared to other countries and territories through continuous learning by using the developments in science and technology" (Emekli, 2011: 29). The tourism learning area is considered as a philosophy that takes into account both the economic and social dimensions of the tourism area. It is stated that this basic philosophy covers the following key actions:

- To achieve learning,
- To identify learning needs and act as a catalyst in such fields,
- To improve the quality of learning, and develop skills,
- To support innovation and continuous change in the sector,
- To encourage information sharing,
- To provide strategic information and guide by raising awareness on key issues,
- To facilitate the process bottom-up,
- To cooperate with relevant stakeholders and bring together the tourism products on the basis of learning actions,
- To ensure and support the placement of qualified people in the sector,
- To exhibit a dynamic approach to increase local demand for tourism education to promote labor (Moles, 2003: 134).

The process of the tourism learning area is based on the operation of the following three dynamic wheels, as shown in Fig. 2-3 and defined as elements that move the tourism sector forward:

1. *Stakeholder Wheel (core center):* It consists of the participation of all broad-sense and regional-based stakeholders in the activities of establishing consistent partnerships and information and cooperation networks and coordination with a holistic perspective from top to bottom and from bottom to top in order to solve the challenges and current problems related to the sector with the tourism learning area. These include Public Authorities, SMEs and Professionals, Social Partners-Non-Governmental Organizations (NGOs), Research and Development (R&D) Organizations, Learning Providers, Expert-Advisory Organizations, and Local Community.

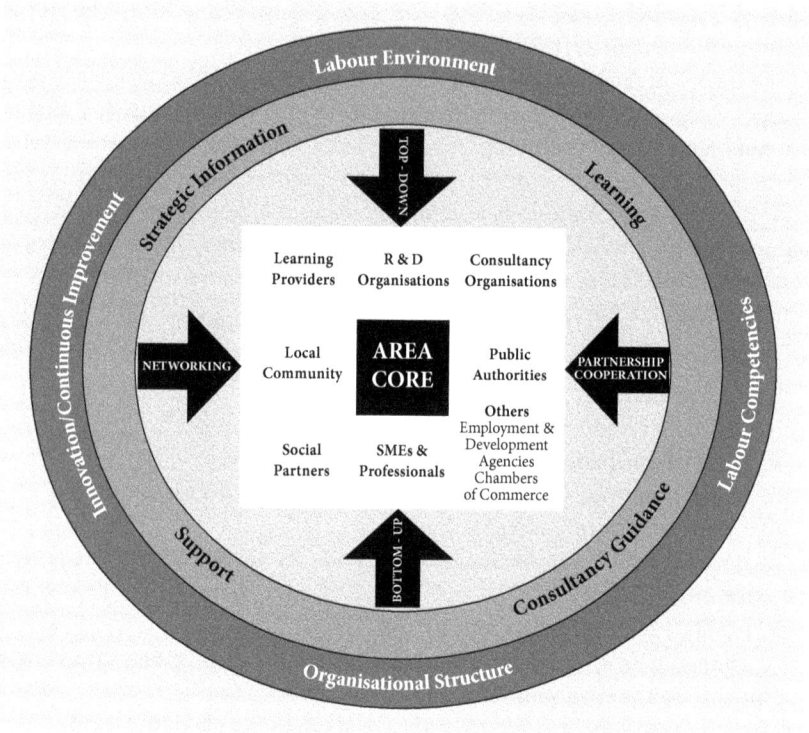

Fig. 2-3: Learning Area Process – Improving Economic and Social Dimensions in the Tourist Sector
Source: European Communities, 2006: XIV

2. *Learning Wheel (inner part):* The tourism learning area closes the gap between education and training systems and learning processes in the industry to improve the transfer of knowledge between the research/education institutions and the industry. This wheel consists of strategic information, learning, consultancy/guidance, and support units. Strategic information presented to the industry is considered necessary to prioritize management decisions and key learning issues. It is also stated that learning processes require guidance, counseling, and mentoring at all levels to transform and execute knowledge under innovation.

3. *Competitive Power-Effective Working Environment Factors (outer part):* When the competitiveness of the enterprises is analyzed in terms of

human resources, it is seen that the improvement of business processes and innovations are integrated with three basic factors. These include the Working Environment, Organizational Structures, and Development of Competitiveness in Transforming Information into Innovation. Therefore, it is stated that the relationship between the individual and the organization, established on condition that a healthy working environment is provided, will result in continuous improvement and innovation in labor competencies and organizational management (European Communities, 2006: 16–17). This study suggests that these three wheels, each being dynamic and interactive, should act together in order to turn "learning," which is the baseline perspective of this study, into "innovation."

It is suggested that the tourism sector is a very important test bed in the development and implementation of innovative ideas and concepts. In this context, all initiatives in the tourism learning area are created to take place together with and in continuous collaboration among public authorities, professional organizations, social partners, and commercial service organizations (public/private) and educational institutions such as vocational education institutes and universities (tourism stakeholders), and locals in order to improve competitiveness through better learning (Moles, 2003: 141). All wide-ranging tourism stakeholders, who are of the greatest importance in the learning tourism area and expected to carry out the learning process interactively with each other, are divided as follows:

A) *Government–Politicians:* National, regional, and local authorities; national, regional, and local tourism information offices; and tourism agencies.
B) *Social Partners:* Federations representing the tourism industry sectors, trade associations, and informal organizations.
C) *Commercial and Industrial Tourism Suppliers:* Hospitality industry; food and beverage industry (Restaurants, cafes, bars, etc.); transportation sector; construction industry; entertainment and activity industry; and tourist guide services.
D) *Commercial Intermediaries in the Tourism Industry:* Tour operators and travel agencies/retailers.
E) *Academic and Scientific Organizations*: Education and training institutions and research and consultancy organizations.
F) *Public, Private, Public-Private Partnerships, Civil Society, and Non-Governmental Organizations (NGOs):* Women, youth, locals, farmers, employees, community dwellers, visitors, and multi-stakeholder networks (European Communities, 2006: 38).

The tourism learning area links its local/regional public and private sector stakeholders to a range of advanced learning opportunities to create innovative, competitive, and sustainable enterprises. It is stated that during the implementation of the learning area approach in the tourism sector, stakeholders should be able to perform the following activities and further improve themselves in these fields:

- To create partnerships, networks, and clusters,
- To provide timely information flow to regional stakeholders and improve its quality,
- To access higher-level learning experience at the destination/regional level,
- To provide quality products and services,
- To exhibit innovative capabilities,
- To increase its performance and success in order to achieve more competitive power,
- To have a more autonomous structure,
- To provide a more flexible workplace,
- To operate in an entrepreneurial way,
- To develop strategies adaptable to globalization and global changes,
- To comprehend how to work in a more sustainable way,
- To increase the quality of life by reaching the highest level in accessing information (European Communities, 2006: 11).

The handbook prepared by the European Commission points out that since the concept of the tourism learning area is very new, yet it is difficult to find application models of this approach. However, example tourism learning areas, 7 of which are geographical areas (destination), and 3 of which are thematic areas, are presented, which were determined in the research process of this handbook, which contains valuable experiences on how to establish a tourism learning area, but are unconfirmed/conditional yet. The map in Fig. 2-4 shows the tourism areas, which are examples but not yet certain on a geographical basis:

⇒ Algarve (Portugal),
⇒ Bergamo (Italy),
⇒ Eastern Hungary (Hungary),
⇒ Pongau (Austria),
⇒ Bodensee – Lake Constance (Germany) (Germany – Austria – Swiss border),
⇒ Gelderland (Germany),
⇒ East Riding (U.K.) (European Communities, 2006: 98–103).

Turning "Learning" to Innovation in Tourism 59

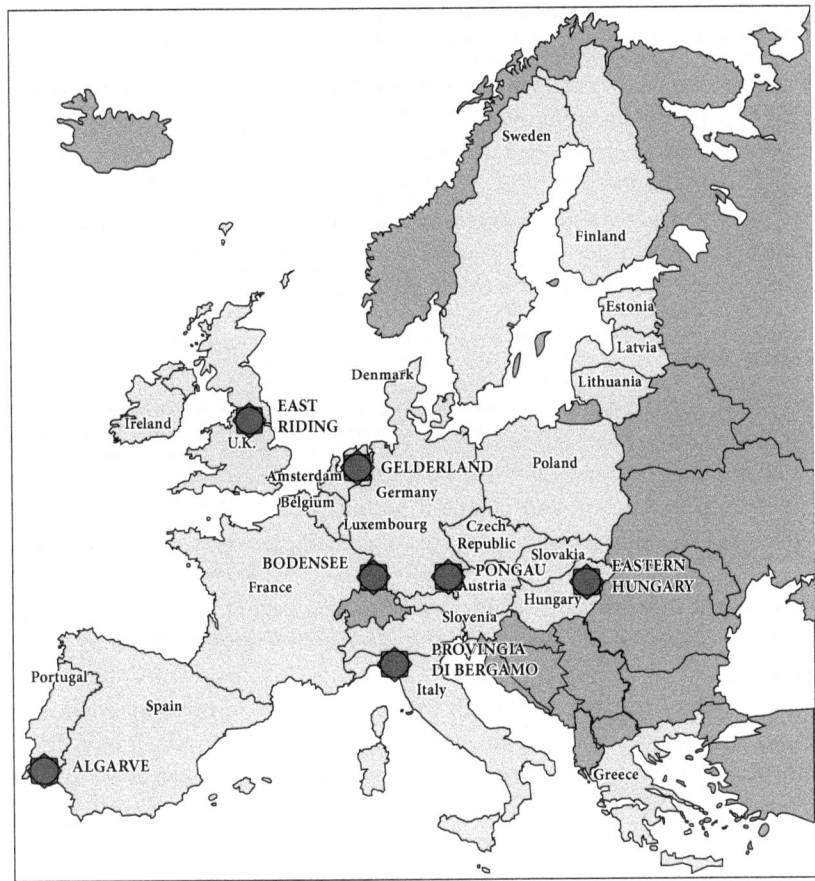

Fig. 2-4: Provisional Learning Area Sites Used to Develop the Handbook
Source: European Communities, 2006: XIV

The tourism learning areas shown as an example at the thematic region level are as follows:

⇒ Rural Tourism – Agricultural Tourism (the Netherlands),
⇒ Spa Rural Tourism (Austria),
⇒ DestiNet Sustainable Production and Consumption (EU) (European Communities, 2006: 103–105).

Finally, the concept of learning area in terms of the tourism sector can be described as an approach that seeks a holistic solution to the basic and current challenges/problems of the tourism sector at the local/regional level, overcomes these difficulties, and improves competitiveness *creating innovation through learning.*. It is concluded that the developments provided by this approach include inter-alia the growth of the local economy, sustainable development of tourism, creating permanent employment, and improving the working environment and career perspectives of local people who were already included in the sector or who newly enter it (Moles, 2003: 134–140).

Conclusion

This study prepared on the basis of the importance of turning "learning" into "innovation" considering current and future progress and requirements of tourism suggests that the theory of the "tourism learning area" is a solution method for current and potential problems of tourism. Considering the problems in the tourism areas, this study, which is based on the realization of the need for a "learning-focused mentality change," is predicated on the effects of the importance of the strategic mind and the aspect of the competitive tendency toward the knowledge, skills, and ideas in tourism in the current age of knowledge-intensive capitalism.

Under global influences, the need for effective use of investments in new technologies and science, knowledge, and learning in local and regional development strategies further emerges over time. In this context, the study aims to present a detailed conceptual analysis of the theory of tourism learning area, the importance of which is recognized, and ignite an idea about future analytical and applied research. It is observed that the theory can offer very useful and effective suggestions on both business and social and human aspects of both international and national tourism. In this regard, the theory of tourism learning area, which learns as a trend that aims innovation, sustainability, and competitiveness in tourism development, can be evaluated within the framework of social philosophy in this study since, thanks to the opportunities it offers with the vision of "lifelong learning," it seems that it will contribute to both tourism development and social development by reducing social division at a local and regional level and gathering all tourism stakeholders (actors) around common socio-economic interests and hopes.

The theory of the tourism learning area supports the sustainable development of tourism thanks to the benefits it provides to both the sector and the community dwellers by creating permanent employment in the area and improving the welfare level of the locals. The theory of tourism learning area, which was presented as a tourism-based regional development tool among the European Union countries for the first time, is a method that aims to increase human potential and performance in seeking solutions to the "skill gap" that is frequently encountered in tourism in the context of low wage, seasonal work, unqualified and unstable labor problems, and to the inadequacy of the level of education and learning in tourism, and the necessity of responsibilities for natural and cultural heritage. To this end, all wide-ranging tourism stakeholders, especially public and private officials, professional organizations, social partners, commercial service institutions (public/private), educational institutions such as vocational education institutes and universities, and locals are actively involved in the process of the tourism learning area in constant cooperation and interaction with each other. In conclusion, it is recommended to conduct studies so that the theory of the tourism learning area can steer the current and future tourism trends as a philosophy that takes into account both economic and social dimensions of the field of tourism, and it can be put it into practice in existing or new tourism areas.

Bibliography

Coşkun, R. (2000), "Transition from Traditional Organization to Learning Organization: Theoretical Framework and Practical Suggestions", *Bilgi*, 2 (1), pp. 109–116.

Çalkavur, E. (2016), *Learning Organization Journey, A Success Story*, Istanbul: Remzi Kitabevi.

Demirkol, Ş. and Çetin, G. (2014), *Tourism Policies and Alternative Management Approaches*, Sakarya: Değişim Aktüel Kitabevi.

Emekli, G. (2011), "Tourism Learning Areas, Cities and Theoretical Approach to Urban Tourism", *Ege Coğrafya Journal*, 20 (2), pp. 27–39.

European Communities (2006: 14). *Innovation In Tourism-How to Create a Tourism Learning Area Handbook- Developing Thematic, Destination Level and Regional Tourism Knowledge Networks*, Luxemburg: Office for Official Publications of the European Communities.

Florida, R. (1995) "Toward the Learning Region", *Futures*, 27 (5), pp. 527–536.

Jerković, D. (2019), "Increasing the Competitiveness and Interdependence of Agriculture and Tourism by Using Learning Organizations, Doctoral Dissertation Summary", *Tourism and Hospitality Management*, 25 (1), pp. 232–237.

Moles, V. J. (2003), Learning Region as an Alternative for Enhancing the Economic and Social Dimension in Tourist Areas: Some Reflections, *Agora XI The Learning Region-Thessaloniki* (pp. 127–150), Belgium: Europe Centre for Development of Vocational Training (CEDEFOP Panaroma Series).

Senge, P. M. (2018), *Fifth Discipline, Learning Organization Art and Practice,* Translated by Ayşegül İldeniz, Ahmet Doğukan, Barış Pala. Istanbul: Yapı Kredi Yayınları.

Shin, H. W., Picken, J. C., and Dess, G. G. (2017), "Revisiting the Learning Organization", *Organizational Dynamics*, 1 (46), pp. 46–56.

Stahl, T. (2003), Regional Development Networks in Europe. *Agora XI the Learning Region-Thessaloniki* (pp. 11–31), Belgium: Europe Centre for Development of Vocational Training (CEDEFOP Panaroma Series).

Watkins, K. E. and Marsick, V. J. (1993), *Sculpting the Learning Organization: Lessons in the Art and Science of Systemic Change*, San Francisco, CA: Jossey-Bass.

V. Rüya EHTİYAR and Aslı ERSOY

Culture and Management Relations in the Context of Intercultural Differentiation in Tourism Industry: A View on Turkish Management Culture

1 Introduction

With the 21st century globalization phenomenon, different cultures have come together between real or virtual borders, and the interdependence of economies, political systems and cultures has increased. The age of advanced technology and communication accompanying this trend has created a kind of "Global Village." Therefore, culture in the business world has become more important than ever before in terms of its impact on various aspects and practices of organizations, as well as increasing the synergy, productivity and well-being of different workforce. In a global business environment that is characterized by increasing mobility across national borders, culture is considered as a variable in determining similarities and differences in terms of organization and management practices among countries, and also offers new perspectives in explaining the nature of organization and management. As a matter of fact, mergers, changes in hands and the restructuring of investments in the international arena have brought along a prediction and necessity about how cultural difference will affect human relations within the organization (Moeller & Svahn, 2004; Weber & Tarba, 2012; Stahl et al., 2017). Tourism is one of the largest multinational economic activities in the world, which is among the top five export industries for many countries. The tourism industry is the most visible face of globalization as it includes the flow of goods, services and people (Reisinger, 2009). With the impact of globalization, employees with different cultural backgrounds work together in the tourism industry, and these differences help to develop new perspectives and innovative ideas to carry out the activities (Malik et al., 2017).

The discussions on culture and management relations brought up the questioning of some theories and models developing in the field of organizations and management, especially in western cultures. As a result of these discussions, it was concluded that organizational and managerial practices as well as organization and management theory are largely limited to culture

(Şişman, 2002). Therefore, it is essential to analyze whether the methods and principles that make the people of a culture productive and direct them to work will show the same effect in a different culture. Since each society has its own culture, it is not possible to evaluate the concepts, principles, processes and practices related to organization and management separately from the cultural context. In this case, there may be some local differences specific to each culture as well as some universal facts (Sargut, 1994). Research (Mor et al., 2013; Stahl et al., 2017) shows that one of the main determinants of managerial understanding and practices is culture. Briefly, the perception of "management" and "manager" concepts, their meanings and management practices may change according to countries and cultures. This situation reveals the necessity of considering the characteristics of culture in understanding and solving administrative problems (Şişman, 1996).

From this point of view, in this study, which deals with differences in a cultural context, evaluations have been made based on the literature on the management of differences in the tourism industry; it has been tried to determine why there is a need to manage the differences in the tourism industry and the features of the Turkish management culture have been examined.

2 Culture and Management Relationship

Many different views are put forward on how best to define culture that has been studied by anthropologists for a long time and what features should be emphasized in this definition (Communal & Senior, 1999). Culture is mental programming that distinguishes a group of people from others (Hofstede, 1980). On the other hand, according to Trompenaars & Hampden-Turner (1997), culture is a set of attitudes, beliefs, behavioral norms and fundamental assumptions shared by a group of people, and a complex whole that affects each member's behavior and their comments on others' behavior. As seen, social culture includes accepted behavioral norms. These norms are formed as a result of historical accumulation, and it forms the cultural heritage of that culture by collecting the characteristics of the society in which it is lived. This heritage distinguishes it from another culture and makes it unique (Nişancı, 2012). The generalizability of the management of organizations is not limited to any society or state. Although management functions are the same in every culture, the way management functions are implemented differs according to cultures (Erdoğan, 2007). Considering the dominant impact of social culture on organizational behavior such as mentality, way of life and working style and communication systems, social trends, which are an important dimension in organizational cultures, should be determined. Such a determination will give businesses important tips on adapting to new requirements, what challenges to face and what changes can be made easily (Erdem, 1996). Features that determine a country's management culture: (1) the hierarchy and the caste system in a society, (2) whether people with higher social hierarchy have more privileges than those who consider hierarchy in socio-economic and political issues, (3) respect for the elderly in the community, (4) the place of men and women in society, (5) belief in fate in society and (6) whether social recognition is based on family and caste history rather than true human quality (Jamil & Dangal, 2009).

From this point of view, the cultural differences in the tourism industry in terms of organizations and management will be tried to be outlined in the following sections.

3 Management in Terms of Cultural Differences in Tourism Industry

Cultural differences can be evaluated within the context of cultural values, social behavior and attitudes, perceptions, needs, expectations, experiences, beliefs, norms, motivations and verbal and nonverbal behavior (Reisinger & Turner, 1997). Managing diversity refers to a variety of management issues and activities, including recruiting and actively utilizing those from different cultures (Cox & Blake, 1991). Managing diversity does not mean controlling differences, but it means ensuring that each individual in the organization realizes their own potential (Thomas, 2001). To achieve this, managing the differences means minimizing potential barriers such as prejudice that can damage the diverse workforce, while maximizing the potential advantages of the difference. The main purpose here is to make the employees more sensitive to individual cultural differences and to adapt them better (Dessler, 2014). In the context of the tourism industry, hotels and restaurants have always been culturally diverse. Provided that organizations manage diversity effectively, diversity in the workforce can be a source of strength for organizations (Christensen-Hughes, 1992). On the other hand, it has become a necessity to manage the differences so that tourism enterprises can gain a competitive advantage through the quality of service in the face of changing workforce and customer profile (Maxwell et al., 2000). There are five core competencies that an international manager must possess in order to manage cultural differences successfully: Cultural awareness, communicative competence, cognitive competence, valuing differences and gaining synergy from differences (Iles, 1995). According to Harris et al.

(2004), the first step in managing cultural differences effectively is to increase the general cultural awareness of individuals. Individuals must first of all understand the concept and characteristics of culture and then be able to evaluate the impact of their cultural backgrounds and the cultural backgrounds of other individuals with whom they interact. Accordingly, a global leader who is sensitive to cultural differences values the differences of individuals and avoids imposing their own attitudes and approaches by considering these factors while communicating with them. Adler and Gundersen (2008) propose three strategies for managing cultural differences effectively: parochial, ethnocentric and synergistic. Accordingly, in organizations that adopt a parochial approach, managers believe that the best way to organize and manage is their own way. In organizations that adopt the ethnocentric approach, managers believe that their own way of organizing jobs is the best way and consider other ways worthless. On the other hand, managers have a synergistic approach when they clearly know the culture and believe that cultural differences have potential advantages and disadvantages. Managers and employees who use the synergistic approach argue that the ways of their own and others are different and none of them is naturally superior to the other.

These perceptions and assumptions have different effects on managing differences in organizations (Thomson, 1998). Accordingly, parochial organizations assume that the effect of culture can be ignored and do not care about cultural differences. This strategy eliminates the possibility of minimizing the negative effects of cultural differences and increasing their positive effects. Ethnocentric organizations assume that cultural differences have a negative effect and minimize cultural difference. Managers advocating this strategy may attempt to choose a culturally homogeneous workforce or to adapt all employees to the behavioral patterns of the dominant culture. Thus, ethnocentric organizations also prevent the opportunity to benefit from the difference. Finally, synergistic organizations see both positive and negative sides of cultural differences. The strategies of synergistic organizations are to minimize potential problems by managing the effects of cultural differences. Synergistic organizations direct their members to recognize cultural differences and use them to create advantages for the organization. The management of cultural differences is very important for the tourism industry, which includes employees with different cultural backgrounds. The aim of organizations that advocate and value cultural differences is to reveal ethnorelativism rather than ethnocentrism in the organization. Because an ethnocentric perspective and a culture that accepts only one way to do things can cause a number of problems for minority workers. In this case, individuals who have differences can feel themselves under extreme pressure to comply with others, become victims of stereotyped attitudes and can be considered inadequate because they are different (Daft, 2008). Valuing cultural differences also improves intercultural relations. Valuing differences means that an individual can go beyond accepting equal treatment for people of different racial and ethnic groups, respecting and enjoying a wide range of cultural, demographic and individual differences (Dubrin, 2012). Employee loyalty and commitment are formed as a result of valuing differences in organizations. Accordingly, organizations that value differences and provide equal opportunities to employees can be successful in recruiting and retaining talented employees (Metcalfe, 2010).

Learning to manage cultural differences is a tool that allows individuals to think broadly, act and be more effective personally and professionally. Cultural differences provide benefits when considered and used as a resource, otherwise they cause costs (Harris et al., 2004). Another effect of the management of cultural differences is related to performance. This is because Dadfar and Gustavson (1992) states that performance will increase as a result of successful management of differences, otherwise it may result in decreased performance and high costs. Robbins and Coulter (2012) suggest that effective management of cultural differences also affects the decision-making processes of managers. Accordingly, managers can make effective decisions by understanding cultural differences, using effective decision-making processes and building an organization that adapts to unexpected situations and the changing environment.

4 Intercultural Dimensions

Some researchers tried to examine different dimensions in order to better understand the concept of culture. Some of the studies on intercultural differences can be listed as follows: Hofstede (1980), Schwartz (1994), Hall (1985), Smith et al. (1996), and Trompenaars & Hampden-Turner (1997). The most prominent and most referenced research in the literature is Geert Hofstede's study of social values in terms of four basic dimensions. Hofstede and Peterson (2000) state that organizational and national cultures and cultural structures that make sense for all nations are visibly different at many points. In the comparative study of Hofstede (1980), which includes 11,600 individuals working in 40 different countries under various institutions to determine the differences between cultures, it is determined that the national culture basically differs in four dimensions. These dimensions will be briefly explained below (Şişman, 2002):

- Power distance,
- Uncertainty avoidance,
- Individualism-collectivism,
- Masculinity-femininity,
- Time orientation.

Power distance refers to the distribution of power in a society, the degree to which differences and inequalities in this distribution are approved and accepted by members of the culture. In cultures where power distance is relatively low, political equality is encouraged among individuals, whereas in cultures with high power distance, individuals see their managers as an autocrat father and are highly dependent on their superiors (Şişman, 2002). ***Uncertainty*** avoidance refers to the tolerance shown by individuals against uncertainty in a society (Erdem, 1996). Countries with high uncertainty tend to establish institutions to avoid risk. In countries where uncertainty is low, individuals often take risks more easily and do not tolerate alternative behaviors and ideas (Oliver & Cravens, 1999). ***Individualism and collectivism*** are seen as an important way of distinguishing national cultures. Individual cultures are not closely tied and are loosely united. In individual cultures, individuals are expected to take care of themselves and their families, and status stems from individual success. Collectivist cultures, on the other hands, attach importance to being members of groups such as social classes, religious communities or extended families for identity and status. In these cultures, people are protected by the group and are expected to act in the interests of the group (Newman & Nolen, 1996). ***Masculinity-Femininity,*** in masculine cultures, individuals emphasize and focus more on independence and success, such as power, wealth and status, while feminine cultures place more emphasis on quality of life and value dependence, relationships and the well-being of other people. In masculine cultures, individuals are encouraged to be ambitious, compete and strive for financial success (Park et al., 2002). ***Time orientation*** is about whether people have long or short orientation. Cultures with a long orientation primarily base family life on a pragmatic arrangement, but attention is paid to true love and care for young children. Also, being assertive is not encouraged and old age is seen as a happy period and starts early. In short-term cultures, individuals experience two groups of norms. The first is to respect social rules and to be seen as a stable individual, and the other is to address urgent need satisfaction, spending and sensitivity to social trends in consumption. In these cultures, old age is seen as an unhappy period, but it starts late (Hofstede & Minkov, 2010).

5 Turkish Management Culture

It is thought that it will be useful to briefly talk about the features of Turkish culture before touching on the features of Turkish management culture that has been shaped around Turkish culture in a historical process. Turkish culture, which forms the basis of the Turkish management model, has also reached the present day, just like the culture passed down from generation to generation. The Turkish nation has a history of about 5000 years, and during this historical period, Turks established many states and spent most of their life in the Central Asian region by prioritizing the formation of many civilizations or cultural areas (Çağlar, 2001). Turkey's historical roots are traced back to the rise and fall of the Ottoman Empire. With the decline of the empire, many Muslim groups living in the old Turkish lands around Southeastern Europe and the Northern Black Sea migrated to their hometowns, creating a western-oriented subculture that still prevails today. Today, Turkish culture is characterized as having the elements of modernity, tradition and Islam, and as a result of the globalization trends experienced worldwide, new lifestyles are created especially among the young population (Kabasakal & Bodur, 2007).

Cüceloğlu (1997) calls Turkish culture as a traditional authoritarian culture. The author explains the basic assumptions on which this culture is based as follows: Man is not able to decide what is good and what is bad; people are a helpless being, so they must submit to their fate; change is considered bad, and it is good to continue traditions and customs, but time planning is not important, because what happens in the future depends on fate, the person's age, position and people with whom he relates can determine his degree of dignity; competition is seen as a shame; expression through indirect statements and implication is preferred over direct and explicit expression, and men and women are not equal.

Sargut (1994) makes the following observations about Turkish culture: Turkish culture gives priority to collective behavior rather than individualism; it is not dominated by the system of male dominant values; organizational power distance is high, uncertainty is high, and it is located in a high context culture; individuals adopt a probabilistic form of speaking rather than speaking precisely and adapt the conversations according to the status of the other person.

Research conducted by Hofstede (1980) is the most widely used research that is the source of other researches. Hofstede describes the features of the Turkish management culture as follows: High power distance, numerous intermediate staff, organizational commitment based on interest, weak competition among employees, employees with low self-confidence, controlled initiative in the delegation of authority and high emphasis on status symbols (Köse & Ünal, 2000).

In a study conducted by Schwartz in 1994, 38 cultures were evaluated. Accordingly, Turkey ranked 12th in conservatism, ranked 5th in hierarchy, ranked 13th in social equity and ranked 16th in harmony (Fikret Paşa, 2000).

Kanungo & Aycan carried out a study covering 7 countries in 1997. As a result of this study, it was determined that China, Turkey, India and Pakistan were identified to have more paternalistic values than Romania, Canada and the United States.

Kabasakal & Bodur carried out a study including 62 countries in 1998. Accordingly, Turkey ranked 4th in collectivism and ranked 10th in power distance, and Turkey was above the world average of 62 countries.

Trompenaars and Hampden-Turner (1998), in a 38-country study, found that employees were committed to their leaders and had an upright hierarchical structure in organizations (Fikret Paşa, 2000).

Briefly, as a general conclusion drawn from the studies on Turkish management culture, it can be put forward that Turkish culture has a high power distance, a high rate of uncertainty avoidance, a collective structure and a feminine culture.

Conclusion

In this study, in order to understand the relationship between culture and management in the context of intercultural differentiation in the tourism industry, it has been tried to create a framework for introduction to related studies by acting on the information in the literature. In the light of the information in the literature, the subject has been examined conceptually and the basic features of the Turkish management culture have been tried to be determined. As a result, it is possible to say that the values of Turkish administration are almost exactly in parallel with the cultural values of the Turkish society and the main characteristic features that have been formed over the years have been transferred to the Turkish administration culture. Considering the impact of culture on the management of the organizations, employees and the way they do business, it can be argued that tourism enterprises and the managers of the enterprises have a great role in valuing the differences and managing the differences successfully. In this sense, it can be stated that managers can both create an appropriate working atmosphere for employees and provide organizational effectiveness and efficiency by accepting and valuing differences. As a matter of fact, in today's world where globalization is intense, it has become almost a necessity to manage the workforce, which has different ideas, cultures, traditions and customs, far from an ethnocentric approach, by valuing and respecting differences effectively. It can be stated that the establishment of an effective communication system is the basis for the successful management of the differences, and this situation may provide some advantages for both organizations and employees.

As a result, businesses and managers who want to work in the international arena such as tourism must understand the relationship between culture and management in the context of intercultural differentiation. Businesses that consciously perceive and manage intercultural differentiation take on the international business identity and thereby increase their performance. In understanding the relationship between culture and management in the context of intercultural differentiation, the biggest task falls on the academicians who conduct research on the subject and the managers who can transfer this information.

Although the research carried out by Hofstede in 1980 is one of the most referenced studies on the relationship between national culture and working values, it is observed that there are valuable but limited studies dealing with Turkish management culture. It is believed that it may be beneficial to carry out new studies that determine the characteristics of Turkish management culture in particular. Such an approach may have beneficial results in terms of revealing whether the Turkish management culture has changed on an annual basis.

In future studies, the relative effects of national culture, when these effects are most effective on organizational dynamics (teamwork, motivation and performance etc.), can be determined. For example, to what extent do centralized human resource management policies in multinational organizations reduce national cultural differences? Indeed, although strong cultures are likely to provide an integrative framework that reduces national differences for employees, there is little information about the interaction between national and strong cultures and their impact on the feedback process. It is also considered that it would be extremely important to develop more specific insights into important principles that affect the intercultural differentiation and management relationship process in future studies. In this context, it can be claimed that more research is needed in different sample groups and different types of work.

Bibliography

Adler, N. J. and Gundersen, A. (2008), *International Dimensions of Organizational Behavior*, Mason, OH: Thomson/South-Western.

Çağlar, İ. (2001), "Yönetim-Kültür Bağlamında Türk Yönetim Modelinin Saptanmasına Yönelik Kavramsal Bir Çalışma", *GÜ İİBF Dergisi*, 3 (3), pp. 125–148.

Christensen-Hughes, J. (1992), "Cultural Diversity: The Lesson of Toronto's Hotels", *Cornell Hotel and Restaurant Administration Quarterly*, 33 (2), pp. 78–87.

Communal, C. and Senior, B. (1999), "National Culture and Management: Messages Conveyed by British, French, German Advertisement for Managerial Appointments", *Leadership and Organizational Development Journal*, 20, pp. 26–35.

Cox, T. H. and Blake, S. (1991), "Managing Cultural Diversity: Implications for Organizational Competitiveness", *Academy of Management Executive*, 5 (3), pp. 45–57.

Cüceloğlu, D. (1997), *Yeniden İnsan İnsana*, Remzi Kitabevi.

Dadfar, H. and Gustavsson, P. (1992), "Competition by Effective Management of Cultural Diversity: The Case of International Construction Projects", *International Studies of Management and Organization*, 22 (4), pp. 81–92.

Daft, R. L. (2008), *Management*, Thomson Learning.

Dessler, G. (2014), *Fundamentals of Human Resource Management*, Pearson.

Dubrin, A. J. (2012), *Essentials of Management*, South-Western Cengage Learning.

Erdem, F. (1996), *İşletme Kültürü*, Üniversite Yayını.

Erdoğan, İ. (2007), *İşletmelerde Davranış*, Kişisel Yayınlar.

Fikret Paşa, S. (2000), "Leadership Influence in a High Power Distance and Collectivist Culture", *Leadership & Organization Development Journal*, 21 (8), pp. 414–426.

Hall, E. T. (1985), *Hidden Differences: Studies in International Communication*, Gruner & Jahr.

Harris, P. R., Moran, R. T., and Moran, S. V. (2004), *Managing Cultural Differences. Global Leadership Strategies for the 21st Century*, Butterworth-Heinemann.

Hofstede, G. (1980), *Culture's Consequences: International Differences in Work Related Values*, Sage.

Hofstede, G. and Minkov, M. (2010), "Long-versus Short Term Orientation: New Perspectives", *Asia Pacific Business Review*, 16 (4), pp. 493–504.

Hofstede, G. and Peterson, M. F. (2000), Culture: National Values and Organizational Practices. In N. M. Ashkanasy, C. P. M. Wilderom, and

M. F. Peterson (Eds.), *Handbook of Organizational Culture and Climate* (pp. 401-416), Sage.

Iles, P. (1995), "Learning to Work with Difference", *Personnel Review*, 24 (6), pp. 44-60.

Jamil, I. and Dangal, R. (2009), "The State of Bureaucratic Representativeness and Administrative Culture in Nepal", *Contemporary South Asia*,17 (2), pp. 193-211.

Kabasakal, H. and Bodur, M. (1998), Leadership, Values and Institutions: The case of Turkey. *Research Paper,* Boğaziçi University.

Kabasakal, H. and Bodur, M. (2007), Leadership and Culture in Turkey: A Multifaceted Phenomenon. In J. S. Chhokar, F. C. Brodbeck, and R. J. House (Eds.), *Culture and Leadership in across The World, The Globe Book of in-depth Studies of 25 Societies* (pp. 835-874), Psychology Press.

Kanungo, R. N. and Aycan, Z. (1997), Organizational Cultures and Human Resource Practices from A Cross-Cultural Perspective, *Symposium Conducted at the Canadian Psychological Association Annual Conference*, Toronto.

Köse, S. and Ünal, A. (2000), Türk Yönetim Kültürü Tarihi Açısından Çağdaş Türk İşletmelerinde Yönetim Değerleri, *Erciyes Üniversitesi 8. Ulusal Yönetim ve Organizasyon Kongresi Bildiriler,* 15-27 Mayıs, Nevşehir.

Malik, R., Madappa, T., and Chitranshi, J. (2017), "Diversity Management in Tourism and Hospitality: An Exploratory Study", *Foresight: The Journal of Futures Studies, Strategic Thinking and Policy*, 19 (3), pp. 323-336.

Maxwell, G., McDougall, M., and Blair, S. (2000), "Managing Diversity in the Hotel Sector: The Emergence of a Service Quality Opportunity", *Managing Service Quality,* 10 (6), pp. 367-73.

Metcalfe, B. D. (2010), Reflections on Difference: Women, Islamic Feminism and Development in the Middle East. In J. Sawad and M. Ozgilbin (Eds.), *Diversity Management in Asia* (pp. 141-60), Edward Elgar.

Moeller, K. and Svahn, S. (2004), "Crossing East-West Boundaries: Knowledge Sharing in Intellectual Business Networks", *Industrial Marketing Management,* 33 (3), pp. 219-228.

Mor, S., Morris, M., and Joh, J. (2013), "Identifying and Training Adaptive Cross-Cultural Management Skills: The Crucial Role of Cultural Metacognition", *Academy of Management Learning & Education,* 12 (3), pp. 453-475.

Newman, K. L. and Nollen, S. D. (1996), "Culture and Congruence: The Fit between Management Practices and National Culture", *Journal of International Business Studies,* 27 (4), pp. 753-779.

Nişancı, Z. N. (2012), "Toplumsal Kültür-Örgüt Kültürü İlişkisi ve Yönetim Üzerine Yansımaları", *Batman Üniversitesi Yaşam Bilimleri Dergisi*, 1 (1), pp. 1279-1293.

Oliver, E. G. and Cravens, K. S. (1999), "Cultural Influences on Managerial Choice: An Empirical Study of Employee Benefit Plans in the United States", *Journal of International Business Studies*, 30 (4), pp. 745-762.

Park, H., Borde, S.F., and Choi, Y. (2002), "Determinants of Insurance Pervasiveness: A Cross-National Analysis", *International Business Review*, 11 (1), pp. 79-96.

Reisinger, Y. (2009), *International Tourism: Cultures and Behavior*, Butterworth-Heinemann.

Reisinger, Y. and Turner, L. (1997), "Cross-Cultural Differences in Tourism: Indonesian Tourists in Australia", *Tourism Management*, 18 (3), pp. 139-147.

Robbins, S. P. and Coulter, M. (2012), *Management*, Prentice-Hall.

Sargut, S. (1994), *Kültürlerarası Farklılaşma ve Yönetim*, Verso Yayınları.

Schwartz, S. H. (1994), "Are There Universal Aspects in the Structure and Contents of Human Values", *Journal of Social Isues*, 50 (4), pp. 19-45.

Şişman, M. (1996), "Yönetim Kuramı ve Kültürlerarası Farklılaşma Açısından Yönetim Uygulamaları", *Eğitim Yönetimi*, 2 (2), pp. 295-308.

Şişman, M. (2002), *Örgütler ve Kültürler*, Pegem Yayıncılık.

Smith, P. B., Dugan, S., and Trompenaars, F. (1996), "National Culture and the Values of Organizational Employees: A Dimensional Analysis across 43 Nations", *Journal of Cross-Cultural Psychology*, 27 (2), pp. 231-264.

Stahl, G. K., Miska, C., Lee, H. J., and De Luque, M. F. S. (2017), "The Upside of Cultural Differences: Towards a More Balanced Treatment of Culture in Cross-Cultural Management Research", *Cross Cultural & Strategic Management*, 24 (1), pp. 2-12.

Thomas, R. R. (2001), "From Affirmative Action to Affirming Diversity, *Harvard Business Review on Managing Diversity*, March-April, pp. 1-31.

Thomson, T. M. (1998), *Management by Objectives*, Jessey-Bass/Pfeiffer.

Trompenaars, F. and Hampden-Turner, C. (1997), *Riding the Waves of Culture*, Nicholas Brealey Publishing.

Trompenaars, F. and Hampden-Turner, C. (1998), *Riding the Waves of Culture: Understanding Cultural Diversity in Global Business*, McGraw-Hill.

Weber, Y. and Tarba, S. Y. (2012), "Mergers and Acquisitions Process: The Use of Corporate Culture Analysis", *Cross-Cultural Management: An International Journal*, 19, pp. 288-303.

Ahmet BAYTOK, Özcan ZORLU and Ali AVAN

Service Culture: A Framework for Hotel Enterprises

1 Introduction

Understanding how businesses think and explaining their inner world is only possible by knowing the culture of the organization and how its elements emerge, can be kept alive and changed.. Organizational culture plays a more active role in the service sector, where human resources are used intensely in production, with its guiding behavior and attitudes. The hospitality industry has a unique culture as compared to other industries. Simultaneous production and consumption in hospitality industry, in other words, the frequency of guest–employee interaction and the fact that the employees interacting with the guests play a vital role in the distribution of services makes the culture important (Huertas-Valdivia, Gallego-Burín & Lloréns-Montes, 2019: 402). In addition, it is not possible for most of the employees in the hospitality establishments to have administrative control over the moment of service delivery to the customers. At this point, the only fact that makes it possible for the management to control the employees is to ensure that they act in line with the organizational values and beliefs adopted. Therefore, good planning and management of the human element in service organizations is only possible with an appropriate organizational culture.

2 The Concept of Culture

The etymological origin of the word culture comes from the Latin word "cultura." Latin cultura stands for "a cultivating, agriculture," figuratively "care, culture, an honoring," and is based on the word colere, expressed as "to tend, guard; to till, cultivate" (www.etymonline.com). The concept of culture, which is a phenomenon derived from the origin of the word by putting a figurative meaning to the idea of cultivating the soil (Morgan, 1998: 140), refers to making the common behaviors and thoughts of a community or group of people who come together according to certain principles meaningful. Culture is used to express human interactions and its consequences as a valid phenomenon for all human communities that come together on a certain basis (Hodgetts & Luthans, 2000: 108). The concept does not include the random behaviors of individuals, who constitute the society, but the behaviors or habits that are common to the society (Erdoğan, 1983: 113). In other words, it expresses symbols and meanings common to members of a group (Alvesson, 2002: 4).

Although the basis of the concept of culture comes from the science of anthropology (Smircich, 1983: 348), there are difficulties in the definition of culture, since it is a common research topic that is of interest to many disciplines and the concept has changed over time. Taylor defined culture as "a complex whole that includes the knowledge, art, morals, traditions and other abilities and habits gained by man as a member of society" (cited in Güvenç, 1985: 22). The culture defined by Hofstede (1981: 24) as "collective thinking program that differentiates a group of people from others." Culture has its own characteristics. As accepted by many researchers (Erdoğan, 1983: 134; Morey & Luthans, 1985: 221; Güvenç, 1996: 101; Vural, 2003: 38), general cultural characteristics are culture is learned, shared by people, transgenerational and cumulative, passed from one generation to the next, symbolic, patterned, and adaptive.

3 The Concept of Organizational Culture

In organizations created by a group of people with certain characteristics, thoughts, philosophies, common behaviors, and usages of people are the main issues of organizational culture. Members of the organizations grow up in a certain community, and their socialization process shapes according to this community culture. Thus, values, norms, ideologies, assumptions, and other cultural elements of relevant community culture directly affect organizational culture. Namely, principally each organization is a projection and subculture of this common culture. Because of this, each organization both embraces the characteristics of common culture and develops an idiosyncratic culture. However, every organization is different from each other and this organization exists because it consists of people with different characteristics. At this point, behavioral norms, expectations, different practices, and organizational values are the main factors of differences (Vanderberghe, 1999: 175).

Although organizational culture considered as a monolithic structure in terms of understanding this concept, today it is widely accepted that more than one culture could exist in organizations. Within this context, the culture that is prevalent in the whole business is called dominant culture and the others are called subcultures. Branches of a multinational company operating in different countries, geographically distributed establishments of a national enterprise, different departments in a company, and even norms, values, and assumptions of working groups are certain examples of subcultures. Each member of the organization voluntarily participates in the organizational culture that symbolizes the unchanged behavioral aspect of the organization.However, the culture existing in organization automatically controls organizational life (Daft, 2000: 314). The organizational culture, which is a basic field of interest as much as strategy, structure and control mechanisms in today's management philosophy, is a characteristic of organizations rather than individuals. Nevertheless, it can be understood and measured with individuals' actions and intellectual behaviors (Hofstede, 1998: 479).

In organizational context, three approaches developed to understand organizational culture conceptually. These are exogenous variables approach, endogenous variables approach, and organizations as metaphor approach. (Smircich, 1983: 339-358). Exogenous variables approach accepts organizational culture as an independent variable and asserts that organizational culture form from individuals' cultural characteristics in which they transferred from common community culture. Contrary to this, endogenous variables approach claims that organizations are the structures forming their own culture due to their idiosyncratic traditions, legends and rites. In organizations as metaphor approach, each organization itself is accepted as a specific culture. To this approach, considering the culture as an endogenous variable is incorrect. Because, the culture spreads everywhere like an air in the room and surrounds the organization. Although each approach has valid assumptions, organizational culture basically consists of these three approaches. Alvesson (1987), a researcher criticizing on this issue, asserts that organizational culture could be a whole consisting all three approaches rather than only one approach by stating that both endogenous variable approach and organizations as metaphor approach consider organizations as cultural assets. In addition, he states that members of any organization become different from others by being characterized with cultural similarities consisting common meanings such as values, norms, beliefs, and symbols.

Organizational culture, which is briefly expressed as "it is just the way it is (here)," can be defined in various ways such as:

a) A set of values and beliefs shared and understood by organization members (Recardo & Jolly, 1997: 5),
b) Philosophy guiding organizational policy toward employees and customers (Robbins, 1987: 357),
c) Beliefs shared by senior management on how to manage themselves and their employees and how to carry out business activities (Lorsch, 1986: 95),

d) Norms, behaviors, expectations, beliefs, assumptions, values, ideologies, and philosophies that hold a community or group together and shared by its members (Kilmann et al., 1986: 89).

Hofstede, Hofstede and Minkov (2010: 6) define culture as "the collective programming of the mind that distinguishes the members of one group or category of people from others" and regards daily practices and shared perceptions as the core of organizational culture. Schein (1992: 12), emphasizing the functions of culture, considers the culture as the basic assumptions that a group has developed in the learning process and is effective in solving problems in order to solve the problems of internal integration and external harmony. According to researcher, these assumptions enable newcomers of organization to perceive, think, and feel appropriately.

Organizational culture is a strong social bond that ties organization members to each other and organization to community. An organization can only exist and achieve its goals if its members share harmonized common goals, standards and values, philosophy, and ideologies (Eren, 1997: 373). For this purpose, culture is the needed power that creates common language, expressions, and concepts to create internal loyalty among the members of the organization, organizes their strengths, status, ascension, and relations among each other and enables them to make sense of all organizational events (Özkara, 1999: 84). Culture plays an important role in organization's compliance with its external environment besides its function to create intraorganizational loyalty. Hence, an organization constantly interacts with external environment and cannot survive if isolated from external world. For the organizations, existence, surviving, and achieving common goals depends on their compliance level with external environment and their ability to reduce environmental uncertainties. From this perspective, culture involves mission, strategies, goals, instruments used for achieving relevant goals, and standards settled for measuring how organization members reach predetermined goals (Özkara, 1999: 85).

Understanding, conceptualizing, and evaluating organizations with a cultural perspective are possible with cultural elements. Elements composing organizational culture can be evaluated in two stages as "norms and shared values" (Beyer & Trice, 1987: 6; Kotter & Heskett, 1992: 5; Daft, 2000: 315), or in three stages as artifacts, espoused values and basic assumptions" (Singh & Hofstede, 1990: 75; Schein, 1992: 17; Schermerhorn et al., 1997: 271).

Artifacts: Artifacts identifying visible aspects of organizational culture are composed of architectural structure, technology and products, clothes, language, myths, heroes, symbols, rituals, rites, and celebrations. Schein

(1992) states that artifacts can easily be observed; however, it is very difficult to render why it is that way and, the underlying assumptions cannot be easily encoded. Senior management use these artifacts during learning, transferring/sharing, and changing organizational culture. Especially language, rites, rituals, symbols, stories, myths, and heroes are accepted as critical elements for learning, transferring, and changing organizational culture. **Language:** Language is one of the most important instruments for sharing culture. Organizations like human communities have also their own language. Many organizations and sub-groups in organizations use language to identify culture or members of subculture. Employees accept organizational culture by learning this language and they help protecting it (Robbins, 1987: 367). **Rites and rituals:** Rites express actions involving special meanings for organizations. Beyer and Trice (1987) state that rites provide more information for understanding culture since language, symbols, jests, physical structures, and institutionalized behaviors are used at preparing and performing rites. Trice and Beyer (1987) classify rites based on their configuration goals and outcomes as transition (to teach organization culture to newcomers), relegation (transferring employees to another department due to certain reasons, to relegate them or discharging), promotion (promoting organizations members' status and social roles with ceremonies), novation (organization development, establishing work teams, quality circles, etc.), reducing conflict (working collectively, establishing committees), and integration (annual meetings, new year parties, and annual picnics). Rites turn into rituals when they become a regular activity within the company. Award ceremonies, annual company picnics, weekly beer parties, introductory meeting of new managers, and to sing organization's anthem together at Japan firms are the apparent examples for rituals (Terzi, 2000: 50). **Symbols:** They are the objects, activities, and events that have special meanings. In fact, rites, rituals, and stories are the symbols and symbolize extensive values of organization (Daft, 2000: 317). Company logos, flags, brand names, clothes, and physical structures can be cited as other examples for symbols. Symbols contain affective messages, which cannot be verbalized. Many basic norms and values of organization cannot be delivered among members without symbols. **Stories and myths:** they are the events about winning or losing, success or failure, occurred in the past of organization. Stories are important since they reflect the changes in organization and provide comprehensive information about relations among employees being in different positions in organization (Martin et al., 1983: 439). The most important stories in organizations are the ones, which are related with foundation of organization or founder of organization (Schermerhorn et al., 1997: 272). The story mentioning one employee's disallowance to IBM chairman Tom Watson to restricted area due to absence of his ID card and telling of this story as the success for applying company rules (Beyer & Trice, 1987: 6; Martin et al., 1983: 439) is a good example in terms of teaching organizational values. In fact, myths are not extraordinary events that cannot be accomplished by humans. The reason for their unusual perception is that they do not comply with the hierarchy in the existing structure as a behavior model. **Heroes:** They represent the basic values of the organization and are role models for cultural managers and employees (Daft, 2000: 317). Heroism is not a myth or legend. Stories about heroes are events that can be accomplished by all employees. For this reason, it is a good motivation tool for employees to develop their abilities and capacities. The institutor of the organization, important characters contributed to organizations, managers and employees from previous periods and today can be heroes.

Espoused Values: Hofstede (2001: 5) defines the concept of value as "the broad tendency to prefer certain states of affairs over others" and sees it as one of the building blocks of the social system and culture. On the organizational level, the value system is important for understanding the behavior of the members of the organization and defining culture. The studies of Hofstede (1980) on the national cultures and cultures of multinational companies in different countries, and the work of Pascale and Athos (1981) in Japanese and American companies provide important data in terms of showing the effect of the value system on behavior at the organizational level. For example, Pascale and Athos in their study stated that there is not much difference between the elements of the visible side of culture in the enterprises of both countries, but the difference is in beliefs, values, and assumptions depending on the country's cultures. Authors have observed that the same or similar behavior produces different meanings and consequences for employers in two countries. The espoused values guide organizational management and employees in decision-making to carry out organizational processes. Ways of thinking such as price leadership, perfect customer satisfaction, and "customer is our blessing" can be evaluated as examples of value. Today, with the effect of technological and social transformation, the change of the structure of the work and the workforce cause differentiation of motivation and control tools in enterprises. Now, internal control comes to the fore instead of external control. Values play a very important role in achieving this (McDonald & Gandz, 1992: 65). For this reason, today some organizations transmit these values, which constitute the basic understanding systems, to their employees with names such as company philosophy, company oath, and golden rules (Vural, 2003: 159). For example, the golden standards of Ritz Carlton Hotels, the quality, service, comfort, and value (QSCV) that constitutes Mcdonalds' vision, and the mission statement of Johnson & Johnson are important examples in this regard.

Basic Underlying Assumptions: The basic assumptions, as indisputable or injured behavior, determine how group members perceive, think, and feel. In reality, assumptions are learning based on adopted values. When we encounter a problem at the first stage, our values govern our behavior. Behavior becomes a value when it solves the problem and is accepted by the members of this group. It is possible for the behavior to become the basic assumption, if it gives the same successful result in the problems encountered in a certain subject. When this happens, when the members of the organization face the same problem, they show the same reaction as the reflex (Schein, 1984: 3). Thus, since assumptions are proven methods in the organizational problem-solving process, they leave the questioning area of the employees and settle in the subconscious in time (Sathe, 1983: 8). Schein (1992) states that the basic assumptions are based on the nature of reality and truth, the nature of time and space, and the nature of human nature, activity, and relationships.

The nature of reality and truth: Assumptions that define neither truth nor reality within society or group. It can be handled in three dimensions: physical, social, and individual reality. Physical facts are objective and can be revealed by scientific experiments. However, in western societies, it is believed that while the physical facts are explained by scientific methods, in other words, they are considered objective, while in eastern societies it can be explained by logical, philosophical, and metaphysical principles.

Nature of time: Time is an important determinant in the regulation of social life. In modern organizations, all activities such as starting work, quitting work, breaks, career development, and retirement are organized according to time. For example, when an employee comes to work late or waste his/her time, it is about making assumptions based on the nature of the time employees feel and react.

Nature of the space: Symbolic meanings of the assumptions about the field are strong. The area has both physical and social meanings. The physical arrangement of the floors where the top management offices are located and the attitudes and behaviors of the staff working here are related to the nature of the physical space. For example, in the headquarters of General Motors, the upper management offices are I-shaped at the end of the 14th floor, and the 14th floor is used when expressing the upper management offices in the common language among all employees (Martin & Siehl, 1983: 56).

The Nature of human: Human nature is perceived differently in different societies. For example, in eastern societies, people are inherently bad. In western societies, human nature is seen as neutral. Being good or bad arises in the life process. At the organizational level, the nature of the human structure is related to how employees and managers are seen, and it is the main determinant in the creation of human-related administrative tools and practices. For example, since human beings were seen as a rational economic creature in the early stages in western societies, monetary rewards were proposed as a motivational tool. The essence of Argyris, Maslow, and McGregor's work is nothing more than exploring how businesses can be managed or made more productive according to human nature.

The nature of human activities: It is about how people will behave depending on their relationship with their environment. Many factors affect individuals' behaviors in organizations. Undoubtedly, the biggest determinant among these factors is the employees themselves. For example, when a problem arises, the efforts of the employees to do something instead of avoiding the problem, to work with other people, to help, to adapt to the environment by developing the employees' own abilities and skills can be evaluated within this scope.

The nature of human relations: Based on each culture, individuals' relationships with each other are key determinants to make the group safer, more comfortable, and efficient. If the values are not shared by the majority of individuals who constitute the group, problems and anarchy arise. At this point, individuals being participatory and individualistic, and role relationships are the main determinant.

The cultural stages that Hofstede (1980) has revealed in his study on national cultures (individualism versus collectivism, power distance, masculinity versus femininity, uncertainty avoidance, and Confucian dynamism (long versus short-term orientation)) have demonstrated that some countries are individualistic (U.S., Canada, and UK), while some countries are collectivist (Japan, Singapore, Turkey, and the Philippines). Similarly, the clan culture in Ouchi's (1981) Theory Z approach has the effects of the samurai–peasant relationship from the Samurai period in Japan.

4 Service and Service Culture

4.1 Service and Service Characteristics

Service, generally defined as "a work or action simultaneously produced and consumed, a performance, social facts or efforts" (Uyguç, 1998: 8), is different from industrial products in terms of its structure (Wilderom, 1991: 7). Service has some differences due to its input, production, and output processes. **Input:** The dependency to customer is much more important in service production process. Customers' active participation to service production significantly heightens the uncertainty of input. **Production:** For the services, the production and consumption is simultaneously performed. In addition, direct interaction between customer and employee play an important role for the production process. **Output:** Service is intangible, compared to industrial goods and products. Industrial goods and products are the objects, but service is a performance, an action, an effort, or a case. Based on this, individuals never own the services; they can only experience them. The fact of service sticks in individuals' mind as an experience. Zeithaml et al. (1985) refers that there are four distinctive characteristics in the nature of service as intangibility, perishability, inseparability, and variability.

Intangibility: It refers inability to touch, test, and see of the services before buying. Services differ based on their intangibility characteristics. For instance, education, transportation, and sportive activity services are most intangible ones. However, these services fundamentally contain some tangible assets as well. Although some assets are tangible, service can only be perceived when it is delivered (Kurtz & Clow, 1998: 10). For example, in a restaurant, customers buy a meal prepared in the kitchen. However, buyers of the meal not only bought a meal, instead they also experienced service atmosphere, cheery service delivery, and qualified service process. Intangibility holds simultaneous production and consumption, variability of quality, and perishability characteristics of services (Kasper, 2002: 1051). Thus, companies try to make physical assets of service product more attractive to make it more tangible. In this context, they inform customers by easing mutual interaction between customer and employee and try to learn customers' expectation about the service. Additionally, they also try to ensure that customers are more knowledgeable about the service, using tools such as advertising, posters, and brochures (Kurtz & Clow, 1998: 11).

Perishability: Perishability refers to inability to stock or keep services as products. As mentioned before, production and consumption processes occur simultaneously in services. For example, if a hotel company has 100 rooms and 60 of them are occupied, the rest 40 rooms mean a loss for the company since they were not sold relevant night. Based on this fact, stocking of rooms and using them according to demand is not possible. An unsold service in a restaurant, an unsold seat in a plane or bus, a vacant seat in a soccer stadium, a vacancy quota of a private school, or vacant seats in auditorium can be other examples of perishability to services. However, this is different for industrial products.

Inseparability: It means simultaneous production and consumption of the services. Normally, for the industrial products, it is possible to produce advance and sell later on. However, this is not valid for the other services. Inseparability is essential for service enterprises due to emphasizing two critical points. First, it is a determiner of service quality. Quality, in terms of service, is the performance of service staff and experience perceived by customers (Siehl, 1992: 18) based on the interaction between customer and employee, in the moment of service delivery (Kurtz & Clow, 1998: 12). Second, it indicates that human factor is the critical for the success of service enterprises. With this structure, inseparability is a feature that guides senior management in creating their administrative applications.

Variability: It means that the services do not always provide the experience at the same level to customer. The variability feature of the service results in the fact that the service is not suitable for automation and standardization depending on the human factor playing a major role in production and causes the difficulty of quality control due to the standardization of the service. Today, some companies try to improve productivity and work efficiency by transforming service production process to standardized procedures and by using machines for production (Kurtz & Clow, 1998: 14). For example, in McDonald's, there are detailed job descriptions that describe what the employee will say, what information and requests will learn in order, as soon as the guest arrives at the cash box for the order (Morgan, 1998: 26). Contrary to this, some enterprises try to heighten the service quality level by personalizing the service production process (Kurtz & Clow, 1998: 14).

4.2 Service Culture

Culture, as the perception of employees about what management believes (Hallowell et al., 2002: 8), is more important for service enterprises than manufacturing firms due to the nature of the service. Because, standardization of service production is impossible and the service has a structure in which human element plays a crucial role at service encounter, and it is not possible to know the behavior of the customer beforehand (Grönroos, 1990: 243). Therefore, the structure and characteristics of the service play a key role in determining the cultural values and assumptions to be created in service businesses. Especially, the intangibility of the service, the distribution process and the contact with the customer in the distribution, the participation of the customer in the delivery of the service, and the fact that the employee is the basic input in production are the determining factors in the development of cultural values in the service enterprises. Accordingly, the culture to be created in service enterprises should have a structure that includes high values such as employee interaction, bilateral relations, and service quality (Kasper, 2002: 1048).

The production and consumption of services take place simultaneously, and the customer–employee interaction plays a key role in service production. The fact that this moment cannot be controlled administratively and instantaneously causes the employee to need a guiding tool about what kind of behavior he/she should behave. This tool is nothing but organizational culture. For this reason, employees in service enterprises are managed not by management but by the organizational culture (Hallowell et al., 2002: 7).

Research on rewarding and control systems has shown that performance cannot be clearly measured if it cannot be monitored directly. The intangibility and separability characteristics of services make it difficult to develop legal rewarding systems and to build meaningful performance criteria for measuring employee performance. This raises the need to use culture-oriented shared values (Siehl, 1992: 19), instead of legal systems in managing and controlling the employee in service businesses.

One of the key issues for customers to evaluate service organizations is the distribution process. This is because the emotional aspect of service quality perception is more dominant. This situation requires service employees to know about the quality and the responsibilities of the customer and service provider in order to provide perpetual and continuous service delivery process. The fact that the employee knows about the quality and the responsibilities of the customer and the service provider can be achieved by creating a group of cultural values that focus on these three elements.

The structure of the services requires service companies to create a customer and employee-oriented cultural structure. The decisive tool in communication for the customer in service organizations is the culture of the organization. A customer understands with culture what he expects from the organization. Culture shapes the value perceived by customers and helps establish their expectations.

Service is an emotional perception rather than a physical product and uncertainty is high. For this reason, the organizational culture guides to the employee how he/she should think during the meeting with the customer (Siehl, 1992: 18). In other words, the employee uses the culture of the organization as a reference for how should he behave in certain situations. In addition, an employee defines an organization with its cultural elements, and knows what the top management finds acceptable and unacceptable with the organizational culture.

The key point in the success of service organizations is the moment of contact (moment of truth) of the customer and the employee. For this reason, an

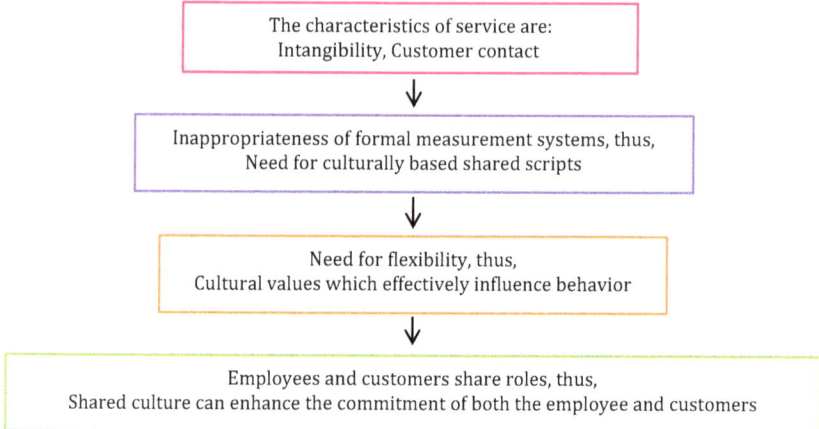

Fig. 4-1: The Relationship between Characteristics of Service and Culture
Source: Siehl, 1992: 20.

organization that wants to be successful has to adopt and develop a customer-oriented service approach. A service organization that wants to become employee and customer oriented is for this (Holden, 1997: 678): it should develop a customer-oriented culture, raise dedicated workforce, should lead with examples, know their customers and the work they do, learn from competitors, and know how to make a profit; it should be different.

Considering the intangibility of the service and the contact with the customer as the main determinant of the culture to be created in service organizations, Siehl (1992) tries to explain the relationship between service and culture as follows (Fig. 4-1). Depending on the characteristics of the service, the legal measurement systems are not suitable for the control and evaluation of the employee, and cultural values compatible with the characteristics of the service are required to control the employee in terms of management. Due to the fact that the employee plays a key role in the production of the service, each service producer acts like a small factory and the employee must have the flexibility to move by the reason of uncertainty of customer demands at the time of consumption. Flexibility cannot be regulated by policies and rules; the legal structure and written rules create stagnation and rigidity, limiting the sphere of activity of the employee. In addition, since customers and employees share roles in service encounter, they must be acquired, educated, socialized, and rewarded as employees do.

Siehl (1992) states that four principles are needed to cultural values designed based on service characteristics that are sharable and explain them as follows: *First principle:* All staff of the service company must be aware that high-quality service is the key source for successful competition, and this belief must be shared by all members of the organization. If employees do not well understand why high-quality service makes positive differences, the service culture will not be created throughout the organization. *Second principle:* Service companies and service staff must be managed as a whole. Thus, it can be possible for the customers to make appropriate assessments and have satisfaction form the service. To control everything is very difficult without shared values/beliefs due to intangibility of the service and active participation of customers to service production. Because attitudes and behaviors of employees can only be affected with the culture. *Third principle:* Service enterprises must manage their customers like their staff. By doing so, participative customers become integrated to company. Additionally, to create loyal customers can only be possible with customer-oriented service concept. *Fourth principle:* Service enterprises must manage their staff like a customer. Researches on this issue reveal that employees managed with the quality deliver high-quality services. This also provides empathy to employees.

Conceptual information about how the service enterprises should have structure in terms of culture and four principle determined by Siehl (1992) emphasize the critical areas for service leaders as designing organization system, achieving managerial responsibilities (vision, mission, and strategies) and allocating cultural values (values, beliefs, and assumptions). Service leaders must consider these critical areas to be able to be successful and to determine the structure of service culture.

Conclusion

Culture, related to making meaningful common behaviors and ideas of a group or human community that was leagued together regarding certain rudiments, can be defined as a personality of a company in an organizational context. Organizational culture also defines as the environment in which employees work. In tourism sector and especially in accommodation enterprises, culture becomes more important due to distinctive characteristics of the services. As it is well known, accommodation enterprises operate at a complex and global scale. Thus, they serve in different cultures, and they have to employ workers who have different cultural backgrounds. This cultural diversity makes it necessary to create an idiosyncratic culture to keep and develop service quality in hotel enterprises. Hotel enterprises are also composed of different departments performing different tasks in terms of organizational structure, including the front office, housekeeping, accounting, maintenance, sales, human resources, and food and beverage. Each of these areas has its own individual daily tasks, but all share common values and expectations.

Hotel enterprises operate in an environment characterized with intensive competition. In hotels, organizational structure and physical amenities, which are determiner of service quality, can be easily imitated. On the other hand, culture is the most difficult factor to imitate in hospitality industry. Thus, culture constitutes the core competency of hospitality enterprises. As emphasized in Michelli's research (2009), investigating the culture of Ritz-Carlton hotels, certain organizational cultures that have some assets (organizational culture elements that are well defined and constructed on service, selection of right employee, appropriate training, empowerment, learning by benchmarking, and using lessons learned for further services), notion, managerial strategy, and philosophy can be a guide for other hotel enterprises for creating an organizational culture. In conclusion, as Thomas Kempis stated, "The higher the building, the deeper the foundation must be laid" (Michelli, 2009: 33).

Bibliography

Alvesson, M. (1987), "Organizations, Culture and Ideology", *International Studies of Management & Organization*, 17 (3), pp. 4-18.

Alvesson, M. (2002), *Understanding Organizational Culture*, Sage.

Beyer, J. M. and Trice, H. M. (1987), "How an Organization's Rites Reveal Its Culture", *Organizational Dynamics*, Spring, 15, pp. 5-24.

Burwash, P. (1997), *Liderliğin Anahtarı, Farklı Hizmet Sektörünün Yeniden Keşfedilen İlkeleri*, Çev.: H. Aydın, İstanbul: Timaş Yayınları.

Daft, R. L. (2000), *Organization Theory and Design* (7th Ed.), South-Western Publishing.

Dönmezer, S. (1982), *Sosyoloji* (8. Baskı), Savaş Yayınları.

Erdem, F. (1996), *İşletme Kültürü*, Friedrich-Naumann Vakfı ve Akdeniz Üniversitesi.

Erdoğan, İ. (1983), *İşletmelerde Davranış*, İstanbul Üniversitesi Yayın No: 3077, EvrimOfset.

Eren, E. (1997), *İşletmelerde Stratejik Yönetim ve İşletme Politikası*, Der Yayınları.

Grönroos, C. (1990), *Service Management and Marketing: Managing the Moment of Truth in Service Competition*, Lexington Books.

Güvenç, B. (1985), *Kültür Konusu ve Sorunlarımız* (2. Baskı), Remzi Kitabevi.

Güvenç, B. (1996), *İnsan ve Kültür* (7. Baskı), İstanbul: Remzi Kitabevi.

Hallowell, R., Bowen, D., and Knoop C. I. (2002), "Four Season Goes to Paris", *Academy of Management Executive*, 16 (4), pp. 7–23.

Hatch, M. J. (1993), "The Dynamics of Organizational Culture", *Academy of Management Review*, 18 (4), pp. 657–693.

Hodgetts, R. M. and Luthans, F. (2000), *International Management, Culture, Strategy and Behavior* (4th Ed.), McGraw-Hill.

Hofstede, G. (1980), *Culture's Consequences: International Differences in Work-Related Values*, Beverly Hills, CA: Sage.

Hofstede, G. (1981), "Culture and Organizations", *International Studies of Management & Organization*, 10 (4), pp. 15–41.

Hofstede, G. (1998), "Attitudes, Values and Organizational Culture: Disentangling The Concepts", *Organization Studies*, 19 (3), pp. 477–492.

Hofstede, G. (2001), *Culture's Consequences: Comparing Values, Behaviors, Institutions, and Organizations Across Nations* (2nd Ed.), Sage.

Hofstede, G., Hofstede, G. J., and Minkov, M. (2010), *Culture and Organizations: Software of the Mind*, McGraw-Hill.

Holden, P.A. (1997), "Success through Service", *Management Decision*, 35 (9), pp. 677–681.

Huertas-Valdivia, I., Gallego-Burín A. R., and Lloréns-Montes, F. J. (2019), "Effects of Different Leadership Styles on Hospitality Workers", *Tourism Management*, 71, pp. 402–420.

Kasper, H. (2002), "Culture and Leadership in Market-Oriented Service Organisations", *European Journal of Marketing*, 36 (9/10), pp. 1047–1057.

Kilmann, R. H., Saxton, M. J., and Serpa, R. (1986), "Issues in Understanding and Changing Culture", *California Management Review*, 28 (2), pp. 87–94.

Kotter, J. P. and Heskett, J. L. (1992), *Corporate Culture and Performance*, The Free Press.

Kurtz, D. L. and Clow, K. E. (1998), *Service Marketing*, John Wiley & Sons.

Lorsch, J. W. (1986), "Managing Culture: The Invisible Barrier to Strategic Change", *California Management Review*, 28 (2), pp. 95–109.

Martin, J. and Siehl, C. (1983), "Organizational Culture and Counterculture: An Uneasy Symbiosis", *Organizational Dynamics*, 12 (2), pp. 52-64.

Martin, J., Feldman, M. S., Hatch, M. J., and Sitkin, S. B. (1983), "The Uniqueness Paradox in Organizational Stories", *Administrative Science Quarterly*, 28, pp. 438-453.

McDonald, P. and Gandz, J. (1992), "Getting Value from Shared Values", *Organizational Dynamics*, 20 (3), pp. 64-77.

Michelli, J. A. (2009), *Yeni Altın Standardı*, (Çev: İlker Gülfidan), İstanbul: Optimist Yayım Dağıtım.

Morey N. C. and Luthans, F. (1985), "Refining the Displacement of Culture and the Use of Scenes and Themes in Organizational Studies", *The Academy of Management Review*, 10 (2), pp. 219-229.

Morgan, G. (1998), *Yönetim ve Örgüt Teorilerinde Metafor*, Çev.: G. Bulut, Mess Yayın No: 280.

Online Etymology Dictionary (2020), "Origin and Meaning of Culture", https://www.etymonline.com/word/culture, (9.10.2020).

Ouchi, W. (1981), *Theory Z: How American Business Can Meet the Japanese Challenge*, Avion Books.

Özkara, B. (1999), *Evrimci ve Devrimci Örgütsel Değişim*, Afyon Kocatepe Üniversitesi.

Pascale, R. T. and Athos, A. G. (1981), *The Art of Japanese Management Application for American Executives*, Simon & Schuster.

Recardo, R. and Jolly, J. (1997), "Organizational Culture and Teams", *SAM Advanced Management Journal*, 62 (2), pp. 4-7.

Robbins, S. P. (1987), *Organization Theory: Structure, Design, and Applications* (2nd Ed.), Prentice Hall.

Sathe, V. (1983), "Implications of Corporate Culture: A Manager's Guide to Action", *Organizational Dynamics*, 12 (2), pp. 5-23.

Schein, E. H. (1984), "Coming to New Awareness of Organizational Culture", *Sloan Management Review*, 25 (2), pp. 3-16.

Schein, E. H. (1992), *Organizational Culture and Leadership* (2nd Ed.), Jossey Bass.

Schermerhorn Jr, J. R., Hunt, J. G., and Osborn, R. N. (1997), *Organizational Behavior* (6th Ed.), John Wiley & Sons.

Siehl, C. (1992), "Cultural Leadership in Service Organizations", *International Journal of Service Industry Management*, 3 (2), pp. 17-22.

Singh, J. and Hofstede, G. (1990), "Managerial Culture and Work-Related Values in India", *Organization Studies*, 11 (1), pp. 75-101.

Smircich, L. (1983), "Concepts of Culture and Organizational Analysis", *Administrative Science Quarterly*, 28, pp. 339-358.

Şişman, A. (2002), *Örgütler ve Kültürler*, Pegem A Yayıncılık.

Terzi, A. R. (2000), *Örgüt Kültürü*, Nobel Yayınları.

Uyguç, N. (1998), *Hizmet Sektöründe Kalite Yönetimi Stratejik Bir Yaklaşım*, Dokuz Eylül Yayınları.

Vanderberghe, C. (1999), "Organizational Culture, Person-Culture Fit, and Turnover: A Replication in The Health Care Industry", *Journal of Organizational Behavior*, 20, pp. 175-184.

Vural, Z. B. A. (2003), *Kurum Kültürü ve Örgütsel İletişim*, İletişim Yayınları.

Wilderom, C. P. M. (1991), "Service Management/Leadership: Different from Management/Leadership in Industrial Organizations?", *International Journal of Service Industry Management*, 2 (1), pp. 6-14.

Zeithaml, V. A., Parasuraman, A., and Berry, L.L. (1985), "Problems and Strategies in Service Marketing", *Journal of Marketing*, 49, pp. 33-46.

Çağdaş ERTAŞ

Culture of Tourism Employees: A Research in Southeastern Anatolia, Turkey

1 Introduction

The main target for businesses operating in the service industry is consumers to display positive attitudes toward the business. Employees shape consumers' attitudes in service businesses such as tourism (Bitner et al., 1990; Heskett et al., 1994; Hartline & Ferrell, 1996; Testa et al., 1998; Dawson et al., 2011; Naseem et al., 2011). For this reason, it can be said that one of the most important factors in the success of tourism enterprises is the employees of the business (Tepeci & Bartlett, 2002). This importance of the employees reveals the importance of their attitudes and behaviors. It is stated that the positive attitudes and behaviors of the employees provide a competitive advantage for enterprises (Lucia-Casademunt et al., 2015). However, the positive attitude and behavior of the employees depend on their working under good conditions within the organization. Because the employee will provide a good service to the customer as long as he is satisfied in the organization (Pienaar & Willemse, 2008). This reveals that the organizational behavior of the employees should be positive (Pienaar & Willemse, 2008; Dick & Metcalfe, 2001). Therefore, the positive organizational behavior of employees emerges as an important factor (Avey et al., 2008). Many factors can be mentioned that affect the organizational behavior of employees. However, perhaps the most important of these factors is the culture of the employees (Magnini et al., 2013). Culture affects individuals' lives, attitudes, beliefs, and especially the working lives of employees (Probst & Lawler, 2006). Researches carried out on tourism industry (Devine et al., 2007; Jones & McCleary, 2007; Sledge et al., 2008; Fisher et al., 2010; Akdeniz & Aytemiz Seymen, 2012; Çalışkan & Özkoç, 2016; Yüksel & Bolat, 2016) showed that the culture of employees influences their organizational behavior. Therefore, tourism enterprises should analyze the culture of their employees well (Türker & Karadağ, 2019).

Low wages, job insecurity, long working hours, limited career opportunities, and seasonality characterize the tourism industry (Dawson et al., 2011). These negativities can lead to undesirable consequences for tourism businesses. For example, an employee who performs a bad service against customers can cause the business to lose customers. Considering organizational behavior, an employee dissatisfied with his job is likely to quit the job (Fernandes et al., 2018). On the other hand, the mentioned negativities in the industry are obstacles in hiring and retaining qualified employees (Fernandes et al., 2018). Considering that a significant part of the jobs in the tourism industry requires teamwork, the negative behavior of the employee can disrupt the motivation of the whole team. Therefore, foreseeing these behaviors and taking precautions in this direction before developing the behavior of the employees is an essential issue for the managers of tourism enterprises. It is possible to predict what kind of behaviors employees can do in the business by knowing their culture (Hofstede, 1980). Culture is the dominant values in a society, and no employee is independent of the culture of the society in which he is a member (Hofstede et al., 2010).

The tourism industry, especially the accommodation industry, operates in a multicultural environment (Sledge et al., 2008). In this respect, it can be said that the tourism industry has a unique culture (Jones, 1999). The fact that tourism businesses operate in a multicultural environment also paves the way for employing people from different cultures (Devine et al., 2007; Sledge et al., 2008). Attitudes and behaviors of employees with different cultures will also be different (Hui et al., 1995; Kirkman & Shapiro, 2001; Lam et al., 2002; Huang & Vliert, 2004; Erdogan & Liden, 2006; Fisher et al., 2010). Different behaviors can make it difficult to manage employees effectively. For this reason, in sectors with high cultural diversity such as tourism, managers have to know what kind of effects culture has (Nazarian et al., 2017). It is important for tourism business managers to know the culture of their employees, especially in terms of managing them (Chen et al., 2012). While the impact of culture on employees' behavior is so obvious, it will be one of the biggest mistakes of today's managers if they ignore the culture of their employees (Taras et al., 2011).

Although there are some researches in the sample of tourism employees (Sledge et al., 2008; Akdeniz & Aytemiz Seymen, 2012; Yüksel & Bolat, 2016; Türker & Karadağ, 2019), it is noteworthy that tourism researchers have not given enough importance to culture factor. This chapter was written both to contribute to the tourism management literature and to draw attention to the importance of the culture in tourism management. In the chapter, firstly, some conceptual and theoretical information is given about the concept of culture and cultural characteristics of tourism employees. Then, empirical research and results carried out in Turkey were shared with the aim of learning about the culture of tourism employees.

2 Concept of Culture and Culture of Tourism Employees

Every society has its own culture. Therefore, there are cultural differences among societies. The theories of cultural differences include Herskovits' five-dimensional model, Ronen-Kraut's smallest space analysis, Ronen-Shenkar's clustering countries, Hall's priority message system, and Hofstede's national culture theory. Although criticized (McSweeney, 2002; McCoy et al., 2005), Hofstede's dimensions of culture are considered to be the most important and most comprehensive study conducted in this field (Fullagar et al., 2003; Lucia-Casademunt et al., 2015). In recent studies on the hospitality industry, Hofstede's cultural dimensions have been taken into consideration (Reisinger & Crotts, 2010). This chapter is also based on the cultural dimensions defined by Hofstede. Hofstede (1980) defined culture as collective mental programming separating a group of people from others. Even though the mental programming in question starts in the family, the future work life is also effective in mental programming. Therefore, the culture of the individual is shaped by the interaction with the social environment (Hofstede et al., 2010).

Each individual has a unique family, environment, work, or social life. This uniqueness is manifested in the mental programming of the individual. Therefore, the mental programming of individuals who grow up in different environments leads to unique cultures for each individual. These differences together make up the basic culture of societies (Hofstede et al., 2010). So, what constitutes the essence of both mental programming and the culture that occurs as a result of mental programming? According to Hofstede et al. (2010), the essence of culture consists of values. Values can be defined as general tendencies to prefer certain relationship states to others (Reimann et al., 2008). The values to which individuals attach importance are considered the culture of the society to which they belong (Meglino & Ravlin, 1998). Therefore, knowing the values of individuals gives information about the fundamental cultural characteristics of the society (Trompenaars & Hampden-Turner, 2011). For this reason, it will be more accurate for research on culture to focus on values. The most comprehensive study that examined the cultural dimensions based on values was made by Hofstede (Posthuma, 2009; Gunkel et al., 2015; Hauff et al., 2015).

Hofstede proposed five dimensions of cultures. "Power distance," "Individualism vs. collectivism," "Masculinity vs. femininity," and "Uncertainty avoidance" dimensions were identified by Hofstede (1980) when studying the cultural dimensions of the countries and in the sample of IBM (International Business Machines) employees in 76 countries, while "long-term orientation vs. short-term orientation" dimension was determined by

Hofstede and Bond (1988) in their study in China. In every society, some individuals are strong, and some are weak. Those who hold power can think that they have every right. These individuals can sometimes even be unjust. While some societies accept such injustices, some can react to them. This is explained by the concept of power distance. Power distance is defined as the extent to which individuals within a country accept that power is distributed unequally (Hofstede et al., 2010). In working life, power distance is concerned with the reaction of subordinates to the orders of their superiors. Employees with high power distances believe that their superiors are stronger than them, do not consider themselves equivalent to them, and fulfill their orders without question. Also, they do not want their managers to consult themselves before making a decision (Hofstede et al., 2010). Low-power distance employees, on the other hand, think that managers are not people to obey and want their managers to consult them before making a decision (Rodrigues, 1998). However, considering the structure of the tourism industry, it is thought that power distance level of the employees may be lower than that of society. Relationships between subordinates and superiors in the industry are more intense (Mia & Patiar, 2002), which may reduce power distances. What effects the power distance level has on employees are summarized in Tab. 5-1 (Hofstede et al., 2010).

Perhaps the most studied dimension of culture is collectivism (Probst & Lawler, 2006). Collectivism can be defined as "the emphasis on collective rather than individual action or identity." In other words, the concept of essence underlies collectivism (Hofstede, 2001). Employees with a collectivist culture have a high level of responsibility toward the group they feel they belong to (Hofstede, 1980). The interests of the group that collectivist individuals belong to precede their interests (Oyserman et al., 2002). Also, employees with a collectivist culture want to establish sincere relationships with themselves and with their managers (Newman & Nollen, 1996). Tourism industry requires teamwork (Polat Üzümcü, 2015). Teamwork, after a while, brings the team members closer and can create the feeling of "we." Therefore, team members prioritize their teams, and consequently, the culture of tourism employees becomes a collectivist culture. Tab. 5-2 bireyci ve çoğulcu çalışanların davranışlarını özet olarak karşılamaktadır (Hofstede et al., 2010).

It is possible to define the concept of masculinity as "a clear differentiation of emotional sex roles." It can be said that this cultural dimension is related to society's perspective of the social and work life (e.g., working and being promoted) of men and women. High income, appreciation by superiors, promotion, and individual success are very important for employees with masculine

Tab. 5-1: Power Distance Level and Working Life

Low Power Distance	High Power Distance
The hierarchy is seen only as inequality of roles.	The hierarchy is seen as the inequality that should be between the highest ranks and the lowest ranks of the business.
It is asked to have a small number of supervisory staff in businesses.	It is asked to have a large number of supervisory staff in businesses.
It is believed that the salary difference between the lowest and the highest levels of the business should be very small.	The wide salary difference between the bottom and the top levels of the business is met normally.
Subordinates expect managers to consult with them on decisions.	Subordinates do not want to participate in decisions, but want instructions.
The ideal boss is considered to be a democrat who can find a cure for everything.	The ideal boss is considered to be a benevolent autocrat or a good father.
Employees do not welcome status symbols and privileges.	Employees see status symbols and privileges normally, strive for status, and want to be privileged.
Manual work and a deskwork (office work) have the same status for employees.	Employees value white-collar jobs more than blue-collar jobs.

Tab. 5-2: Collectivism/Individualism and Working Life

Individualism	Collectivism
Employees work for the personal interests of their employers.	Employees work for the interests of the group to which they belong.
It is considered that hiring and promotion decisions should be made only according to the rules and skills.	It is considered that hiring and promotion decisions should be made by considering the group to which the employee belongs.
It is considered that management is management of individuals.	It is considered that management is management of groups.
Employees think the task is more important than relationships.	Employees think that relationships are more important than duty.

culture (Newman & Nollen, 1996). These employees do not care about good relationships with management and colleagues. Besides, masculine cultures attach importance to working independently from other employees (Hauff et al., 2015). Considering that the tourism industry requires teamwork, having a masculine culture in the industry may have undesirable consequences for

Tab. 5-3: Masculinity/Feminity and Working Life

Masculinity	Feminity
It is thought that managers should be decisive and aggressive.	It is thought that managers should be predictive and accommodating.
Conflicts between employees are resolved by ensuring that the strongest win.	Conflicts between employees are resolved through compromise.
Businesses are asked to be large.	Businesses are asked to be small.
Working for employees is the goal in life.	Working for employees is a tool for living.
Employees prefer to earn more money, rather than more free time.	Employees prefer to have more free time rather than more money.
Employees think that making a career is mandatory for men, and optional for women.	Employees think that making a career is optional for both women and men.
Executive women are few.	Executive women are many.
Humanization of work by job content enrichment.	Humanization of work by contact and cooperation.

the success of the business. Another feature of employees with masculine culture is that they do not want women to work (Schippers, 2007). The tourism industry is noteworthy with the high proportion of female employees. This high proportion of female employment in the industry may be a sign that tourism employees do not tend to show features of masculinity. How masculinity and femininity are reflected in business life was summarized in Tab. 5-3 (Hofstede et al., 2010).

The dimension of uncertainty avoidance refers to a society's level of taking measures in the face of future uncertainty (Harvey, 1997). Job security is important for employees with high levels of uncertainty avoidance (Hofstede, 1984). These employees are more unhappy than those with low levels of uncertainty avoidance (Hofstede et al., 2010). Because the level of avoiding uncertainty increases, stress will increase (Hofstede, 1980). Employees with anxiety about dismissal and uncertainty regarding their working conditions are likely to have an intention to quit if they perceive alternative employment opportunities. Those who work in businesses that operate periodically, such as the tourism industry, then develop the opposite behavior (Bozkurt & Demirel, 2019). When the perception of job insecurity in the tourism employees increases, they can increase their performance in their work in order not to lose their current job instead of leaving (Staufenbiel & König, 2010). This situation is likely to increase the uncertainty level of the employees. The anxiety of employees about uncertainty will result in stress, dissatisfaction, and insecurity in working life (Gunkel et al., 2015). Long-term orientation, which is associated with uncertainty avoidance and is Hofstede's last cultural dimension, describes "whether an individual is focused on the present or the future" (Ayoun & Moreo, 2008). In the tourism industry, employees are usually young individuals who do not bear the responsibility of the family. The lack of such responsibility leads to a high turnover rate. One of the basic requirements of avoiding uncertainty is the desire for long-term employment (Hofstede et al., 2010). However, the high turnover rate in the tourism industry (Akova et al., 2015) might be since the employees are not future-oriented or long-term oriented. This, in turn, decreases the ratio of employees avoiding uncertainty. However, the crisis experienced by the Turkish tourism industry in recent years has caused many hotels to be closed or to dismiss a significant number of their employees. This situation causes tourism employees to feel uncertain about their future, and this feeling causes them to consider the future rather than today. Finally, time leftover from working life is important for long-term oriented employees (Hofstede et al., 2010). Considering the long working hours in the tourism industry (Dawson et al., 2011), the fact that employees have a long-term oriented culture may cause them to be unhappy at work in the long run. Tab. 5-4 summarizes how uncertainty avoidance level and long- /short-term orientation reflect on working life (Hofstede et al., 2010).

Hofstede defined Turkish culture as having high power distance, feminine, collectivist, avoiding uncertainty (Hofstede, 1980), and long-term oriented (Hofstede et al., 2010). However, as it can be understood from the findings of a limited number of studies (Akdeniz & Aytemiz Seymen, 2012; Çalışkan & Özkoç, 2016; Türker & Karadağ, 2019) on the tourism industry, especially some cultural dimensions of the tourism employees can differ from the cultural dimensions that Hofstede defined for Turkey. For example, in the research conducted by the author of this chapter (see Ertaş, 2017), it was determined that the employees in 5-star hotels operating in İstanbul, İzmir, Ankara, and Antalya have a low power distance. On the other hand, Akdeniz and Aytemiz Seymen (2012) found that those working in hotels in Istanbul have a masculine culture rather than a feminine culture. Türker and Karadağ (2019) found that tourism employees have a medium level long-term orientation rather than the high long-term orientation that Hofstede has specified for Turkish culture. They stated that the levels of uncertainty avoidance of tourism employees differ from region to region, although they are in the same country. The researchers found that hotel employees in Trabzon have moderate uncertainties, but hotel employees in Şanlıurfa avoid

Tab. 5-4: Uncertainty Avoidance Level – Short- /Long-Term Orientation and Working Life

Low Uncertainty Avoidance	High Uncertainty Avoidance
Individuals often change jobs.	The frequency of changing jobs is low.
It is easier to establish a work-life balance.	It is more difficult to establish a work-life balance.
Individuals work hard only when they need it.	Individuals have an emotional need to engage. There is an inner urge to work hard.
Time is a framework for orientation.	Time is money.
Uncertainty and chaos are tolerated.	Official rules are needed.
Individuals act relatively independently of the rules.	Individuals act under existing rules.
Success motivates employees.	Job security motivates employees.
It is thought that there is not much need for the rules.	Even if it does not work in practice, there is an emotional need for rules.
Short-term Orientation	**Long-term Orientation**
The core business values for employees are freedom, rights, success, and thinking for oneself.	The core business values for employees are learning, honesty, compliance, responsibility, and self-discipline.
Leisure time is important for employees.	Leisure time is not important for employees.
Meritocracy prevails (management is in the hands of a certain audience); it is wanted that rewards are given according to abilities.	Employees do not want broad social and economic differences.
Employee loyalty varies according to business needs.	Employee loyalty is high.
Employees prioritize abstract rationality.	Employees prioritize common sense.

high levels of uncertainty. According to the researchers, terrorist incidents in the Southeastern Anatolia Region, socio-economic underdevelopment of the region, and limited job opportunities increase the uncertainty level of the employees (Türker & Karadağ, 2019). There are also researches that the cultural dimensions of Hofstede will not be valid for tourism employees in other countries. For example, according to Hofstede et al. (2010), the United Kingdom has a high level of masculinity. However, Nazarian et al. (2017) found that hotel employees in London have low masculinity. These findings indicate that the unique culture of the tourism industry is reflected in the tourism employees.

3 Research Method

The population of this research consists of employees who have been working in the tourism business certified hotels in Şanlıurfa, Mardin, and Diyarbakır. The main reason for taking hotel employees as the population is that one of the most intensively employed areas in the tourism industry is hotels. Also, the fact that the accommodation establishments are the heart of the tourism industry (Rogersen & Kotze, 2011; Fernandes et al., 2018) has been effective in the selection of hotels as the application area. The survey was used as a data collection tool in the research. The questionnaire consists of two parts. In the first part, there were five questions to learn about employees' gender, marital status, educational background, age, and monthly personal income. In the second part of the questionnaire, the scale for learning the culture of the employees was included. The "Cultural Values Scale" developed by Yoo and Donthu (2002) was used to measure the culture of the employees. The scale consists of five dimensions (collectivism, power distance, masculinity, uncertainty avoidance, and long-term orientation) and 26 items. Scale items were rated according to Likert scale (from 1: Strongly Disagree to 5: Strongly Agree).

Research data were collected between August 1 and December 10, 2019, via face-to-face interviews by the author of this chapter. A purposeful sampling method was used in data collection. The main reason for this method to be preferred is to reach only people entering the research population. There are a total of 54 tourism business certified hotels in Şanlıurfa, Mardin, and Diyarbakır (General Directorate of Investments and Enterprises, Republic of Turkey Ministry of Culture and Tourism, 2019). There is no information about the total number of employees in these hotels. For this reason, the population of the research could not be determined exactly. As a result of the data collection process, 420 usable questionnaires were obtained. 110 of the questionnaires were obtained from Mardin, 145 are from Diyarbakır, and 165 are from Şanlıurfa. The database of the research consisted of 420 participants. According to Child (2006), if the sample size is at least five times the number of items, reliable results will be obtained. Also, according to Comrey and Lee (2016), 100 weak, 200 medium, 300 good, 500 very good, and 1000 perfect for sample size. Therefore, it can be said that the sample used in this research is of sufficient size.

Data analysis was done through SPSS 23.0 statistics package program. Frequency and percentage distributions were used to determine the demographic characteristics of the participants. Cronbach's Alpha coefficient was taken as a basis to test the reliability of the scale used in the research, and the explanatory factor analysis was used to test the construct validity of the scale.

4 Results

Demographic characteristics of participants are as follows: 44.5 % are female, 55.5 % are male; 38.5 % are married and 61.5 % are single. 58.5 % have a monthly income of minimum wage (2020 TL) or below, 15.0 % have a monthly income of between minimum wage and 3000 TL; 10.5 % have a monthly income of between 3001 TL and 4000 TL, 9.0 % have a monthly income of between 4001 TL and 5000 TL, and 7.0 % have a monthly income of 5001 TL or above. 4.5 % are primary school graduates, 19.0 % are secondary school graduates, 39.0 % are high school graduates, 19.0 % have an associate degree, 15.0 % have a bachelor's degree, and 3.5 % are graduates. Also, the age of the participants is in the 18–50 range, with a mean of 28 years.

Cronbach's Alpha overall reliability coefficient of the culture scale was determined to be .895. According to the social sciences literature, an alpha reliability coefficient greater than .70 is sufficient for the reliability of the scales (Baum & Wally, 2003). It can be said that the culture scale used in this research is highly reliable (Hair et al., 2010). If the corrected item-total correlation value of an item is below .250, that item should be removed from the scale because it impairs the summability feature of the scale. In this context, corrected item-total correlations of the culture scale were also examined. Corrected item-total correlation values were found to be between .287 and .580. It was determined that the reliability coefficient of the scale decreased when items were removed. This value indicates that each item in the scale contributes to the scale at acceptable limits (Hair et al., 2010). For this reason, no item was removed from the scale.

In line with the main purpose of the research, explanatory factor analysis was applied to the data set of the culture scale. The following conditions were taken into account when applying the explanatory factor analysis: at least 50 % of the total variance should be explained (Tinsley & Tinsley, 1987), factor loadings should be at least .500, communalities of the items should be at least .500 (Hair et al., 2010), the difference of factor loadings of the items loading on two factors should be at least .100, the principal component analysis should be used (Osborne & Costello, 2004), and Varimax should be preferred (Kaiser, 1958). In the factor analysis (Tab. 5-5), the Bartlett Sphericity test performed on the data of culture scale was found to be significant (X^2: 3599,051 s.d .: 253 p <0.001), which indicates that the data set has a multivariate normal distribution. Also, the result of the Kaiser-Meyer-Olkin (KMO) Test for Sampling Adequacy (.866) shows that the sample size is sufficient for factor analysis (Hair et al., 2010). As a result of the first-factor analysis of the data set of the culture

Tab. 5-5: Culture of Hotel Employees

Factors and Items	Communalities	Factor Loading	Explained Variance	Mean	Alpha
Collectivism			21,831 %	3,420	0,950
Individuals should sacrifice self-interest for the group that they belong to.	0,866	0,919			
Individuals should stick with the group even through difficulties.	0,872	0,918			
Group welfare is more important than individual rewards.	0,839	0,888			
Individuals should pursue their goals after considering the welfare of the group.	0,810	0,867			
Group success is more important than individual success.	0,791	0,828			
Group loyalty should be encouraged even if individual goals suffer.	0,716	0,818			
Power Distance			17,398 %	2,764	0,927
People in higher positions should avoid social interaction with people in lower positions.	0,829	0,882			
People in higher positions should not delegate important tasks to people in lower positions	0,811	0,868			
People in lower positions should not disagree with decisions by people in higher positions.	0,757	0,842			
People in higher positions should not ask the opinions of people in lower positions too frequently.	0,779	0,822			
People in higher positions should make most decisions without consulting people in lower positions.	0,727	0,762			

(continued on next page)

Tab. 5-5: Continued

Factors and Items	Communalities	Factor Loading	Explained Variance	Mean	Alpha
Masculinity			13,595 %	2,941	0,902
Solving difficult problems usually requires an active forcible approach, which is typical of men.	0,830	0,852			
Men usually solve problems with logical analysis; women usually solve problems with intuition.	0,789	0,848			
It is more important for men to have a professional career than it is for women.	0,785	0,844			
There are some jobs that a man can always do better than a woman.	0,714	0,746			
Uncertainty Avoidance			10,872 %	3,626	0,791
Instructions for operations are important.	0,653	0,749			
It is important to closely follow instructions and procedures.	0,665	0,709			
It is important to have instructions spelled out in detail so that I always know what I'm expected to do.	0,695	0,685			
Standardized work procedures are helpful.	0,539	0,685			
Long-term Orientation			10,610 %	3,739	0,774
Working hard for success in the future	0,660	0,764			
Personal steadiness and stability	0,741	0,737			
Long-term planning	0,579	0,731			
Going on resolutely in spite of opposition (Persistence)	0,645	0,728			

Principal Components Analysis with Varimax Rotation - Total variance explained: 74.306 %; KMO Sample Adequacy: 0.866 - Bartlett Sphericity Test: X^2: 3599.051 s.d .: 253 p <0.001; Overall Average: 3.289 – Alpha for the Whole Scale: 0.892

Response categories: (1) Strongly Disagree (2) Disagree (3) Undecided (4) Agree (5) Strongly Agree scale, the communalities of the 9th item (*Rules and regulations are important because they inform me of what is expected of me*) were found to be .263, the 21st item (*Careful management of money*) .468, and the 25th item *(Giving up today's fun for success in the future)* .468. Therefore, these items were excluded from the analysis. As a result of the factor analysis applied to the remaining 23 items, all preliminary assumptions were met, and thus, five factors emerged.

According to factor analysis, the contribution rate of the five factors to total variance is 74.306 %. In other words, the scale used in this research explains about 75.00 % of the culture of the hotel employees considered within the scope of this study. Examination of the factors in terms of their contribution to total variance will show that collectivism contributes by 21.831 %, power distance 17.398 %, masculinity 13.595 %, uncertainty avoidance 10.872 %, and long-term orientation 10.610 %. It can also be inferred that the highest average was obtained in long-term orientation (3.739), and uncertainty avoidance (3.626), while the lowest average was obtained in power distance (2.764).

Conclusion and Discussion

The abstract feature of the service increases the importance of employees in tourism businesses. Because, the attitudes and behaviors of employees in tourism businesses are an important determinant of consumers' positive perceptions. However, employees must have positive organizational behaviors to create positive perceptions of consumers (Heskett et al., 1994). One of the most important factors that determine the organizational behavior of employees is culture. Research conducted in the sample of tourism employees (Devine et al., 2007; Jones & McCleary, 2007; Sledge et al., 2008; Fisher et al., 2010; Akdeniz Ay, 2015; Çalışkan & Özkoç, 2016; Yüksel & Bolat, 2016) supports this importance. For this reason, it is important to know the culture of tourism employees. The present study was conducted to determine the culture of tourism employees in a sample of hotel employees.

According to the present study, hotel employees attach importance to collectivism, avoid uncertainty, are long-term oriented, have a medium level masculine, and low power distanced culture. The low power distance may be attributed to many reasons. One of the reasons may be the change in Turkish people's views on power resulting from the increase in the average duration of education in Turkey. Another reason could be the prolongation of Turkish people's experience of democracy and the developments caused by changes in communication and transportation technologies. Besides, considering that the hotel industry in Turkey mainly accommodates foreigners from many different countries, it can be thought that the employees interact intensively with different cultures, and this interaction contributes to the decrease of the power distance of the employees (Ertaş, 2017). The long-term orientation and uncertainty avoidance of the hotel employees to be high may be considered normal, considering the political and economic conditions of the region. Terrorist incidents in the region, the socio-economic backwardness, and limited job opportunities increase the concerns of the employees (Türker & Karadağ, 2019). The high level of collectivism can be explained by the social structure in the region. The most prominent feature of the social structure in the region is the commitment of the local people to institutions such as tribes. This commitment led to collective identities rather than individual identities in the region (Kaya, 2013). Findings of the present study support the research findings of Çalışkan and Özkoç (2016). However, some researchers found some different results. For example, Akdeniz and Aytemiz Seymen (2012) found a high level of masculinity in their study on hotel staff. Türker and Karadağ (2019) found that hotel employees in Şanlıurfa have a medium level of long-term orientation and avoidance of uncertainty. It is interesting that although both studies were conducted in the same region they found different determinations in the mentioned dimensions. The reason for these differences can be examined with new research. Bu araştırmada ayrıca Mardin ve Diyarbakır'daki otel çalışanları da bulunmaktadır. Güneydoğuya doğru gidildikçe terör olaylarının artıyor olması, bu araştırma katılımcılarının uzun dönem odaklılık ve belirsizlikten kaçınma düzeylerini Türker ve Karadağ'ın bulgularının aksine az da olsa artırmış olabilir.

The fact that the employees have a collectivist culture shows that the hotel employees place the interests of the group and the business above their interests. Hotel managers should make an effort to keep this cultural trend. In this sense, they can strengthen the relations between employees by taking measures to increase group loyalty among employees. These measures will also strengthen the spirit of teamwork of hotel employees. The medium level of masculinity of the hotel employees is a sign of danger. Employees with a masculine culture are far from teamwork (Hofstede et al., 2010). The high level of uncertainty avoidance and long-term orientation of the employees may cause hotel employees to experience negative situations such as job dissatisfaction, job stress, and future insecurity. Hotel managers need to take precautions to decrease their uncertainty avoidance levels of employees. In this context, employees should be guaranteed job security, and job descriptions should be clearly defined. A low power distance indicates that hotel employees agree with decisions, consider themselves equal with other employees, and superiors are not autocratic. It is advised to hotel managers not to spoil this cultural level. If tourism managers want to effectively manage employees, they can start by learning employees' cultures. Otherwise, trying to manage them by ignoring the culture of the employees may lead to wrong practices.

Although the present research tried to determine the cultures of tourism employees, it has some limitations. The first limitation of the research is related to its sample. Research should be conducted on the culture of employees in different tourism businesses. On the other hand, conducting similar studies with employees in hotels in different tourism destinations will make it easier to comment on the culture of tourism employees. By comparing the culture of tourism employees in different countries, research can be conducted on how regional differences affect the culture of employees. Also, in this research, no hotel classification (four-star, five-star, etc.) was made. Future research can achieve different results, taking this into account. This study only attempted to identify the culture of hotel employees. For this reason, it has not been investigated whether the culture of employees varies according to their demographic characteristics. It should also be investigated whether the tourism industry has any impact on the culture of its employees. In the present study, no questions were asked about the departments where the hotel employees work or how long they have been working in the tourism industry or in the same hotel. These issues should also be considered in similar research in the future. Thus, it will be possible to see the big picture of the culture of tourism employees. Finally, this study has no purpose to test the work of Hofstede. It is possible that the culture of societies may have changed slightly, over the 40 years since Hofstede's work.

Bibliography

Akdeniz Ay, D. (2015), *Interrelationship between National Culture, Organizational Culture, Perception of Organizational Politics and Intention to Join a Union* (Unpublished Doctoral Dissertation), Balıkesir University, Turkey.

Akdenlz, D. and Aytemiz Seymen, O. (2012), "Diagnosing National and Organizational Culture Differences: A Research in Hotel Enterprises", *Nevşehir University Journal of Social Sciences*, 2, pp. 198–217.

Akova, O., Tanrıverdi, H., and Kahraman, O. C. (2015), "Risk Factors Effecting the Labour Turnover Rate in Hotel Establishments", *Suleyman Demirel University Visionary Journal*, 6 (12), pp. 87–107.

Avey, J. B., Wernsing, T. S., and Luthans, F. (2008), "Can Positive Employees Help Positive Organizational Change? Impact of Psychological Capital and

Emotions on Relevant Attitudes and Behaviors", *The Journal of Applied Behavioral Science*, 44 (1), pp. 48-70.

Ayoun, B. M. and Moreo, P. J. (2008), "The Influence of the Cultural Dimension of Uncertainty Avoidance on Business Strategy Development: A Cross-National Study of Hotel Managers", *International Journal of Hospitality Management*, 27, pp. 65-75.

Baum, J. R. and Wally, S. (2003), "Strategic Decision Speed and Firm Performance", *Strategic Management Journal*, 24, pp. 1107-1129.

Bitner, M. J., Booms, B. H., and Tetreault, M. S. (1990), "The Service Encounter: Diagnosing Favorable and Unfavorable Incidents", *Journal of Marketing*, 54 (1), pp. 71-84.

Bozkurt, H. Ö. and Demirel, Z. H. (2019), "Job Security as the Predictor of Turnover Intention in Hotel Businesses: Mediating Role of Job Embeddedness", *Business & Management Studies: An International Journal*, 7 (4), pp. 1383-1404.

Chen, R. X. Y., Cheung, C., and Law, R. (2012), "A Review of the Literature on Culture in Hotel Management Research: What is the Future?", *International Journal of Hospitality Management*, 31 (1), pp. 52-65.

Child, D. (2006), *The Essentials of Factor Analysis*, Bloomsbury Academic.

Comrey, A. L. and Lee, H. B. (2016), *A First Course in Factor Analysis*, Psychology Press.

Çalışkan, N. and Özkoç, A. G. (2016), "Determination of National Culture Dimensions Affecting Paternalistic Leadership Perception in Organizations", *Journal of Yasar University*, 11, pp. 240-250.

Dawson, M., Abbott, J., and Shoemaker, S. (2011), "The Hospitality Culture Scale: A Measure Organizational Culture and Personal Attributes", *International Journal of Hospitality Management*, 30, pp. 290-300.

Devine, F., Baum, T., Hearns, N., and Devine, A. (2007), "Managing Cultural Diversity: Opportunities and Challenges for Northern Ireland Hoteliers", *International Journal of Contemporary Hospitality Management*, 19 (2), pp. 120-132.

Dick, G. and Metcalfe, B. (2001), "Managerial Factors and Organisational Commitment-A Comparative Study of Police Officers and Civilian Staff", *International Journal of Public Sector Management*, 14 (2), pp. 111-128.

Erdogan, B. and Liden, R. C. (2006), "Collectivism as a Moderator of Responses to Organizational Justice: Implications for Leader-Member Exchange and Ingratiation", *Journal of Organizational Behavior*, 27 (1), pp. 1-17.

Ertaş, Ç. (2017), *The Mediation Role of Organizational Culture in the Effect of Cultural Values on Organizational Justice: Case of Hotel Employees* (Unpublished Doctoral Dissertation), Mersin University, Turkey.

Fernandes, A., Alturas, B., and Laureano, R. (2018), "Validation of the Hospitality Culture Scale in the Context of Hotel Industry", *Tourism & Management Studies*, 14 (1), pp. 43–52.

Fisher, R., McPhail, R., and Menghetti, G. (2010), "Linking Employee Attitudes and Behaviors with Business Performance: A Comparative Analysis of Hotels in Mexico and China", *International Journal of Hospitality Management*, 29 (3), pp. 397–404.

Fullagar, C. J., Sumer, H. C., Sverke, M., and Slick, R. (2003), "Managerial Sex-Role Stereotyping: A Cross Cultural Analysis", *International Journal of Cross Cultural Management*, 3 (1), pp. 93–107.

General Directorate of Investments and Enterprises, Republic of Turkey Ministry of Culture and Tourism (2019), "Tourism Facilities", http://yigm.kulturturizm.gov.tr/TR-9579/turizm-tesisleri.html, (30.07.2018).

Gunkel, M., Schlaegel, C., Rossteutscher, T., and Wolff, B. (2015), "The Human Aspect of Cross-Border Acquisition Outcomes: The Role of Management Practices, Employee Emotions, and National Culture", *International Business Review*, 24, pp. 394–408.

Hair, J. F., Black, W. C., Babin, B. J., and Anderson, R. E. (2010), *Multivariate Data Analysis*, Upper Saddle River, NJ: Prentice Hall.

Hartline, M. D. and Ferrell, O. C. (1996), "The Management of Customer-Contact Service Employees: An Empirical Investigation", *Journal of Marketing*, 60 (4), pp. 52–70.

Harvey, F. (1997), "National Cultural Differences in Theory and Practice: Evaluating Hofstede's National Cultural Framework", *Information Technology & People*, 10 (2), pp. 132–146.

Hauff, S., Nicole, F. R., and Tressin, T. (2015), "Situational Job Characteristics and Job Satisfaction: The Moderating Role of National Culture", *International Business Review*, 24, pp. 710–723.

Heskett, J. L., Jones, T. O., Loveman, G. W., Sasser, W. E., and Schlesinger, L. A. (1994), "Putting the Service-Profit Chain to Work", *Harvard Business Review*, 72 (2), pp. 164–174.

Hofstede, G. (1980), "Culture and Organizations", *International Studies of Management & Organization*, 10 (4), pp. 15–41.

Hofstede, G. (1984), "Cultural Dimensions in Management and Planning", *Asia Pacific Journal of Management*, 1 (2), pp. 81–99.

Hofstede, G. (2001), *Cultures Consequences: Comparing Values, Behaviors, Institutions and Organizations Across Nations*, Beverly Hills, CA: Sage.

Hofstede, G. and Bond, M. H. (1988), "The Confucius Connection: From Cultural Roots to Economic Growth", *Organizational Dynamics*, 16, pp. 4-21.

Hofstede, G., Hofstede, G. J., and Minkov, M. (2010), *Cultures and Organizations: Software of the Mind*, Berkshire, England: McGraw-Hill.

Huang, Xu and Vliert, E. Van de (2004), "Job Level and National Culture as Joint Roots of Job Satisfaction", *Applied Psychology*, 53 (3), pp. 329-348.

Hui, C. H., Yee, C., and Eastman, K. L. (1995), "The Relationship Between Individualism-Collectivism and Job Satisfaction", *Applied Psychology*, 44 (3), pp. 276-282.

Jones, D. L. and Mccleary, K. W. (2007), "Expectations of Working Relationships in International Buyer-Seller Relationships: Development of a Relationship Continuum Scale", *Asia Pacific Journal of Tourism Research*, 12 (3), pp. 181-202.

Jones, P. (1999), "Operational Issues and Trends in the Hospitality Industry", *International Journal of Hospitality Management*, 18 (4), pp. 427-442.

Kaya, M. (2013), *Transformation of Tribes Within the Modernization Process: The Familial and Tribal Associations in Şanlıurfa* (Unpublished Master's Dissertation), Selçuk University, Turkey.

Kaiser, H. F. (1958), "The Varimax Criterion for Analytic Rotation in Factor Analysis", *Psychometrika*, 23 (3), pp. 187-200.

Kirkman, B. L. and Shapiro, D. L. (2001), "The Impact of Cultural Values on Job Satisfaction and Organizational Commitment in Self-Managing Work Teams: The Mediating Role of Employee Resistance", *The Academy of Management Journal*, 44 (3), pp. 557-569.

Lam, S. S. K., Schaubroeck, J., and Aryee, S. (2002), "Relationship Between Organizational Justice and Employee Work Outcomes: A Cross-National Study", *Journal of Organizational Behavior*, 23, pp. 1-18.

Lucia-Casademunt, A. M., García-Cabrera, A. M., and Cuéllar-Molina, D. G. (2015), "National Culture, Work-Life Balance and Employee Well-Being in European Tourism Firms: The Moderating Effect of Uncertainty Avoidance Values", *Tourism & Management Studies*, 11 (1), pp. 62-69.

Magnini, Vincent P., Hyun, S., BeomCheol, K., and Uysal, M. (2013), "The Influences of Collectivism in Hospitality Work Settings", *International Journal of Contemporary Hospitality Management*, 25 (6), pp. 844-864.

McCoy, S., Galletta, D. F., and King, W. R. (2005), "Integrating National Culture Into IS Research: The Need for Current Individual Level

Measures", *Communications of the Association for Information Systems*, 15 (1), pp. 211–224.

McSweeney, B. (2002), "Hofstede's Model of National Cultural Differences and Their Consequences: A Triumph of Faith-a Failure of Analysis", *Human Relations*, 55 (1), pp. 89–118.

Meglino, B. M. and Ravlin, E. C. (1998), "Individual Values in Organizations: Concepts, Controversies, and Research", *Journal of Management*, 24 (3), pp. 351–389.

Mia, L. and Patiar, A. (2002), "The Interactive Effect of Superior-Subordinate Relationship and Budget Participation on Managerial Performance in the Hotel Industry: An Exploratory Study", *Journal of Hospitality & Tourism Research*, 26 (3), pp. 235–257.

Naseem, A., Sheikh, S. E., and Malik, K. P. (2011), "Impact of Employee Satisfaction on Success of Organization: Relation Between Customer Experience and Employee Satisfaction", *International Journal of Multidisciplinary Sciences and Engineering*, 2 (5), pp. 41–46.

Nazarian, A., Atkinson, P., and Foroudi, P. (2017), "Influence of National Culture and Balanced Organizational Culture on the Hotel Industry's Performance", *International Journal of Hospitality Management*, 63, pp. 22–32.

Newman, K. L. and Nollen, S. D. (1996), "Culture and Congruence: The Fit between Management Practices and National Culture", *Journal of International Business Studies*, 27 (4), pp. 753–779.

Osborne, J. W. and Costello, A. B. (2004), "Sample Size and Subject to Item Ratio in Principal Components Analysis", *Practical Assessment, Research, and Evaluation*, 9, pp. 1–9.

Oyserman, D., Coon, H. M., and Kemmelmeier, M. (2002), "Rethinking Individualism and Collectivism: Evaluation of Theoretical Assumptions and Meta-Analyses", *Psychological Bulletin*, 128 (1), pp. 3–72.

Picnaar, J. and Willemse, S. A. (2008), "Burnout, Engagement, Coping and General Health of Service Employees in the Hospitality Industry", *Tourism Management*, 29 (6), pp. 1053–1063.

Polat Üzümcü, T. (2015), "Perceptions of Tourism Training for Hotel Managers: A Study on Managers of Hotel That is in Kocaeli", *Kocaeli University Journal of Social Sciences Institutes*, 30, pp. 123–150.

Posthuma, R. A. (2009), "National Culture and Union Membership: A Cultural-Cognitive Perspective", *Relations Industrielles/Industrial Relations*, 64 (3), pp. 507–529.

Probst, T. M. and Lawler, J. (2006), "Cultural Values as Moderators of Employee Reactions to Job Insecurity: The Role of Individualism and Collectivism", *Applied Psychology: An International Review*, 55 (2), pp. 234-254.

Reimann, M., Lünemann, U. F., and Chase, R. B. (2008), "Uncertainty Avoidance as a Moderator of the Relationship Between Perceived Service Quality and Customer Satisfaction", *Journal of Service Research*, 11 (1), pp. 63-73.

Reisinger, Y. and Crotts, J. C. (2010), "Applying Hofstede's National Culture Measures in Tourism Research: Illuminating Issues of Divergence and Convergence", *Journal of Travel Research*, 49 (2), pp. 153-164.

Rodrigues, C. A. (1998), "Cultural Classifications of Societies and How They Affect Cross-Cultural Management", *Cross Cultural Management*, 5 (3), pp. 31-41.

Rogerson, J. M. and Kotze, N. (2011), "Market Segmentation and the Changing South African Hotel Industry (1990 to 2010)", *African Journal of Business Management*, 5 (35), pp. 13523-13533.

Schippers, M. (2007), "Recovering the Feminine Other: Masculinity, Femininity, and Gender Hegemony", *Theory and Society*, 36 (1), pp. 85-102.

Sledge, S., Miles, A. K., and Coppage, S. (2008), "What Role Does Culture Play? A Look at Motivation and Job Satisfaction Among Hotel Workers in Brazil", *The International Journal of Human Resource Management*, 19 (9), pp. 1667-1682.

Staufenbiel, T. and König, C. J. (2010), "A Model for the Effects of Job Insecurity on Performance, Turnover Intention, and Absenteeism", *Journal of Occupational and Organizational Psychology*, 83 (1), pp. 101-117.

Taras, V., Steel, P., and Kirkman, B. L. (2011), "Three Decades of Research on National Culture in the Workplace: Do the Differences Still Make a Difference?", *Organizational Dynamics*, 40 (3), pp. 189-198.

Tepeci, M. and Bartlett, A. L. B. (2002), "The Hospitality Industry Culture Profile: A Measure of Individual Values, Organizational Culture, and Person-Organization Fit as Predictors of Job Satisfaction and Behavioral Intentions", *International Journal of Hospitality Management*, 21 (2), pp. 151-170.

Testa, M. R., Skaruppa, C., and Pietrzak, D. (1998), "Linking Job Satisfaction and Customer Satisfaction in the Cruise Industry: Implications for Hospitality and Travel Organizations", *Journal of Hospitality & Tourism Research*, 22 (1), pp. 4-14.

Tinsley, H. E. and Tinsley, D. J. (1987), "Uses of Factor Analysis in Counseling Psychology Research", *Journal of Counseling Psychology*, 34, pp. 414–424.

Trompenaars, F. and Hampden-Turner, C. (2011), *Riding the Waves of Culture: Understanding Diversity in Global Business*. Nicholas Brealey International.

Türker, N. and Karadağ, D. (2019), "Cultural Differences: A Study in Trabzon and Şanlıurfa in the Context of Hofstede's Cultural Dimensions", *Journal of Economy Culture and Society*, 60 (1), pp. 1–25.

Yoo, B. and Donthu, N. (2002), "The Effects of Marketing Education and Individual Cultural Values on Marketing Ethics of Students", *Journal of Marketing Education*, 24 (2), pp. 92–103.

Yüksel, M. and Bolat, T. (2016), "The Relationships Between Organizational Politics, Hofstede's Organizational Cultural Dimensions, Job Attitude and Job Outcomes", *Eskişehir Osmangazi University Journal of Economics and Administrative Sciences*, 11 (3), pp. 173–204.

Tuğba GÜRÇAYLILAR-YENİDOĞAN, Hatice BARUT and Ayça ÇETİNKAYA

Linking Strategy to Business Model Renewal in the Hospitality Industry: The Case of Land of Legends Theme Park in Antalya

1 Introduction

Bringing innovation to services is never too old, dating back only to the late 1990s (Camisón & Monfort-Mir, 2012). Because of intangible, perishable, and inseparable nature of services, innovation is so difficult to succeed in experience-based products of tourism services. Because of that strategy studies have recently engaged in innovation behavior of tourism organizations. This newly maturing field in the theory and practice of corporate strategy and innovation argues that tourism businesses may differ from their competitors by expanding horizontal and/or vertical scope of the business activities. In this regard, diversification rather than reducing business risks and uncertainties allows organizations to maintain their competitive edge in the marketplace. Such a strategic decision helps improve diversity in and quality of services provided. According to Dundas and Richardson (1980), diversification relates to the extent of market differentiation when firms operate in more than one target market. Firms can differentiate their markets by making additional substantial improvements to the physical product or building a value-based marketing strategy (Furrer, 2016). In case, corporate diversification serves to differentiation in competition by addressing the market segments in a dissimilar fashion, this strategy brings innovation to the performance of service offerings with distinguished features (Sundbo & Gallouj, 2000).

The emerging trend of traveling has increasingly caused shifts in consumer behavior toward independent holidays so that tourists start relying more on direct online booking in order to organize the activities of accommodation and transportation. While the demand for highly standardized generic packages of all-inclusive holiday keeps coming down, the new age of tourism is looking for new solutions to move demand curve upward. As a result, diversification referring as the expansion behavior in products, markets, and different industries has become an important strategic move for tourism organizations in response to the competitive business environment (Yan et al., 2017). They have started diversifying their product markets by creating new brands or purchasing existing ones (Lee & Jang, 2007). Thereby, they have developed capabilities in differentiating service offerings for specific market niches having different life styles or travel purposes.

Undoubtedly, diversification in different market segments allows a wider range of resources and capabilities to be exploited that supports innovation in early stages (Hitt et al., 1997; Kolluru & Mukhopadhaya, 2017). However, continuity in success of a highly diversified multi-business firm depends on the relatedness (in terms of marketing or production) between business units (Porter, 1987; Davis et al., 1992). In other words, positive synergy between business functions extending along the full coverage of the firm scope is the mystery of success. Strong synergies can be accomplished with the knowledge and capability spillovers. Cross-unit integration of physical space, distribution channels, advertising, branding, purchasing, and training induces spillover effects. Ansoff (1965) and Rumelt (1974) argue that functional synergy is extensively practiced in production (shared production facilities or vertical integration) and marketing (undifferentiated markets and marketing programs). If diversified tourism businesses benefit from common production and marketing opportunities, they will likely to be succeeded in competition.

Despite examples of strategy in practice, corporate diversification in the hospitality industry is still in an early stage of the research agenda. Accordingly, the promising venue for future studies is to explain how successful innovations through diversification breed distinctive capabilities, which in turn results in strong competitive position of an organization. To respond this call, this study here aims to exemplify how portfolio diversification in the hospitality industry fosters business model innovation that leads to competitive differentiation. Conducting case study research, this study reveals the secret behind the growth attacks of Rixos hotel chain with its new brand "Land of Legends."

2 Diversification Strategy of Rixos Hotels: The Birth of a New Brand *Land of Legends*

Corporate growth strategy is closely related to the decision an organization takes on how to expand its business scope (Furrer, 2016). On the one hand, some organization with small size and limited resources seem to be pursued a highly concentrated strategy when they penetrate a specific market segment in growth. On the other hand, some others follow specialization strategy by offering the same product to different markets (product specialization) or different products to the same market (market specialization) and hence gradually move away from concentrated growth. And in the end, the others display a more and more diversified structure by serving with a wide variety of products in different market combinations toward the full coverage of the core business, and/or entering new business areas. Thus as an organization becomes a multi-business company, it may be broader or narrower in scope. As a strategic business unit contains a single business (/product), or a line of related businesses (/products) with differentiated market segments, diversification differs consistently in composition of a firm's portfolio of business units (Bartlett & Ghoshal, 1993; Goold & Luchs, 1996; Puranam & Vanneste, 2016). As a result, diversification is not a yes/no decision, but rather a matter of degree that falls somewhere in the firm scope (Furrer, 2016: 3).

Product diversification in tourism supports the value-creation by adding new ones to the existing product portfolio (Benur & Bramwell, 2015). A familiar way to diversify the primary tourism product is to create a new bundle or combination of the tourism service package. For instance, a hotel serving mass tourists can diversify to include sophisticated amusement services (Milman et al., 2010). Spatial closeness between service stations of diversified tourism products increases the percent of repeat guests in existing market and allows hotels to attract new tourist segments. The Land of Legends (TLOL), founded by Rixos hotel chain and started operating in May 2016, is the first theme park hotel of Turkey as one of the world's leading tourism destinations. TLOL is located in the Belek district, known as golfer's paradise, 40 km away from Antalya city center. Under this new brand of Rixos Hotels, recreation and entertainment activities are involved in a blended service package of accommodation, retail, and food services. At a fantasy atmosphere, TLOL offers guest experience in performing dragon-themed mystical characters and shows. With a wide variety of products, the tourism company can meet a combined demand for tourists interested in both coastal resort tourism and city tourism. Hence, it acts as the candidate catalyst of economic development in the Belek region, which is the center of the tourism resorts around. In 2019, TLOL was awarded with the most successful outdoor entertainment center in Turkey by Tureks International Fair Organization (ATRAX) (Tureks, 2019). As well, this new-fashioned amusement park stands in third place in Europe's Best Waterparks. In fear of being ordinary, TLOL is constantly seeking innovation to keep the current demand alive. A strong evidence of this is the adventure playground with endlessly varied activities, which was added into the structure in 2018. Daily visitor capacity reached 9000 visitors after being included Dry Fun Park next to the Wet Park. New projects constantly continue to be improved for differentiating itself from other hotels.

Rixos Hotel Group, the creator of the brand TLOL, is one of the well-known national and international hotel chains operating in Turkey, Azerbaijan, Egypt, Kazakhstan, Russia, Switzerland, and the UAE (Rixos, 2020). Established in 2000, the tourism company provides first class services and exclusive accommodation with 25 hotels, 17 of which are operated as beach resorts. Although the Rixos hotel group originally started to serve for mass tourism catering for sun, sea, and sand, but then it quickly diversify to fulfill its new mandate of being Turkey's Largest Park for accommodation and entertainment. As an amusement facility, the new structure of the tourism place features a five-legged business portfolio, including theme park, accommodation, F&B, shopping, and events (see Tab. 6-1). Different businesses of the tourism company are complementary to meet touristic needs. Under the family branding strategy, TLOL aims to foster the development of operational synergy through diversification by conducting co-marketing activities for products across different businesses located in the same geographic area.

Unlike resort hotels applying all-inclusive system in Turkey, Kingdom Hotel in TLOL serves half board service including bed and breakfast. All rooms are fully equipped with smart technology products such as PlayStation, X-box, and Dual view TVs. The room prices of theme park hotels are higher than un-themed, less themed, and city hotels. Consistently, average room prices per night range between 450 and 500 dollars in the Kingdom Hotel with seven different room types. The hotel with 401 rooms and 1100 beds is the most expensive hotel in the Belek region, yet has a 100% of occupancy rate from the beginning of April to the end of September. The Land of Legends Kingdom Hotel is the first kid-themed hotel in Turkey (see Fig. 6-1). The rooms of the hotel are furnished with wallpapers and colorful designs suitable for children's imagination. Small character toys and kids furniture are also featured in children's accommodation and entertainment units. Carpets in the rooms are 100 % wool and antibacterial; furniture and lighting systems are particularly kid-friendly. Each floor is painted with a different color for children not to forget the room floor. Balloon pools, special playgrounds, spas, painting workshops, and a library are dedicated to the use by children. Free and themed candy stands for kids are available in the hotel lobby. Franko Dragon, who is the founder and artistic director of Dragone Entertainment Group specialized in the creation of large-scale theater shows, adds spirit to the concept of a "happy holiday" built in TLOL by developing 12 characters for children and setting up a show team of 100 members. Plans are also made to create cartoon characters.

TLOL offers a wide range of F&B outlets with different experiences. Hotel customers (for meals except breakfast) and other visitors can get service from

Linking Strategy to Business Model Renewal

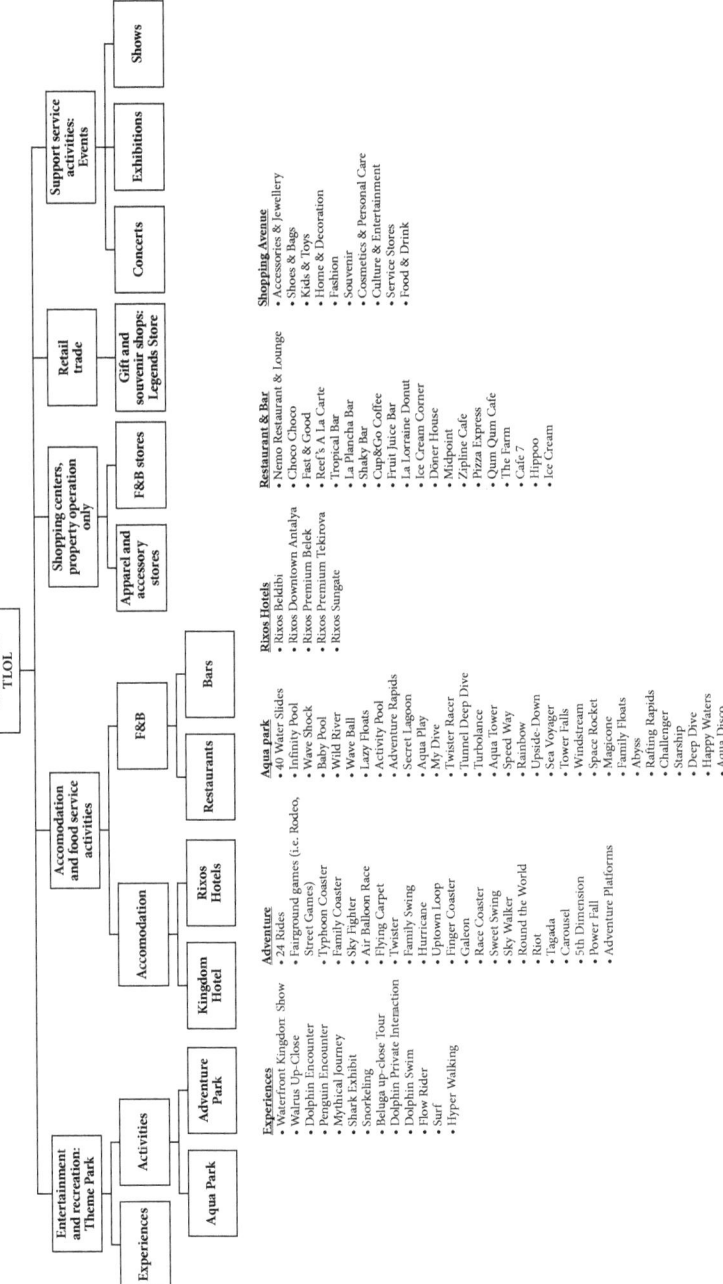

Tab. 6-1: Business Portfolio of TLOL

Fig. 6-1: Kid-Friendly Rooms and Theatrical Shows at TLOL
Source: The Land of Legends, 2020

71 restaurants in the shopping mall hosting stores of well-known food and clothing brands. At the same time, here embodies specialist restaurants in the excellence of world's cuisines. For example, Nemo is an aquarium-themed restaurant located inside the facility area that creates a memorable experience with the stunning views of the underwater (Nemo Restaurants, 2020). Five restaurants, 5 bars, 3 kiosks, and 3 carts compose the F&B outlets under the umbrella of TLOL branding. Located around the theme park and shopping mall, the gift shops sell various kinds of themed products bearing the logo and characters of TLOL. Efforts to increase service stations are made to ensure that expenditures are kept within the property. In the facility 180.000 m² are available for use. The total area with a planned new section will reach 300.000 m². Despite its large land capacity near the city center, the company without meeting facilities does not seek to expand toward the meetings, incentives, conferences, and exhibitions (MICE) segment. However, as an exception, TLOL hosts the gala of groups staying at neighboring hotels on the lawn of the facility.

There are four lands in TLOL theme park: Adventure Rapids, Secret Lagoon, Waterfront Kingdom, and Dry Park. In these sections within the theme park, a wide variety of activities are carried out including water games, adrenaline activities, fairground games and experiences. In general, this place is divided into two as wet and dry park. "Legends of Aqua" was awarded the best waterpark project award at the World Waterpark Association (WWA) in 2017

(Blooloop, 2017). In the theme park, experiences with the discovery of special animals such as sharks, penguins, and seals and activities like scuba diving and snorkeling are offered safely and enjoyably. For the guests who want to spend time with amusement park toys, there are funfair toys equipped with the latest technology. Adventure activities are also designed for those who want to experience excitement.

3 Business Model Renewal of Rixos Hotels: Broadening Revenue Stream to the New Segments *Upper Income Tourists, Excursionists and Competitor's Customers*

The business model addresses fundamentally critical questions to ask about configuring a business system (Keeley et al., 2013). It describes the rationale of what an organization offers and to whom it offers that. It also clarifies how an organization earns money. And in end, the concept elucidates the statement of an organization's revenue generation. A company can create a profitable and sustainable revenue stream by offering differentiated value propositions adjustable to the particular needs of each market segment. Despite some difficulties of defining what business model innovation actually entails, it obviously requires new ways of doing business in a different perceived added value and thus diversifying existing businesses to make more money. Although 70 % of its main business and operations still consist of resort hotel management and all-inclusive concept, Rixos Hotels becomes increasingly more noticeable example of doing tourism in the new age of Antalya by accomplishing business model renewal through diversification.

The management of Rixos Hotels, which is currently pursuing expansion strategy in international market with the similar value proposition, set a new direction in its growth strategy by creating the Land of Legends brand in 2016. It is a specific type of diversification strategy for offering additional product and services to existing customers and creating a completely different and modular service package for new customer markets. This is a holiday structure that combines accommodation, shopping, entertainment, and experience with rest. Additional service packages offered in TLOL increased the repeat customer rate of Rixos Hotels by 5%. The loosely coupled structure of the organization enables to put a modular service system into the practice of hospitality. Thus, guests of the rival hotels in the region have become one of the target market segments. Another attractive market segment of TLOL consists of the local people who make daily visits. Other hotels at the coastal regions of Antalya almost do not get a share from the local market niche. In efforts to satisfy specific needs of

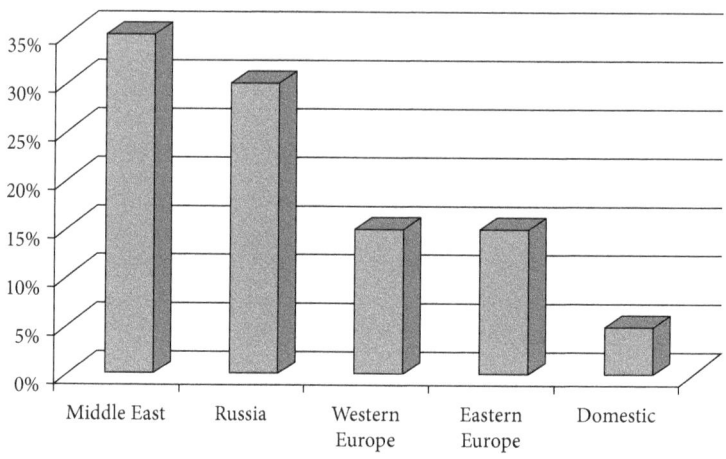

Fig. 6-2: Customer Profile of TLOL
Source: The compiled data from face-to-face interviews with managers

excursionists through local tourism marketing TLOL seems to increase market share from the diversified coverage of products (see Fig. 6-2).

The theme park is free to use for all customers staying in the chain-branded family hotels. However, other Rixos hotel guests, except those staying at the Kingdom hotel, cannot benefit from the theme park services on the first and last day of their stay. By this way, the desire to visit the theme park is expected to extend the overnight stays in the hotel. Besides the services included at the accommodation package, specialist restaurants, F&B stands (kids food corner with cotton candy, popcorn etc.), TLOL branded souvenir shops, and some parts and products of the theme park (e.g., fast track service, cabanas, lockers, beach boxes, blue sky balloon, hyper walk, float rider, private park tour, videos & photos, fairground games, and dolphinarium activities) are extra charged. Other customers (e.g., excursionists and other tourists staying at rival hotels) benefit from the theme park by paying for the theme park ticket in variable pricing according to the season. All in all the total revenue generated by whole facilities in the theme park is divided into the 4 major accounts (see Fig. 6-3).

The value proposition of the facility is more diverse and richer than the competitors' offer, in turn helps retaining its prominent position as the differentiator in the industry. TLOL with a fantastical world of fun and adventure breeds at lower risk of being drifted in price competition. The major sources of product differentiation in TLOL are as follows:

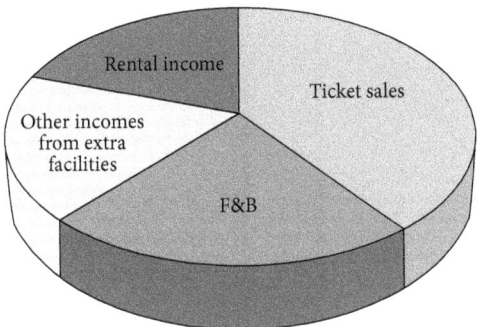

Fig. 6-3: The Revenue Streams of the Theme Park in the Four Different Accounts
Source: The compiled data from face-to-face interviews with managers

- Differentiating as being the first in customers' minds: *Turkey's first kid-themed hotel*
- Differentiating at high prices: *High-priced sales across the region and country with the half board accommodation policy*
- Differentiating with distinctive features: *"Distinguish itself" in design and uniqueness – a hotel for families with children* in many ways, such as the architecture, landscaping, room designs, staff clothes, and large-scale theater shows.
- Differentiating as the newest: *Rooms equipped with the latest technologies* like dual view flat screen TVs.

In conclusion, the new business model initiated by Rixos Group with the TLOL brand is a successful outcome of business diversification and hence differentiation in competition. The tourism company supports its new business model with network innovations. One of the commercial partners includes Dubai-based Emaar, which is one of the world's largest real estate corporations. With the vast experience in the sector of shopping malls and retail stores Emaar controls the management of the shopping units in TLOL. The other partner, Franco Dragone, is responsible for the architecture and the creativity of entertainment world throughout the facility.

Conclusion

Depending on the rapid growth in the children population that reach 2.5 billion around the world and 25 million in Turkey, the increasing impact of

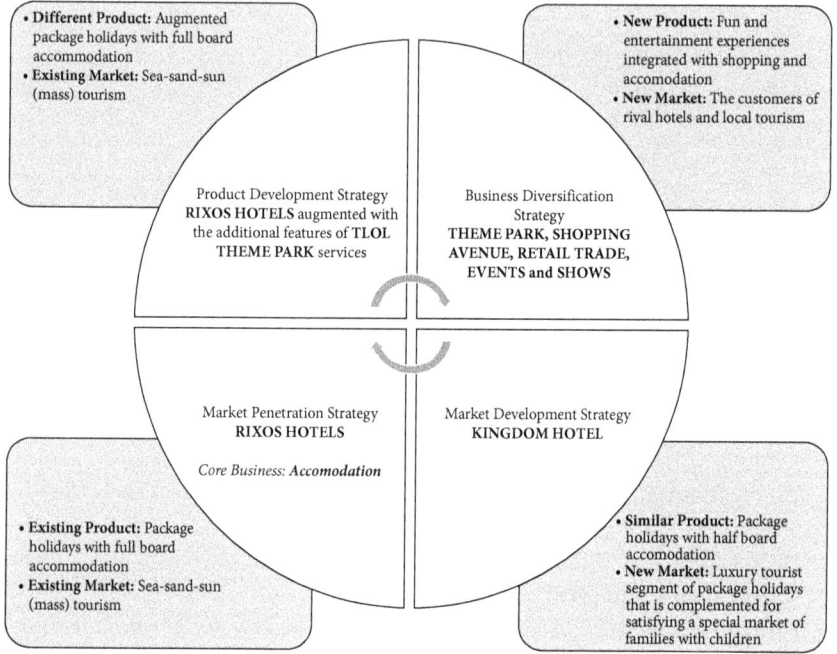

Fig. 6-4: Linking Growth Strategy to Business Model Renewal – The Case of Rixos Tourism Group

Source: Prepared by authors

children on consumption decisions has resulted in spreading out the concept of child-friendly tourism (Tuna, 2018). As the child population increases in tourism sector, changing family structures are evaluated a business opportunity to go beyond the triangle of sea-sand-sun tourism and all-inclusive resort hotel management by Rixos Hotels known as one of the world's fastest growing companies. By adding the Land of Legends brand next to the Rixos brand, the giant tourism company aims to follow a hybrid strategy in growth (see Fig. 6-4). With the rooms suitable for child concept, mystical shows designed by Franco Drogan, theme park, gourmet tastes, and shopping opportunities, the new venture of Rixos Hotels offers different or combined products to different customer groups. In turn, it distinguishes itself from the competition by addressing to each market segment in a dissimilar fashion: Adults, Children, Local visitors, and the Guests staying at other hotels. Whereas the augmented product creates a new and great customer experience, the tourism company shifts toward a new business model with a better pattern of cash flow. Thanks to contribute building "happy city," the facility is attracting a considerable number of visitors from Europe, Russia, the Middle East and local market.

In summary, this chapter exemplifies transformation of Rixos Hotels toward a more diversified company with new product markets addressed as never before. Rixos Group seems to successfully expand its services into new markets by making brand differentiation. In a theoretical sense, the case of TLOL includes lessons learned from growth patterns of an organization that help ascertaining new ways to convert its offers and other value sources into cash. A diversified growth strategy supports developing new markets in the core business, creating a distinguished brand value for new businesses, increasing market share in the core business and/or expanding the market coverage by opening new markets.

Bibliography

Ansoff, H. I. (1965), *Corporate Strategy*, New York: McGraw-Hill.

Bartlett, C. A. and Ghoshal, S. (1993), "Beyond the M-form: Toward a Managerial Theory of the Firm", *Strategic Management Journal*, 14 (2), pp. 23–46.

Benur, A. M. and Bramwell, B. (2015), "Tourism Product Development and Product Diversification in Destinations", *Tourism Management*, 50, pp. 213–224.

Blooloop (2017), "The World Waterpark Association's WWA Show", https://blooloop.com/features/world-waterpark-association-wwa-2017, (12.4.2020).

Camisón, C. and Monfort-Mir, V. M. (2012), "Measuring Innovation in Tourism from the Schumpeterian and the Dynamic-Capabilities Perspectives", *Tourism Management*, 33 (4), pp. 776–789.

Davis, P. S., Robinson, R. B., Pearce, J. A., and Seung Ho, P. (1992), "Business Unit Relatedness and Performance: A Look At The Pulp And Paper Industry", *Strategic Management Journal*, 13 (5), pp. 349–361.

de la Peña, M. R., Núñez-Serrano, J. A., Turrión, J., and Velázquez, F. J. (2016), "Are Innovations Relevant for Consumers in the Hospitality Industry? A Hedonic Approach for Cuban Hotels", *Tourism Management*, 55 (August), pp. 184–196.

Dundas, K. N. M. and Richardson, P. R. (1980), "Corporate Strategy and the Concept of Market Failure", *Strategic Management Journal*, 1 (2), pp. 177–188.

Furrer, O. (2016), *Corporate Level Strategy: Theory and Applications*, London: Routledge.

Goold, M. and Luchs, K. S. (1996), *Managing the Multibusiness Company: Strategic Issues for Diversified Groups*, New York, Cengage Learning EMEA.

Hitt, M. A., Hoskisson, R. E., and Kim, H. (1997), "International Diversification: Effects on Innovation and Firm Performance in Product-diversified Firms", *Academy of Management Journal*, 40 (4), pp. 767–798.

Keeley, L., Pikkel, R., Quinn, B., and Helen, W. (2013), *Ten Types of Innovation: The Discipline of Building Breakthroughs*, New Jersey, John Wiley & Sons.

Kolluru, S. and Pundarik, M. (2017), "Empirical Studies on Innovation Performance in the Manufacturing and Service Sectors since 1995: A Systematic Review", *Economic Papers: A Journal of Applied Economics and Policy*, 36 (2), pp. 223–248.

Lee, M. J. and Jang, S. S. (2007), "Market Diversification and Financial Performance and Stability: A Study of Hotel Companies", *International Journal of Hospitality Management*, 26 (2), pp. 362–375.

Milman, A., Okumus, F., and Dickson, D. (2010), "The Contribution of Theme Parks and Attractions to the Social and Economic Sustainability of Destinations", *Worldwide Hospitality and Tourism Themes*, 2 (3), pp. 338–345.

Muharrem, T. (2018), "Temalı Parklar ve Çocuk Dostu Turizm", *Sivas Interdisipliner Turizm Araştırmaları Dergisi*, 1, pp. 47–53.

Nemo Restaurants (2020), "Experience", http://nemorestaurants.com, (15.1.2020).

Porter, M. E. (1987), "From Competitive Advantage to Corporate Strategy", *Harvard Business Review*, 65 (3), pp. 43–59.

Puranam, P. and Vanneste, B. (2016), *Corporate Strategy: Tools for Analysis and Decision-making*, Cambridge, Cambridge University Press.

Rixos (2020), "Hotels", https://www.rixos.com, (4.1.2020).

Rumelt, R. P. (1974), *Strategy, Structure and Economic Performance*, Cambridge, MA: Harvard University Press.

Sundbo, J. and Faïz, G. (2000), "Innovation as a Loosely Coupled System in Services", *International Journal of Services Technology and Management*, 1 (1), pp. 15–36.

Sundbo, J. and Gallouj, F. (2000), "Innovation as a Loosely Coupled System in Services", International Journal of Services Technology and Management, 1 (1), pp. 15–36.

The Land of Legends (2020), "Rooms", https://thelandoflegendsthemepark.com/en, (8.2.2020).

Tuna, M. (2018), "Temalı Parklar ve Çocuk Dostu Turizm", Sivas İnterdisipliner Turizm Araştırmaları Dergisi, 1, pp. 47–53.

Tureks International Fair Organization (ATRAX) (2019), "Attraction Star Awards", http://tureksfuar.com.tr/en, (14.3.2020).

Yang, Y., Yang, C., and Grace, L.-T. (2017), "Product Diversification and Property Performance in the Urban Lodging Market: The Relationship and its Moderators", *Tourism Management*, 59, pp. 363–375.

Mustafa GÜLAYDIN and Gonca AYTAŞ

Investigation of Emotional Labor Behaviors in Hotel Enterprises

1 Introduction

The hospitality businesses, travel businesses, food and beverage businesses and recreation businesses operating in the tourism sector meet the needs of tourists such as accommodation, transportation, food and beverage, entertainment and recreation and it is noticed that given the labor-intensive nature of tourism, service enterprises benefit from the human factor the most in terms of service delivery. In tourism businesses the most significant production factor and the key component in making a difference is humans. Because the touristic product produced in tourism businesses has an abstract feature and is consumed at the place where it is produced (Kaplan, 2010: 1). It is recognized that as the capacity of accommodation, transportation, food and beverage and recreation enterprises increase, the number of employees increases proportionally and as a result, large organizational structures are generated. Although automation systems are used in tourism businesses, the human aspect holds the most crucial place in service delivery. Therefore, it is of great prominence to manage employee behavior in businesses operating in the tourism sector.

In a globally competitive environment, there is a need to trace the increase in efficiency and assuring this increase in order for businesses to survive. The most essential parts of efficiency are individual and organizational performance variables, and the main actor of these variables is the human constituent (Turunç & Çelik, 2010: 210). When evaluated from this point of view, managing employee behavior is major for situations such as organizational success, creating a competitive and superior advantage, productivity and increasing the quality of service provided in tourism businesses. Production and consumption come to pass simultaneously in the tourism sector and reach the maximum level of employee and customer interaction during service delivery. This situation increases the emphasis on behavioral management in tourism businesses one more time.

Emotional labor is expressed as employees managing their emotions while interacting with the customer and their effort during this management (Hochschild, 1983). These behaviors that the organization wants to be indicated emerge as surface acting, genuine acting and deep acting (Ashforth & Humphrey, 1993: 88–94). Even if the employee does not feel at that moment, the behavior expected by the organization is interpreted as surface acting. However, the effort of the employee to feel the actual emotions desired by the organization is called deep acting. On the other hand, genuine acting describes the action of unveiling the natural and real feelings that the employee feels. Employees in tourism businesses adopt these three behaviors during service delivery and while interacting with the customer, and they keep their emotions under control according to these behaviors. Because emotional labor is the reverberation of the right emotions desired and anticipated by the organization to the customers, regardless of the emotions felt by the employee (Yürür & Ünlü, 2011: 85–86).

This study intends to discover the emotional labor behaviors adopted and exhibited by the employees of tourism businesses. However, it also hopes to ascertain whether there is a difference between the individual characteristics of the employees and the emotional labor behaviors they manifest. It is considered that this study may constitute a resource for managers both in terms of more operative organization and human resource management and more effectual business management.

2 Emotional Labor Concept

Even though the phenomenon of emotion was not as widely used as it is today, it is an area that has been studied since the late 1800s, and various opinions have been put forward. Investigating the feelings and emotional states of mankind and estimating the effects of these situations has always been a subject of interest in the literature. This is why it is possible to come across various explanations of what is felt and emotions in terms of biological, psychological, social, and cultural aspects. Especially since the 1980s, the existence of emotions in working life, the forms of expression, their roles in organizational success have been increasingly discussed (Seçer, 2005: 814).

Güney (2015: 266) specifies emotions as "positive or negative responses of individuals to social interactions." For this reason, management of the employees' emotions by organizations is critical for organizational success. The phenomenon of labor first appeared systematically, especially in the period after the Industrial Revolution and is entitled as the financial and moral meaning of the time spent by the employee at work (Yüksel, 2014: 258). Based on definitions, the concept of labor may be depicted as the effort made by the employees both physically and mentally in order to achieve success toward the goals and objectives of the organization. In addition to the mental and physical efforts added by the employees, the concept of emotional labor, a new concept of labor, was first used, and added by Hochschild (1983).

The early studies on the concept of emotional labor were conducted by Goffman (1959). Goffman observed the behaviors desired from the employees and the organizations' supervision about these behaviors (Oğuz & Özkul, 2016: 132). Influenced by Goffman, American sociologist Arlie Russell Hochschild first addressed the concept of emotional labor in his book called as *The Managed Heart* in 1983 (Köse et al., 2011: 168). In addition to Hochschild's (1983: 6) explanation of emotional labor concept as "managing emotions in order to exhibit facial expressions and body movements that can be observed by everyone," it was also indicated that emotional labor is a value that can change as it is carried out in exchange for a certain fee.

When the necessary literature is examined, it is possible to come across many different definitions regarding the concept of emotional labor. For example, Ashforth and Humphrey (1993: 90) delineated emotional labor as "the behavior to expose the expected emotions during service delivery." Also, Morris and Feldman (1996: 987) outlined the concept of emotional labor as "the effort, planning and control required from the employees to express the emotions desired by the organization in interpersonal relations." On the other hand, Grandey (2000: 95) described emotional labor as "intensifying emotions, suppressing emotions or playacting in order to organize emotional expressions." Additionally, Johnson (2007: 3) depicted the concept of emotional labor as "revealing emotions by the service personnel in line with the organization's wishes and expectations."

The quality of the service provided is judged by the mutual relations between the employee who provides the service and the customer who purchases the service in the jobs where the service is sold. In this regard, organizations, with the service provided, undertake that the smiling face and respect will be shown to the customers by the employees during the service delivery. In this context, organizations specify the rules that employees must follow during service delivery in order to guarantee service quality and supervise them for their implementation (Eroğlu, 2014: 147–148).

Since emotional labor is regarded as a labor process that operates under the control of the organization, its impact on employees is generally considered negative. In order to establish the relationship required by the organization with the customer, the employee can often enter into emotions different from their real feelings and attempt to disclose behavior in this direction. Therefore, these may have negative effects on the employees over time, resulting in job stress, job dissatisfaction, and emotional fatigue. However, rather than defining emotional labor under organizational pressure, if emotional labor is characterized as a process that takes place in collaboration with the organization and employees, approaches that try to eliminate its negative consequences can be obtained (Karaman, 2017: 54).

In order for the employees to have a high-quality emotional labor, organizations should pick the employees who will directly communicate with the customer from the highly motivated ones with strong social relationships, doing their job lovingly. It is possible to provide organizational rewards, job rotation and promotion in order to increase the quality of emotional labor. For instance, sending the successful employees of the hotel business to holiday abroad, assigning the staff working in the front office department to the accounting department at certain times, promoting the personnel who have worked in the housekeeping department in the hotel business for many years and who are successful in their jobs to the housekeeping chief may contribute to increasing the quality of emotional labor behavior to be exhibited by the employees. This may help contribute to the high emotional labor quality of employees and perception of this situation by the customers, and to lead the creation of service that satisfies the customers (Kılıç & Baş, 2015: 23).

Emotional labor has a noteworthy role in exhibiting organizational and managerial behavior. Especially in service businesses, communication skills, interpersonal relationships and emotional efforts of employees are prominent concepts in terms of the survival of businesses and organizations. Hence, many organizations find out the rules regarding the behaviors they expect and demand from their employees and aim to achieve organizational goals (Steinberg & Figart, 1999; Diefendorf & Richard, 2008; Quoted from Basım ve Beğenirbaş, 2012: 80). It is an influential piece that the emotional labor should take place under the will and control of organizational management (Karaman, 2017: 54). Businesses, which can control and manage the emotions of the employees correctly, are successful in achieving their goals and objectives, and this is achieved through emotional labor. In this context, emotional labor is consequential for businesses, especially tourism businesses.

3 Theories of Emotional Labor

The first research on the concept of emotional labor was performed by Hochschild (1983) and then it has become a topic that has been explored by many researchers. Although the subject has been researched by many researchers, the concept of emotional labor and related approaches introduced by Ashforth and

Humprey (1993), Morris and Feldman (1996) and Grandey (2000) contributed to the literature greatly and developed approaches that were adopted more than any others.

The concept of emotional labor was first used and introduced by the American sociologist Arlie Russel Hochschild in his book called *The Managed Heart* based on the emotional requirements of a cabin crew on board. Hochschild (Hochschild, 1983: 6-7) describes emotional labor as "managing emotions in order to exhibit facial expressions and body movements that could be observed by everyone." Also, it was affirmed that emotional labor is a value that can change as it is carried out in exchange for a certain fee.

Hochschild (1983) listed the conditions necessary for the formation of emotional labor. These are as follows:

- As a requirement of the job, the employee should communicate face to face or by voice,
- The employee should create an emotional state in the customer,
- The employer should allow controlling the employee's emotional activities.

Hochschild (1983) announced that there are two kinds of ways to express emotional labor behavior. The first is "surface acting" and the second is "deep acting." Surface acting is a mismatch between the internal emotional and the external behavior in the communication process in which case the employee behaves as desired and expected by the organization but without feeling the real emotion. On the contrary, deep acting is the effort of the employee to feel the emotions under the role expected from him by the organization and to exhibit the behavior by feeling the emotion as desired and expected from him by controlling his emotions (Hochschild, 1983: 35-42; Eroğlu, 2010: 20-21).

Hochschild likened the employees who are in direct contact and communication with customers to the theater actors and the customers to the viewers. From this point of view, it is demanded that actors convey the role required by the game to those who watch it sincerely, intimately and naturally, and to take on the desired identity. This is how viewers believe that the game is authentic and not contrived, and the game becomes a reflection of reality on the stage rather than being a game. Likewise, service sector employees establish a control and audit mechanism over their real feelings while transferring the appropriate emotion to the other party. While these behaviors of the players are formulated as "acting," the behavior of service sector employees is called "emotional labor" (Eroğlu, 2010: 20).

Hochschild indicated that the employees should feel the emotions and live the emotions desired by the organization and exhibit these emotions in the best way during the service delivery as production and consumption are simultaneous in the service sector and the employee establishes a one-to-one relationship with the service-receiving customers. Besides, it was also argued that due to the growing number of businesses operating in the service sector, more responsibility is attached to the employee emotions, and thus criticized that employees' feelings were turned into commodities by purchasing them for a fee in order for employees to establish more one-to-one and face-to-face and audibly relationships with customers, and to maximize customer satisfaction (Hochschild, 1983: 118–119; Ashforth & Humphrey, 1993: 93).

Hochschild (1983) is the researcher who introduced the concept of emotional labor first, conducting the first study on it and proclaimed that emotional labor is exhibited in two dimensions as surface acting and deep acting. In the following years, this concept was developed by adding different dimensions to the concept of emotional labor by different researchers (Keleş, 2014: 54).

Unlike Hochschild, Ashforth and Humpery (1993: 90) ascertained emotional labor as "exhibiting behavior that feels and shows the appropriate emotion." Because, according to Ashforth and Humpery, what matters is rather than what emotion is actually felt or what behavior is felt, it is what behavior is illustrated to make the appropriate emotion felt and perceived (Eroğlu, 2010: 20).

In addition to the "surface acting" and "deep acting" dimensions of emotional labor put forward by Hochschild (1983), Ashforth and Humpery (1993) added the dimension of "genuine acting." Genuine acting is uncovering the behavior expected and desired from the employee without any difficulty or any sense of necessity, as it comes from within according to the owner's wishes. This behavioral dimension of emotional labor transpires when the employee pinpoints the rules set by the organization. In this dimension of emotional labor, the employee reveals emotion rather than acting, but living, and truly experiencing behavior (Ashforth & Humphrey, 1993: 94–98). For example, if a hotel employee has an accident at the workplace, his colleagues feel sad and worried about that person. Moreover, if the person's work accident is caused by the negligence of the business, the reaction of the employees to the business reflects their real and sincere feelings.

Based on these four facets, it may be commented that emotional labor has a substantial place in service businesses, and the fact that the employee adopts the "genuine acting dimension" of emotional labor during service delivery will contribute to the positive opinion of the customer when evaluating the quality of service.

The third approach developed for emotional labor is by Morris and Feldman (1996). According to Morris and Feldman, today, the most relevant part of the work done by many employees is the implementation of the rules set by the organization. Furthermore, according to the definition of Morris and Feldman (1996: 987), emotional labor is "the effort, planning and control for the manifestation of emotions as a result of the rules set by the organization in interindividual relations." To this approach, even if the emotion that the employee feels at that moment and the feeling desired to be exhibited is similar to each other, the employee needs to do much more to fully demonstrate the emotional behavior desired by the organization. Whether the employee exhibits deep acting, surface acting or genuine acting, the use of emotions by the employees in a workplace means that they exhibit emotional labor. The serious thing here is how the employee reflects the emotion outside. Even if the employees really feel the emotions they have to exhibit, the emotions of the employees to show to the other side with this behavior are not left to the initiative of the employee and the emotions are assessed by the rules of emotional behavior. Therefore, the ability of the employee to showcase the emotions as desired and demanded by the organization is also a work of labor (Köse et al., 2011: 170).

According to Morris and Feldman's (1996) theory, emotional labor takes place under four dimensions. These dimensions are voiced as "Frequency of emotional exhibition," "Paying attention to unveil rules," "Diversity of emotions to be exhibited" and "Emotional dissonance."

4 Dimensions of Emotional Labor Behavior

There are many different dimensions related to emotional labor. This research was handled in three dimensions, including the surface acting and deep acting dimensions revealed by Hochschild (1983) and the genuine acting dimension revealed by Ashforth and Humphrey (1993). After reviewing the necessary literature, emotional labor behavior in this study has been discussed in three dimensions since it was established that several studies were performed in tourism businesses with these dimensions.

Surface acting is designated as "the behavior exhibited when there is a mismatch between the emotions felt by the employee and the emotions desired from the employees by the organization" (Ashfoth & Humphrey, 1993: 92–93; Hochschild, 1983: 37). In other words, surface acting is that employees show the emotions desired by the organization by changing their external appearance (facial expressions, tone of voice and body language), which is different from their real feelings (Onay, 2011: 592). Surface acting means that the employee behaves "by wearing a mask" even though he does not feel the emotions, reflecting the emotions required by his job rather than the emotions he really feels and plays a role like an actor (Çoruk, 2014: 81).

Employees who exhibit surface acting act following the rules of emotional behavior resolved and created by the organization rather than presenting their sincere and inner emotions (Nylander et al., 2011: 471). For example, although an employee in the guest affairs department of a hotel business feels depressed, the employee seems to be happy with the customers and greets them with a smile, and he exhibits the behavior as if he feels the emotions desired by the organization.

According to Grandey (2000: 100), surface acting has been a method frequently requested by organizations due to the demands of customers and service buyers to see controlled and regulated emotional expressions even when the employee feels and lives differently than the emotions he/she has to exhibit (Eroğlu, 2014: 149). As an example, in service businesses, the employees are warned that they parade the emotions desired by the business with the notices on corridor walls, offices and exit doors saying "Please smile! You are taking to the podium."

In addition, Hochschild examined the emotional labor processes of cabin crew members by attending courses at Delta Airline's cabin crew training center. According to the research data, it was mentioned that cabin crew candidates were warned to smile constantly during the training and the candidates complained that these smiles did not belong to them (Hochschild, 1983: 8).

Deep acting is the second dimension of emotional labor and is defined as the "effort by the employee to feel the emotions actually that the employee is asked to exhibit" (Ashfoth & Humphrey, 1993: 93). Deep acting takes place when the employee really strives to feel the emotions required by the job. In this dimension of emotional labor, an example of deep acting could be that a waiter working in the hotel business gets angry at the endless requests of the customers and is much aware of this feeling. However, in order to do his job well, he tries to replace this feeling with positive emotions. To put it differently, deep acting means replacing the emotion that one feels inside with a voluntary intervention with the emotion suitable for the job (Başbuğ et al., 2010: 256).

Deep acting turns up when an individual tries to feel the emotions that he/she should exhibit per the emotional behavior rules settled by the organization. Here emotions are actively promoted, suppressed or shaped. According to Hochschild deep acting emerges in two ways (Yürür & Ünlü, 2011: 86): Stimulating the individual's feelings by suppressing or trying to rouse a feeling and managing the individual's imagination by trying to revive ideas, thoughts and memories in order to feel the emotion related to the subject (e.g., thinking that marriage gives happiness).

The different side of the deep acting dimension from the surface acting dimension of emotional labor is that in this dimension, not only the behavior of the employee but also his emotions are harmonized with the code of conduct (Altın Gülova et al., 2013: 45). In other words, in deep acting, the business demands the employees to change their emotions as well as their behavior.

Ashforth and Humphrey (1993: 94) describe the genuine acting as "the sincere behavior by which employees do not need to act in their behavior while doing their job and can show the emotions they actually feel without any effort". According to the authors, the genuine acting is the feelings and behaviors that the organization expects the employee to exhibit, sometimes feeling and exhibiting without the need for any strain. Employees exhibit the emotions they feel as they do without having to take any role in exhibiting these emotions (Biçkes et al., 2014: 100). For example, that a bartender serving in the hotel business to show a sincere concern to relieve a sad customer expresses his true and sincere feelings (Chu & Murrmann, 2006: 1182; Basım & Beğenirbaş, 2012: 79).

In the study of Diefendorff et al. (2005), they specified that the genuine acting was demonstrated more than surface acting behavior or deep acting in the enterprises. As Ashforth and Humphrey (1993) and Diefendorff and Gosserand (2003) suggest that the uncovering of genuine acting should be considered as a third type of behavior along with surface acting and deep acting of emotional labor.

Expressing emotions naturally, in other words, genuine acting accounts for exhibiting the required behaviors and emotions required and expected by the organization while interacting with the customer during the service delivery, not because it is requested by the organization, but because it feels natural from within. Although the employee does not feel real emotions while exposing surface acting and deep acting, employees must make an effort to completely change their emotions or to suppress their real emotions so as to comply with the mandatory exposition rules. On the contrary, there is no need for such an effort in the genuine acting dimension of emotional labor (Bıyık & Aydoğan, 2014: 162). Employees showing the emotions they naturally feel, as in the surface acting dimension and the deep acting dimension, do not have a necessity and obligation. Unlike, the employee reflects the real emotions he feels inside to the outside environment (Basım & Beğenirbaş, 2012: 79).

When evaluated in general, it may be noticed that the concept of emotional labor has become an eminent medium for service businesses because production and consumption ensue simultaneously in service enterprises, and the effect of the enterprise in controlling the behavior of the employee during service delivery is minimal. The emotional labor concept clarified as laying out the desired and expected emotions by the employee during the service delivery constitute the emotions that the employee must exhibit during service delivery. These feelings are expressed by the employees as surface acting, genuine acting and deep acting.

5 Research Method

5.1 Population and Sampling

In order to spot the emotional labor behaviors that the employees of five-star hotel businesses adopt and possess a tendency to exhibit them and to evaluate whether there is a difference between the individual characteristics of the employees and the emotional labor behaviors, they make spectacle outline of the fundamental purposes of this research. The population of the research consists of the employees of five-star hotel businesses operating in Afyonkarahisar. The reason why the scope of the research includes the five-star hotel businesses is that the management nature and organization of such businesses operate more professionally than other types of hotel businesses (one-star, two-star, three-star and four-star) (Pelit, 2011: 125).

According to the data by Afyonkarahisar Provincial Culture and Tourism Directorate, there is a total of 10 five-star hotel businesses operating in the city of Afyonkarahisar, 7 operating in the center and 3 in Sandıklı district, and it was estimated that the total number of rooms of these facilities is 2986 (www.afyonkulturturizm.gov.tr). An exact number of employees in these businesses has not been reached, but considering the information by the Ministry of Culture and Tourism as the number of personnel per room in Turkey is 1.18 (Agaoğlu, 1992: 114), 2986*1,18=3523 employees has been reached. Thus, this number reveals that the population is less than N <10,000 (limited population). Since the number of employees working in five-star hotel businesses in Afyonkarahisar is less than ten thousand, limited populations (N <10,000) and the sampling volume calculation formula proposed for quantitative studies were used (Özdamar, 2001: 257; Ural & Kılıç, 2018: 41–43) and the minimum sample number was calculated as 346.

In order to reach the targeted sample size, a total of 600 questionnaires were distributed to five-star hotel businesses in Afyonkarahisar province between June and August 2018. As a result, the number of returned surveys is 570, and

26 surveys were excluded due to incomplete and incorrect filling. Because of the incomplete and incorrect filling, 26 questionnaires were removed, and 544 questionnaires were included in the analysis.

5.2 Data Collection and Analysis

Since this research owns an applied nature, a literature review was conducted on the emotional labor behaviors exhibited by the employees in tourism businesses, and the data of the study were collected with a questionnaire. In the questionnaire, 10 closed-ended questions are included aiming to calculate some of the individual characteristics of the employees (gender, age, marital status, educational status and education level in the field of tourism) and the characteristics of the business (the department, the term of employment in the industry, the term of employment in the business, the income and the type of employment in the business). However, in order to gauge the emotional labor behavior of employees, Hospitality Emotional Labor Scale (HELS), including 19 expressions, was employed, which was used by Kızanlıklı (2014) and developed by Chu and Murmann (2006) from the work of Brotheridge and Lee (2003), DeLay (1999), Grandey (1999), Kruml and Geddes (2000) for the hospitality industry (Chu & Murrmann, 2006: 1188). The 11 expressions of the Emotional Labor Scale consist of expressions of the emotional contradiction dimension representing the types of "surface acting" and "genuine acting." Also, 8 of the Emotional Labor Scale are expressions of the emotional effort dimension that represents the "deep acting" type of emotional labor. The expressions in the scale are rated with Likert type ranging from (1) Totally Disagree to (5) Totally Agree.

The data obtained in the study were analyzed through the statistical package program utilized for social sciences. Attitudes of employees toward emotional labor behaviors are described by calculating the arithmetic mean and standard deviation values. Explanatory factor analysis was applied to the scales to verify whether the scale used in the research has the same factors (size), and then scales have been tested. Additionally, Cronbach's Alpha value was calculated to test the reliability of the scales for internal consistency. Furthermore, t-test (independent samples t-test) was employed for samples independent of parametric tests in order to regulate whether employees' attitudes toward emotional labor behavior differ significantly based on their individual characteristics, and the study was statistically analyzed by one (unrelated) factor variance analysis (One Way Anova), and the results of the single factor variance analysis are given together with the Tukey Test results.

6 Findings and Discussion

When the findings regarding the business characteristics and individual characteristics of the employees of the five star hotel businesses participating in the research are examined, it may be noticed that the majority of the individuals (52.8 %) who participated in the study were male and 23–27 years old (34.0 %) and married (59.7 %). It was also labeled that the majority of the participants (37.3 %) worked permanently in the tourism sector for 2–4 years and in the business, currently working are 79.8 %. This situation can be interpreted as the fact that city hotels operating year-round offer permanent opportunities to employees in the tourism sector compared to the resort hotels. It was fixed that the majority of individuals consisted of high school (46.3 %) graduates and people who did not receive education in the field of tourism were 49.8 %. In this context, it has been observed that almost half of the employees in hotel businesses are composed of those who were not trained in tourism. The reason for this situation is that the income obtained in the field of tourism is low along with excessive and irregular working hours and absence of overtime payments. Thus, it can be interpreted that people who are trained in the field of tourism tend to move to different sectors. When the income levels are analyzed, it is found that the majority of the employees are composed of individuals with income between 1601 and 2500 TL (52.8 %). As for the department in the hotel businesses, it was detected that 25.9 % works in the food and beverage department, 23.2 % works in housekeeping department, 13.8 % works in front office department, 7.7 % works in fitness and spa department, 7.0 % works in sales and marketing department, 5.7 % works in the guest relations department and 4.8 % works in the human resources department. It has been measured that (11.9 %) of employees work in the "other" (security, accounting, technical service, etc.) category since the number of employees is not high depending on the employment intensity of the departments in hotel enterprises.

Cronbach's Alpha values with explanatory factor analysis to ascertain the construct validity and reliability levels of the emotional labor scale used in this research are given in Tab. 7-1. However, as a result of the normality test as to whether the data shows the normal distribution, it was obtained that Skewness and Kurtosis values of the scale expressions used in the research are between -1.0 and +1.0 and according to Hair et al. (2014: 34), it was evaluated that the data showed the normal distribution.

Tab. 7-1 shows there is a relationship between the variables taken for factor analysis as a result of the Bartlett test (p =0.000<0.05) conducted for the Emotional Labor Scale. As a result of the test, it was found out that

Emotional Labor Behaviors in Hotel Enterprises 141

Tab. 7-1: Findings by Explanatory Factor Analysis and Reliability Level of Emotional Labor Scale

Expressions	Factors		
	1st (Deep)	2nd (Surface)	3rd (Genuine)
I fake a good mood when interacting with customers.	-0,154	**0,573**	0,271
The emotions I display when dealing with customers are different from what I really feel.	-0,088	**0,819**	0,056
I wear fake facial expressions while expressing real feelings about my job.	-0,040	**0,730**	0,023
My behavior and what I really feel differ from each other.	-0,017	**0,830**	0,015
I put on an act in order to deal with customers.	0,003	**0,732**	-0,036
I am in quite artificial interaction with the customers.	0,021	**0,798**	-0,070
I display emotions I am not actually feeling.	0,046	**0,803**	-0,155
I have to cover up my true feelings when dealing with the customers.	0,014	**0,782**	-0,072
The emotions I show to the customers match with what I truly feel.	0,317	-0,055	**0,809**
I actually feel Emotions that I have to display to do my job well.	0,449	-0,025	**0,732**
I reflect the feelings I feel to customers in the same way.	0,395	0,033	**0,782**
I try to reflect on my feelings by changing them into the way I need to show my customers.	**0,698**	-0,143	0,106
While interacting with the customers, I try to create some feelings in myself to present the image that the hotel wants.	**0,743**	-0,098	0,125
While getting ready for work I think of good things.	**0,828**	0,048	0,076
While helping customers, I try to convince myself to get rid of what I actually feel.	**0,714**	-0,033	0,302
When getting ready to go to school, I tell myself that I am going to have a good day.	**0,825**	0,034	0,106
I try to experience the emotions I have to show while interacting with customers.	**0,818**	0,029	0,222
I try to do my best actually to experience the emotions that I must show to the customers.	**0,829**	0,048	0,206

(continued on next page)

Tab. 7-1: Continued

Expressions	Factors		
	1st (Deep)	2nd (Surface)	3rd (Genuine)
I have to concentrate more on my behavior when I want to display an emotion I actually feel.	**0,651**	-0,077	0,247
Eigenvalues	5,190	4,698	2,204
Variance Explanation Rate (%)	27,317	24,728	11,600
Total Variance Explanation Rate (%)	63,644		
Reliability Level for Factors (α)	0,910	0,894	0,857
General Reliability Level (α)	0,868		

1st Factor: Deep Acting, 2nd Factor: Surface Acting, 3rd Factor: Genuine Acting; α: Cronbach's Alpha Value; Kaiser-Meyer-Olkin Value (KMO)= 0,909; Bartlett's Test: X^2= 5993,898; p= 0,000

(KMO = 0,909> 0,60) the sample size is sufficient for factor analysis application. In factor analysis, in all samples with more than 100 participants and the analyses with factor loads of over 0.50, results are statistically significant (Coşkun et al., 2017: 277).

Although the factor load value of 0.45 or higher is a good measure for selection, this limit value can be reduced to 0.30 for fewer expressions in practice (Büyüköztürk, 2003: 124). In the factor analysis application, the varimax method was chosen, and it was assured that the structure of the relationship between the factors remained the same. According to the result of factor analysis, it is seen that factor loads vary between 0.573 and 0.830. However, it was discovered in the factor analysis that the eigenvalue of the scale is greater than 1 and is gathered under 3 dimensions (factor), which explains 63,644 % of the total explained variance. Surveying the study in terms of dimensions, it is recognized that the "deep acting" dimension explained 27.317 % of the total variance, "surface acting" dimension explained 24.728 % of the total variance, and the "genuine acting" dimension explained 11.600 % of the total variance. According to the explained variance value, Emotional Labor Scale was found to be a valid measurement tool. When the Cronbach's Alpha value regarding the Emotional Labor Scale and its sub-dimensions is studied, it may be spotted that the overall reliability level of the Emotional Labor Scale was α = 0.868. Cronbach's Alpha coefficient of the surface acting dimension was found to be α = 0,894 while the Cronbach's Alpha coefficient of the genuine acting dimension was α = 0,857. Finally, the Cronbach's Alpha coefficient of

Tab. 7-2: Descriptive Statistics for Emotional Labor Scale

Scale		X̄	s.d.
Emotional Labor	*Surface Acting*	2,66	0,99
	Genuine Acting	3,46	1,11
	Deep Acting	3,82	0,89

X̄: *Arithmetic mean; s.d.: Standard deviation*

deep acting dimension was calculated to be α = 0.910, and it was computed to be quite reliable.

Arithmetic averages and standard deviation values related to emotional labor behaviors are presented In Tab. 7-2. As the data regarding the emotional labor behaviors of the employees are evaluated, it may be realized that it is the deep acting with the highest mean ($\bar{X}=3,82$). The Genuine acting has the second highest mean ($\bar{X}=3,46$). On the other hand, the surface acting dimension holds the lowest mean ($\bar{X}=2,66$). This state may be interpreted that hotel business employees are in favor of exhibiting the deep acting behavior most and secondly sincere acting whereas they are intended to showcase surface acting the least.

Similar to the studies carried out in the tourism sector by Kızanlıklı (2014) and Kılıç et al. (2016), it was decided that employees front to exhibit deep acting the most, secondly genuine acting and surface acting the least. Whereas, in the study conducted by Kurt (2013), it was figured out that the employees of hotel enterprises face to exhibit surface acting the most, and they inclined to exhibit deep acting the least. However, in the studies carried out by Pala Morkoç (2014) and Ayana (2016), it was indicated that hotel business employees are interested in manifesting genuine acting the most, the second behavior is deep acting and the behavior they are in support of unveiling the least is the surface acting.

The business characteristics and individual characteristics of the participants and the emotional labor behaviors they exhibited were compared and the findings obtained are given in the tables. Tab. 7-3 shows the results of variance analysis for comparing the emotional labor behaviors of the five-star hotel business employees with the age variable. According to the results, it was found that surface acting (p = 0,169; p> 0,05) and deep acting (p = 0,123; p> 0,05) did not differ significantly by the age of the employees. On the other hand, it was obtained that genuine acting (p = 0.028; p<0.05) varied by age variable.

Tukey test was carried out to appraise which age groups differ in the genuine acting reflected by the employees. According to this result, it was ended that employees aged 22 and under and aged 43 and above monitored more genuine acting than those aged 38–42. This situation may take place since those who are 22 years old and under may have just started work in the hotel business or since it may be the first job they take or they may be more enthusiastic than middle-aged employees. Nevertheless, it may be explained since with the sincerity of working in the sector for many years, employees working at the age of 43 and over may dispose to exhibit more genuine acting compared to the middle-aged employees. Similar results were found in the study conducted by Yakar (2015) in the related literature in the tourism sector. That is, there is no significant relationship between the surface acting and deep acting dimensions of the emotional labor exhibited by the employees and the age variable. However, in the genuine acting dimension, it has been identified that employees between the ages of 41 and 45 show more genuine acting compared to those who are between the ages of 26 and 30.

Tab. 7-3 reveals that surface acting ($p = 0.132$; $p > 0.05$) and genuine acting ($p=0.243$; $p>0.05$), which are the emotional labor behaviors exhibited by the employees of five-star hotel businesses, did not differ by the education levels of the employees in the field of tourism. Whereas, it was concluded as a result of variance analysis that the deep acting ($p = 0.008$; $p<0.05$) revealed by the employees varies according to the education level of the employees in the field of tourism. As a result of the Tukey test, performed to find out at what level the deep acting differs by the employees trained in the field of tourism, it was perceived that employees who have received tourism education at "other" (e.g., course) level tend to behave more deep acting than those who were trained at associate degree level.

Likewise, according to the result of the t-test conducted to compare the surface acting ($p=0,237$; $p>0,05$), genuine acting ($p=0,878$; $p>0,05$) and deep acting ($p=0,368$; $p>0,05$) and the gender variable, it was detected that there was no significant difference. It was also gauged as a result of analysis of variance that there was no difference between the surface acting ($p=0,080$; $p>0,05$), genuine acting ($p=0,121$; $p>0,05$) and deep acting ($p=0,838$; $p>0,05$) and the marital status variable. It was computed as a result of analysis of variance that there was no significant difference between the variables of surface acting, genuine acting and deep acting of the participants and education level of the employees, their departments, the term of employment in the tourism sector and the term of employment in the business.

Tab. 7-3: Emotional Labor Behaviors of Participants and Analysis of Variance for the Comparison of Age and Tourism Education Level Variables

Variables			n	X̄	s.d.	F	P	Tukey HSD
Surface Acting	Age group	22 and younger	87	3,26	0,92	1,563	0,169	-
		Between 23–27	185	3,22	1,03			
		Between 28–32	86	3,30	1,01			
		Between 33–37	89	3,41	0,97			
		Between 38–42	54	3,50	0,92			
		43 and older	43	3,60	0,95			
Genuine Acting	Age group	22 and younger[a]	87	3,68	1,04	2,537	0,028*	a>b c>b
		Between 23–27	185	3,43	1,06			
		Between 28–32	86	3,40	1,21			
		Between 33–37	89	3,43	1,08			
		Between 38–42[b]	54	3,12	1,19			
		43 and older[c]	43	3,79	1,08			
Deep Acting	Age group	22 and younger	87	3,91	0,90	1,741	0,123	-
		Between 23–27	185	3,75	0,90			
		Between 28–32	86	3,99	0,79			
		Between 33–37	89	3,80	0,89			
		Between 38–42	54	3,61	1,02			
		43 and older	43	3,95	0,86			
Variables			n	X̄	s.d.	F	P	Tukey HSD
Surface Acting	Education Level in Tourism	No Education	271	3,43	0,96	1,776	0,132	-
		Tourism Vocational High School	145	3,25	0,93			
		Tourism Associate Degree	30	3,13	0,94			
		Tourism undergraduate program and postgraduate program	64	3,15	1,12			
		Other (course, etc.)	34	3,34	1,19			
Genuine Acting	Education Level in Tourism	No Education	271	3,46	1,14	1,369	0,243	-
		Tourism Vocational High School	145	3,39	1,10			
		Tourism Associate Degree	30	3,20	0,86			
		Tourism undergraduate program and postgraduate program	64	3,58	1,13			
		Other (course, etc.)	34	3,76	1,01			

(continued on next page)

Tab. 7-3: Continued

Variables			n	x̄	s.d.	F	P	Tukey HSD
Deep Acting	Education Level in Tourism	No Education	271	3,83	0,93	3,519	0,008**	b>a
		Tourism Vocational High School	145	3,81	0,86			
		Tourism Associate Degree [a]	30	3,40	0,80			
		Tourism undergraduate program and postgraduate program	64	3,82	0,79			
		Other (course, etc.) [b]	34	4,23	0,90			

**: $p<0,001$; *: $p<0,05$

Similar results were obtained in the studies conducted on this subject in the tourism sector. For instance, Kızanlıklı (2014), Yakar (2015) and Karakaş (2015) concluded that surface acting and genuine acting dimensions of emotional labor do not differ by gender. Additionally, Keleş (2014) precipitated that the genuine acting dimension did not differ by the gender variable. It was ended as a result of the studies conducted by Yakar (2015) and Ayana (2016) and Karakaş (2015) that emotional labor behavior of employees did not change according to their marital status. Moreover, it was found in the studies conducted by Ayana (2016) and Karakaş (2015) that emotional labor behaviors of employees did not differ according to the department variable. Similar to the previous studies, it was assessed as a result of the studies carried out by Kızanlıklı (2014) and Yakar (2015) and Karakaş (2015) that the emotional labor behaviors presented by the employees do not differ by their educational status. Studies conducted by Ayana (2016) and Karakaş (2015) show that the emotional labor behavior of employees did not differ according to their term of employment in the sector.

Contrary to the findings obtained in this study, some studies in the field of tourism have attained different results. For example, male employees tend to exhibit more surface, genuine, and deep acting compared to female employees (Ayana, 2016). In another study, it was articulated that female employees exhibit more surface, genuine and deep acting than male employees (Işık, 2015). Şat et al. (2015) put forward in their study that there was a significant difference between the deep acting and genuine acting of behavioral dimensions indicated by the employees and marital status. Additionally, married employees showed more genuine acting than single employees. As for the deep acting, they found that single employees paraded more deep acting compared to married employees. It was also learned that high school graduate employees have exhibited more surface, deep and genuine acting than those who have graduate degrees (Ayana, 2016) whereas it was recognized that the employees with primary and high school education have shown more deep acting than those who have been trained at associate and undergraduate level, and the employees who received education at primary level showed more sincere acting than those who received an education at associate and undergraduate level (Keleş, 2014). Yakar (2015) found that those working in the food and beverage department exhibit more genuine acting than those working in other departments. Similar to the study by Yakar (2015), Keleş (2014) figured out that the employees working in the front office, food and beverage and housekeeping departments showed more deep acting than those working in other (security etc.) departments. Kızanlıklı (2014) detected that the employees with 1–3 years, 3–5 years and 7–9 years work experience showed more surface acting than those working for 10 years or more while it was ended that those working in the tourism sector for 10 years or more showed more genuine acting than those working in the tourism sector for 1–3 years.

It may be noticed in Tab. 7-4 that surface acting ($p=0.626$; $p>0.05$) and deep acting ($p= 0.213$; $p>0.05$) did not differ significantly by the income level of the employees. However, it was resolved as a result of analysis of variance that genuine acting ($p =0.006$; $p<0.05$) showed a significant difference by the income levels of the employees. Tukey test was carried out to show which income levels differed among the employees. As a result of the test, it was understood that the employees with income levels of 1600 TL and less turn to disclose more genuine acting than the employees with an income of 3401 TL and more. As the reason for their low tendency to expose genuine acting, it may be suggested the employees with an income of 3401 TL and above are generally composed of middle and upper level managers in hotel businesses so these people may have difficulty in showcasing their true feelings to both employee management and management of customers-related problems.

In the meantime, it was found according to the variance analysis results that surface acting ($p =0,000$; $p<0.001$) showed a significant difference according to the type of employment in the business whereas genuine acting ($p=0.182$; $p>0.05$) and deep acting ($p=0.908$; $p>0.05$) did not differ significantly by the type of employment in the business. Tukey test was performed to detect in which groups of employees the surface acting differed. According to the Tukey test results, it has been found that the employees working permanently in the business have a tendency to exhibit more surface acting than the employees

Tab. 7-4: Analysis of Variance for Comparison of Emotional Labor Behaviors of Participants and Income Level and Type of employment in the business Variables

Variables			n	$\bar{X}$	s.d.	F	P	Tukey HSD
Surface Acting	Income rate	1600 TL and less	154	3,25	1,00	0,584	0,626	-
		Between 1601–2500 TL	287	3,35	1,02			
		Between 2501–3400 TL	79	3,35	0,90			
		3401 TL and more	24	3,48	0,73			
Genuine Acting	Income rate	1600 TL and less [a]	154	3,66	0,98	**4,232**	**0,006***	a>b
		Between 1601–2500 TL	287	3,43	1,13			
		Between 2501–3400 TL	79	3,37	1,16			
		3401 TL and more [b]	24	2,86	1,14			
Deep Acting	Income rate	1600 TL and less	154	3,86	0,84	1,501	0,213	-
		Between 1601–2500 TL	287	3,86	0,90			
		Between 2501–3400 TL	79	3,68	0,97			
		3401 TL and more	24	3,59	0,91			
Variables			n	$\bar{X}$	s.d.	F	P	Tukey HSD
Surface Acting	Type of employment in the business	Permanent [a]	434	3,41	0,99	**8,408**	**0,000****	a>b a>c
		Seasonal [b]	83	3,11	0,91			
		Other (Internship etc.) [c]	27	2,73	0,95			
Genuine Acting	Type of employment in the business	Permanent	434	3,44	1,12	1,706	0,182	-
		Seasonal	83	3,45	1,04			
		Other (Internship etc.)	27	3,85	0,98			
Deep Acting	Type of employment in the business	Permanent	434	3,83	0,91	0,097	0,908	-
		Seasonal	83	3,79	0,80			
		Other (Internship etc.)	27	3,86	0,94			

**: $p < 0,001$; *: $p < 0,05$

working in seasonal and other (internship etc.) groups. It may be interpreted that this state may stem from the fact that the individuals who work permanently in the hotel businesses have a longer working period in the business than those who work seasonally and other (internship etc.) category. In contrast, it was calculated in the study conducted by Yakar (2015) that employees' emotional labor behaviors did not differ by type of employment in the business.

Conclusion and Suggestions

This research aims to point out the emotional labor behaviors that the employees of five-star hotel businesses adopt and tend to exhibit and to measure whether there is a difference between the individual characteristics of the employees and the emotional labor behaviors they lay out, and the sample group of study consists of 544 employees with different individual characteristics.

As the findings regarding the emotional labor behaviors of the employees are examined, it has been found that the highest $\bar{X}=3.82$ arithmetic mean was in "deep acting" dimension, then $\bar{X}= 3.46$ arithmetic mean was in "genuine acting" dimension. On the other side, among the dimensions, the lowest $\bar{X}=2.66$ arithmetic mean appeared in "surface acting." This situation can be expressed as the tendency of the employees in hotel businesses to exhibit deep acting and genuine acting is high while their tendency to exhibit surface acting is low.

According to the results of the t test and variance analysis to specify the differences between the emotional labor behaviors and individual characteristics of the employees, it was detected that the emotional labor behaviors of the employees showed no significant difference in terms of gender, marital status, educational status, department, term of employment in the tourism sector and the term of employment in the business. However, it was reached that the emotional labor behavior of the employees (surface acting, genuine acting and deep acting) showed a significant difference by age, education level in the field of tourism, income level and the type of employment in the business.

The suggestions that can be presented regarding the emotional labor behaviors of the employees are as follows:

- Training support should be provided to employees. For instance, pieces of training on emotion management should be supplied to help control the emotions of the employees and to enable the employees to array the emotions desired and expected by the business and it may contribute to businesses.
- Employees' feeling that they are supported by managers and their co-workers will contribute to the reduction in the surface acting that employees will

flourish and as a result of the formation of a sense of responsibility, it may contribute to the effort of the employee to illustrate the emotions desired by the company, that is, to exhibit deep acting. After employees go into the tendency of exhibiting deep acting and try to feel the emotions desired by the business during service delivery, after a while this may turn into genuine acting and thus it may contribute to an increase in the service quality, to ensure customer satisfaction and to establish loyal customers. As a result, the company may gain a competitive advantage over its competitors.

- The tourism businesses involve a labor-intensive structure, and the service is consumed where it is produced; the control over the attitudes and behaviors uncovered by the employees during the service delivery is at the minimum level and the behaviors of the employees directly affect customers' perception of service quality make the support of managers and colleagues critical with regards to the businesses. The support of managers and colleagues can positively affect the emotional labor behaviors that employees may exhibit. Hence, they can contribute to improved service quality and ensure customer satisfaction and loyalty.

Bibliography

Ağaoğlu, O. K. (1992), "İşgücünü Verimli Kullanma Tekniklerinin Turizm Sektörüne Uygulanması", *Verimlilik Dergisi, Milli Prodüktivite Yayını*, 457, pp. 1–203.

Altın Gülova, A., Palamutçuoğlu, B. T., and Palamutçuoğlu, A. T. (2013), "Duygusal Emek ile İşe Bağlılık Arasındaki İlişkide Amir Desteğinin Rolü: Üniversitede Öğrenci İşleri Personeline Yönelik Bir Araştırma", *Dokuz Eylül Üniversitesi İktisadi ve İdari Bilimler Fakültesi Dergisi*, 28 (2), pp. 41–74.

Ashforth, B. E. and Humphrey, R.H. (1993), "Emotional Labor in Service Roles: The Influence of Identity", *Academy of Management Review*, 18 (1), pp. 88–115.

Ayana, Ç. (2016), *Duygusal Zekâ ve Duygusal Emeğin Örgütsel Sinizm Algısına Etkisi: Beş Yıldızlı Otel İşletmelerinde Bir Uygulama* (Yayınlanmamış YL Tezi), Afyon Kocatepe Üniversitesi, Afyonkarahisar.

Basım, H. N. and Beğenirbaş, M. (2012), "Çalışma Yaşamında Duygusal Emek: Bir Ölçek Uyarlama Çalışması", *Celal Bayar Üniversitesi İktisadi ve İdari Bilimler Fakültesi Yönetim ve Ekonomi Dergisi*, 19 (1), pp. 77–90.

Başbuğ, G., Ballı, E., and Oktuğ, Z. (2010), "Duygusal Emeğin İş Memnuniyetine Etkisi: Çağrı Merkezi Çalışanlarına Yönelik Bir Çalışma", *Sosyal Siyaset Konferansları Dergisi*, 58, pp. 253–274.

Bıyık, Y. and Aydoğan, E. (2014), "Duygusal Emek ile Örgütsel Vatandaşlık Davranışı İlişkisi: Bir Araştırma", *Gazi Üniversitesi İktisadi ve İdari Bilimler Fakültesi Dergisi,* 16 (3), pp. 159–180.

Biçkes, D. M., Yılmaz, C., Demirtaş, Ö., and Uğur, A. (2014), "Duygusal Emek ile İş Tatmini Arasındaki İlişkide Psikolojik Sermayenin Aracılık Rolü: Bir Alan Çalışması", *Eskişehir Osmangazi Üniversitesi İktisadi ve İdari Bilimler Fakültesi Dergisi,* 9 (2), pp. 97–121.

Brotheridge, C. M. and Lee, R. T. (2003), "Development and Validation of the Emotional Labour Scale", *Journal of Occupational and Organizational Psychology,* 76 (3), pp. 365–379.

Büyüköztürk, Ş. (2003), *Sosyal Bilimler İçin Veri Analizi El Kitabı,* Ankara: Pegem Akademi Yayıncılık.

Chu, K. H. L. and Murrmann, S. K. (2006), "Development and Validation of the Hospitality Emotional Labor Scale", *Tourism Management,* 27 (6), pp. 1181–1191.

Çoruk, A. (2014), "Yükseköğretim Kurumlarında Görev Yapan İdari Personelin Duygusal Emek Davranışları", *Ondokuz Mayıs Üniversitesi Eğitim Fakültesi Dergisi,* 33 (1), pp. 79–93.

Çoşkun, R., Altunışık, R., and Yıldırım, E. (2017), *Sosyal Bilimlerde Araştırma Yöntemleri SPSS Uygulamalı* (5. Baskı). Sakarya: Sakarya Yayıncılık.

Delay, M. T. (1999). *Emotional Labour in the Workplace* (Unpublished Master Thesis), University of Arkansas, Little Rock.

Diefendorff, J. M. and Gosserand, R. H. (2003), "Understanding the Emotional Labor Process: A Control Theory Perspective", *Journal of Organizational Behavior,* 24 (8), pp. 945–959.

Diefendorff, J. M. and Richard, E. M. (2008), Not All Emotional Display Rules are Created Equal Distinguihing between Prescriptive and Contextual Display Rules. In N. M. Ashkanasy and L. C. Cooper (Eds.), *Research Companion to Emotion in Organizations,* United Kingdom: Edward Elgar.

Diefendorff, J. M., Croyle, M. H., and Gosserand, R. H. (2005), "The Dimensionality and Antecedents of Emotional Labor Strategies", *Journal of Vocational Behavior,* 66 (2), pp. 339–357.

Eroğlu, E. (2010), "Örgütsel İletişimin İşgörenlerin Duygu Gösterimlerinin Yönetimine Olan Etkisi", *Selçuk Üniversitesi İletişim Fakültesi Akademik Dergisi,* 6 (3), pp. 18–33.

Eroğlu, Ş. G. (2014), "Örgütlerde Duygusal Emek ve Tükenmişlik İlişkisi Üzerine Bir Araştırma", *Pamukkale Üniversitesi Sosyal Bilimler Enstitüsü Dergisi,* (19), pp. 147–160.

Goffman, E. (1959), *The Presentation of Self in Everyday Life*, New York: Doubleday Anchor.

Grandey, A. A. (1999), *The Effects of Emotional Labor: Employee Attitudes, Stress and Performance* (Unpublished Doctoral Dissertation),Colorado State University, Colorado.

Grandey, A. A. (2000), "Emotion Regulation in the Workplace: A New Way to Conceptualize Emotional Labor", *Journal of Occupational Health Psychology*, 5 (1), pp. 95-110.

Güney, S. (2015), *Sosyal Psikoloji* (4. Basım), Ankara: Nobel Akademik Yayıncılık.

Hair, J. F., Black, W. C., Babin, B. J., and Anderson, R. E. (2014), *Pearson New International Edition (Multivariate Data Analysis)*, England: Pearson Education.

Hochschild, A. R. (1983), *The Managed Heart: The Commercialization of Human Feeling*. Berkeley: University of California Press.

Işık, Z. (2015), *Erzurum Palandöken Kış Turizm Merkezindeki Konaklama İşletmelerinde Çalışan Personellerde Duygusal Emek ve İş-Aile Çatışması İlişkisi* (Yayınlanmamış YL Tezi), Atatürk Üniversitesi, Erzurum.

Johnson, H. A. M. (2007), *Service with a Smile: Antecedents and Consequences of Emotional Labor Strategies* (Unpublished Doctoral Dissertation), University of South Florida.

Kaplan, M. (2010), *Otel İşletmelerinde Etiksel İklim ve Örgütsel Destek Algılamalarının Örgütsel Bağlılık Üzerindeki Etkisi (Kapadokya Örneği)* (Yayınlanmamış Doktora Tezi), Selçuk Üniversitesi, Konya.

Karakaş, A. (2015), *Duygusal Emeğin İş Davranışlarına Etkisi: Otel İşletmesi Çalışanları Üzerine Bir Araştırma* (Yayınlanmamış Doktora Tezi), Dicle Üniversitesi, Diyarbakır.

Karaman, N. (2017), "Çalışma Yaşamında Duygusal Emek", *İş ve Hayat Dergisi*, 3 (5), pp. 30-57.

Keleş, Y. (2014), *Örgütsel Adaletin Duygusal Emek Üzerindeki Etkisi: Antalya'daki Beş Yıldızlı Otel İşletmelerinde Bir Araştırma* (Yayınlanmamış Doktora Tezi), Gazi Üniversitesi, Ankara.

Kılıç, B. and Baş, M. (2015), "Engelli Turistlerin Duygusal Emek Algısının Müşteri Memnuniyetine Etkisi", *Uluslararası Sosyal ve Ekonomik Bilimler Dergisi*, 5 (1), pp. 15-24.

Kılıç. G., Ergen F. D., and Gülaydın, M. (2016), "Duygusal Emeğin, Örgütsel Yabancılaşma ve İşten Ayrılma Niyeti Üzerine Etkisi: Afyonkarahisar'daki Beş Yıldızlı Otel İşletmelerinde Bir Araştırma", *17. Ulusal Turizm Kongresi*

Bildiri Kitabı, Muğla Sıtkı Koçman Üniversitesi Turizm Fakültesi, Muğla, pp. 882-892.

Kızanlıklı, M. (2014), *Otel İşletmelerinde Duygusal Emek Öncüllerini Belirlenmesine Yönelik Bir Araştırma* (Yayınlanmamış Doktora Tezi), Gazi Üniversitesi, Ankara.

Köse, S., Oral, L., and Türesin, H. (2011), "Duygusal Emek Davranışlarının İşgörenlerin Tükenmişlik Düzeyleri ile İlişkisi Üzerine Sağlık Sektöründe Bir Araştırma", *İşletme Fakültesi Dergisi*, 12 (2), pp. 165-185.

Kruml, S. M. and Geddes, D. (2000), "Exploring the dimensions of emotional labor", *Management Communication Quarterly*, 14 (1), pp. 8-49.

Kurt, Z. (2013), *Duygusal Emek Faktörünün Yabancılaşmaya Etkisi: İstanbul'daki Otel İşletmeleri ve Seyahat Acentalarına Yönelik Bir Araştırma* (Yayınlanmamış YL Tezi), Çanakkale Onsekiz Mart Üniversitesi, Çanakkale.

Morris, A. J. and Feldman D.C. (1996), "The Dimensions, Antecedents and Consequences of Emotional Labor", *Academy of Management Review*, 21 (4), pp. 986-1010.

Nylander, P. A., Lindberg, O. and Bruhn, A. (2011), "Emotional Labour and Emotional Strain among Swedish Prison Officers", *European Journal of Criminology*, 8 (6), pp. 469-483.

Oğuz, H. and Özkul, M. (2016), "Duygusal Emek Sürecine Yön Veren Sosyolojik Faktörler Üzerine Bir Araştırma: Batı Akdeniz Uygulaması", *Süleyman Demirel Üniversitesi Vizyoner Dergisi*, 7 (16), pp. 130-154.

Onay, M. (2011), "Çalışanın Sahip Olduğu Duygusal Zekasının ve Duygusal Emeğinin, Görev Performansı ve Bağlamsal Performans Üzerindeki Etkisi", *Ege Akademik Bakış Dergisi*, 11 (4), pp. 587-600.

Özdamar, K. (2001), *SPSS ile Biyoistatistik* (4. Baskı), Eskişehir: Kaan Yayınları.

Pala Morkoç, T. (2014), *Otel Çalışanlarında Duygusal Emek ve İş-Aile Çatışması İlişkisi: İzmir Şehir Otelleri Örneği* (Yayınlanmamış Doktora Tezi), Dokuz Eylül Üniversitesi, İzmir.

Pelit, E. (2011), "Otel İşletmelerinde Operasyonel Risk Yönetimi: Ankara'daki Dört ve Beş Yıldızlı Otel İşletmelerinde Bir Araştırma", *Business and Economics Research Journals*, 2 (2), pp. 117-137.

Seçer, H. Ş. (2005), "Çalışma Yaşamında Duygular ve Duygusal Emek: Sosyoloji, Psikoloji ve Örgüt Teorisi Açısından Bir Değerlendirme", *Sosyal Siyaset Konferansları Dergisi*, 7 (50), pp. 814-834.

Steinberg, R. J. and Figart, D. M. (1999), "Emotional Demands at Work: A Job Content Analysis", *The Annals of the American Academy of Political and Social Science*, 561, pp. 177-191.

Şat, A., Amil, O., and Özdevecioğlu, M. (2015), "Duygusal Zekâ ve Duygusal Emek Düzeylerinin Bazı Demografik Değişkenler Açısından İncelenmesi: Özel Okul Öğretmenleri ile Bir Araştırma", *Sosyal Bilimler Enstitüsü Dergisi*,2 (39), pp. 1–20.

Turunç, Ö. and Çelik, M. (2010), "Algılanan Örgütsel Desteğin Çalışanların İş-Aile, Aile-İş Çatışması, Örgütsel Özdeşleşme ve İşten Ayrılma Niyetine Etkisi: Savunma Sektöründe Bir Araştırma", *Atatürk Üniversitesi Sosyal Bilimler Enstitüsü Dergisi*,14 (1), pp. 209–232.

Ural, A. and Kılıç, İ. (2018), *Bilimsel Araştırma Süreci ve Spss ile Veri Analizi* (5. Baskı). Ankara: Detay Yayıncılık.

Yakar, S. (2015), *Turizm İşletmelerinde Duygusal Emek ve Tükenmişlik İlişkisi: Otel İşletmelerine Yönelik Bir Araştırma* (Yayınlanmamış YL Tezi), Dokuz Eylül Üniversitesi, İzmir.

Yüksel, H. (2014), "Emek Kavramının Ortaya Çıkışında Rol Oynayan Tarihi Dönüm Noktalarının Süreç Merkezli Değerlendirilmesi", *Süleyman Demirel Üniversitesi İktisadi ve İdari Bilimler Fakültesi Dergisi*, 2 (19), pp. 257–273.

Yürür, S. and Ünlü, O. (2011), "Duygusal Emek, Duygusal Tükenme ve İşten Ayrılma Niyeti İlişkisi", *"İş, Güç" Endüstri İlişkileri ve İnsan Kaynakları Dergisi*, 13 (2), pp. 81–104.

http://www.afyonkulturturizm.gov.tr/TR-158749/kultur-turizm-istatistikleri. html (01.06.2018).

Esra ERENLER TEKMEN

Women in Management in the Tourism Industry and the Glass Ceiling Problem (An Invisible Barrier in Workplace)

1 Introduction

Have you ever felt stuck at some point in your career? Have you ever failed to get the promotions you thought you deserved? What about the words "women cannot be managers," "women are so emotional," "you have a child, you can't spend enough time for work" – have you ever heard them? If you are reading these lines and your answer is yes, you may have hit the "glass ceiling" created consciously or unconsciously.

Globally, there has been an unprecedented change in the participation of the labor force in the world economy. The percentage of women in business world and executive positions has increased, though it varies from one country to another. However, the rate of increase is slow and lower than that of men. The increase in the number of women working in business world has not yet led to a major shift in the gender composition of senior leadership and decision-making positions in companies (ILO, 2019). Even though there have been significant improvements along with changes in employment policies aimed for gender equality, women continue to struggle to be employed in the boards and in senior management positions and to be treated fairly in company practices (Sharma & Kaur, 2019). When it comes to senior management, the situation is not pleasant at all for women. Despite the limited data on the subject to analyze the situation, it is possible to draw a general picture of the ratio of women in management positions by using the ILO data and other sources.

According to the ILO survey report on "Women in Business and Management" covering 70 countries and 13,000 businesses, more than 78 % of the businesses in the survey have male CEOs. As the size of the business increases, female CEO ratio drops from 26 % in small businesses to 20 % in medium businesses, and 16 % in large businesses 4.8 % of the companies listed in Fortune 500 in 2018 have female CEOs (ILO, 2019: 120). There are no women executives in more than 25 % of the Fortune 500 companies. Catalyst, the global organization that conducts research and provides solutions for women in leadership, reported that the CEOs of 5,8 % of the Standards and Poor's 500 companies are women. Further only 21.2 % of board seats were held by women (Catalyst, 2020).

According to Turkish Statistical Institute (TurkStat) data in 2019, the population of Turkey is 83 million and half (49.8 %) of the population are women. When the situation of women in business life is reviewed, according to TurkStat data, the female employment rate has increased from 25.3 % in 2002 to 29.4 % in 2018. The number of working women rose significantly. The ratio of working women which was 27.9 % in 2002 has increased to 32.2 % according to the data in 2018. Female employment is highest in the service sector. The rate of women employed in the service sector has increased from 26.3 % in 2002 to 57.9 % in 2018. On the other hand, according to the 2019 local election results, 3 out of 30 metropolitan mayors are women. Female rectors are employed in 19 (9.69 %) of 196 universities. The rate of female deans is 20.5 % (HEI, January 2020). 5.5 % of chiefs of police and 10.7 % of police commissioners are women (General Directorate of Security, January 2020). 11.38 % of the senior managers in the bureaucracy are women. The ratio of women in senior decision-making positions in the bureaucracy is rather low (as cited in, Ministry of Family, Labor and Social Services, 2020). According to TurkStat household labor force survey results, the rate of women in senior and mid-level executive positions has increased from 14.4 % in 2012 to 17.3 % in 2017 (www.turizmgunlugu.com). Even in "female-oriented" sectors where more female managers are expected, the ratio is rather low. Even though there has been a proportional rise both in the world and in Turkey, compared to men and the population, the data indicate that the number of female employees working in senior positions is rather insufficient and support the claim that there is a glass ceiling that prevents women from rising. The vast majority of the studies conducted by academicians and international organizations reveal that women are discriminated in their jobs, wages, working conditions, promotions, and social relations at work due to their gender (Albayrak, 2019: 105). Even though women hold various executive positions, very few of them reach the highest positions (Zel, 2002). Compared to men, women are usually employed in mid-level and lower leadership positions (Mohammadkhani & Dariush, 2016) and are kept from strategical management positions.

In parallel to women's more involvement in business life compared to the past, the number of women working in the tourism industry is also increasing gradually (Öktem, Kubat & Kızıltan, 2018). Tourism is a large and global industry that consists of many interrelated segments such as accommodation, transportation, food & beverage, and entertainment (Knutsan & Schmidgall, 1999). In addition, one job in the core tourism industry creates additional (indirect) jobs in the tourism-related economy. For example for hotel segment, if there is an average of one employee for each hotel room, there are workers indirectly dependent on each person working in hotels, such as travel agency staff, guides, taxi and bus drivers, food and beverage suppliers, laundry workers, shop staff for souvenirs, airport employees, and others. In general, tourism can offer good opportunities to women to work in occupations directly or indirectly in several other sectors (Bazazo et al., 2017). Considering the size of the sector and its potential for employment creation, it does not seem surprising that the number of women who choose to make a career in this sector is increasing.

However, the tourism sector, where women are employed predominantly, is one of the sectors where gender discrimination is experienced. Women who face many obstacles in nearly every sector work under a "glass ceiling" that they cannot pass over in the tourism sector, either. In this sector, men are predominantly involved in senior management positions, and women are subject to discrimination and separate treatment due to their gender (Aydın-Tükeltürk & Şahin-Perçin, 2008). Women can find places mostly in mid-level and staff positions in this industry (Öktem, Kubat & Kızıltan, 2018). The aforementioned discrimination may be implicit or explicit.

2 The Glass Ceiling Phenomenon

The term "glass ceiling" was first used in a special report about the women of business world by Hymowitz and Scehellhardt in Wall Street Journal in 1986. In this chapter, the glass ceiling describes a business life in which women's reaching for the summit is hindered by organizational traditions and prejudices (Jackson, 2001).

As the name suggests, the glass ceiling refers to invisible, artificial barriers that prevent qualified individuals in business from advancing in their organizations and reaching their full potential (Knutson & Schmidgall, 1999). The glass ceiling is a thin, transparent, but strong barrier that prevents women from getting employed in senior management positions (Morrison & Von Glinow, 1990). It is the situation where women do not get the promotion they expect after they have risen to a certain point as a manager due to reasons which are not known. Although men encounter certain problems in promotion, the so-called glass-ceiling problem is used for different practices mostly for female employees (Baxter & Wright, 2000). In this context, especially when it comes to career advancement, it points out to disadvantages experienced by women (Sharma & Kaur, 2019). In a holistic definition, the glass ceiling is an invisible and artificial barrier that hinders their career advancement and keep them stuck at a certain level, prevent them from reaching the upper levels of the organizational hierarchy no matter what their qualifications and how their performances are (ILO, 2019), preventing women's vertical mobility (Baxter & Wright, 2000). However, it should be remarked that the aforementioned barrier is not a one based on the person's inability to work on a higher-level job. This barrier is an invisible one in which women's achievements and personal traits are ignored and not understood (Jackson, 2001). It is called "glass" because it's usually an invisible barrier and a woman may not know its existence until she "hits" it. In other words, it's not an explicit practice of discriminating against women – though specific policies, practices, and attitudes may exist that produce this barrier without the intention to discriminate (Levis, 2019). It means that there is a glass ceiling between the management level of that company and female workers if it is thought that a female employee expecting a career advancement cannot perform a job only because she is a woman and if her promotion is prevented without any concrete evaluation (Ergeneli, 2014: 262). In organizations where the glass ceiling exists, female employees who are qualified, adequately equipped, well educated, prepared, and willing cannot reach their career goals or get promoted even if there are no objective reasons (ILO, 2019). According to Bazazo et al. (2017), many women in some steps of their career from all over the world encounter a "glass ceiling" above mid-level positions that hinder them from reaching higher management positions, whereas men can rise faster to executive positions through "glass escalators."

Taking studies, statistics, and articles about the glass ceiling into account, it can be said that the existence of the glass ceiling is not controversial. Current, female employees earn less than men and experience difficulties in reaching senior management positions, and negative attitudes toward women continue. But why?

In the following section, the factors that hinder women's career development and cause the glass ceiling will be addressed.

3 Factors Leading to the Formation of the Glass Ceiling

It is not easy to explain why the glass ceiling exists. The answer depends on who you ask and it may vary, depending on the differences in perception that the person you ask is a man or woman. However, one might say that the barriers which make it hard for women to reach senior executive positions are universal (Mohammadkhani & Dariush, 2016).

On the subject, structural-centered theories argue that structural, systematic organizational practices are responsible for the glass ceiling. Bias-centered theories argue that the glass ceiling exists because consciously or unconsciously male organizational leaders are biased against the promotion of women to the top levels of the organization (Hull & Umansky, 1997). Cultural-centered theories of the glass ceiling suggest that if male organizational leaders don't give support socially to women, it will end in in glass ceiling (Cohen et al., 2020).

Actually, there are many factors that make it harder for women to move up to top positions than for men. Individual, organizational, and social barriers surrounded by culture and the society are among these factors. Some of these factors are formed by male managers and some by female ones, and some of them concern individuals' own preferences and choices (Öktem, Kubat & Kızıltan, 2018). According to Morgan (2015), the glass ceiling barriers are artificial and they can be understood by comparing them to natural ones (as cited in Sharma & Kaur, 2019).

3.1 Individual Barriers

In a sense, individual barriers refer to the barriers that one puts on oneself. The concern about harming the family life, the belief that says, "I cannot be promoted, anyway,", and the lack of self-confidence brought upon by these are some of the individual barriers that one puts on oneself (Şivye, 2004). The importance given to the family life is one of the greatest challenges in this regard. Thinking about the negative effects of their careers on the family life, the role conflicts between the roles that they have learned during socialization, and the expectations of the business life are shaping the expectations about business life of women who have grown up in prejudice in socialization process (Crampton & Mishra, 1999). This situation, which is also known as multi-role taking, has its roots in the fact that women have to play their roles related to their working lives, as well as their private lives (Özyer & Azizoğlu, 2014). In this type of role conflicts, women tend to prefer their families over their careers more than men (Jakson, 2001). On the other hand, the fact that women sometimes get stuck between their jobs and families leads the decision-makers to believe that women cannot bear certain responsibilities introduced by their career advancement (Mızrahi & Aracı, 2010). Also, some women assist in the conservation of the glass ceiling through internalizing the proposition that males are superior managers or that the male need for promotion (due to their social role of "breadwinners") is more important or superior to the female need for promotion or women hold the belief that they are less deserving the organizational rewards – including salary and promotion – than males for a given level of work (Cohen et al., 2020).

The fact that women find it hard to achieve work-family balance in the tourism industry is largely because of the structure of the sector. Long and busy working hours, having to work on weekends and holidays, stressful and tiring duties such as making business agreements, arranging business travels to attend exhibitions, managing crises in the workplace, and problem-solving may lead to the notion that management is not suitable especially for women married and with children (Aydın-Tükeltürk & Şahin-Perçin, 2008).

3.2 Organizational Barriers

Organizational factors are also thought to be effective in women not being adequately involved in senior executive positions. Some of these include organizational culture and organizational policies, lack of a mentor, lack of participation in informal communication networks, lack of equal opportunities, and the perception of women in business life.

Organizational Culture and Policies: Organizational culture and policies are considered to be one of the important barriers to women's advancement. Organizational culture has a supportive or reducing effect in glass ceiling since it shapes the behavior, attitudes, and treatment toward the employees, the glass ceiling. Women's advancing in their careers and getting equal opportunities change according to the policies of the organization and organizational culture. Since organizations tend to reinforce the system of values regarding the dominant gender, policies and practices that reflect organizational culture maintain the male status quo. In the business life, most work environments have a male-dominated structure (Ergeneli & Akçamete, 2004). The fact that women have stepped into the business life later makes it difficult for them to find a place for themselves in the male-dominated organizational culture. The male dominance in the organization and especially in senior management levels strengthens men's impact as decision-makers and makes it difficult for women to get promoted (Badjo & Dickson, 2001). Organizational practices are effected by this situation, as well (Bazazo et al., 2017). In male-oriented organizational cultures, female managers are not considered suitable for some jobs and tasks, receive less value than they are worth, and, in some cases, do not even have the opportunity to prove themselves. In fact, the negative impact of organizational culture and practices on women begins at employment. While men are placed in positions that are thought to be more strategic, such as R&D, operations, and finance, women are employed in support departments (such as human resources) that do not offer the opportunity to advance to senior management. Employing women in such areas prevents them from developing their talents and visibility by senior management and damages their career advancement (ILO, 2019). According to Bergman et al. (2002), it is important for women to comply with predetermined rules that reflect organizational culture for a positive image in male-dominant companies.

On the other hand, it should be noted that even if a company has a formal, written policy that claims to support women, this policy may be weakened with certain attitudes toward women and especially women managers (Knutson & Schmidgall, 1999). In this context, some managers can give male employees jobs that they can prove themselves more than female employees and let them benefit from development opportunities more than women (Taşkın & Çetin, 2012).

The glass-ceiling phenomenon is also addressed in terms of promotion and wage inequality (Baroudi & Igbaria, 1995; Sumner & Niederman, 2003). According to studies, in most organizations, women's salaries are generally lower than men's. Nevertheless, women accept lower salaries, anticipating that they can prove their worth in time and get a fairer salary (Knutson & Schmidgall, 1999). However, it should be emphasized that it is hard to distinguish how much of the differences in wages and promotions are related to discrimination and how much is related to the choices and preferences of women and men (Sharma & Kaur, 2019).

Women encounter various difficulties in maintaining effective mentorship relations, planning their careers, and networking, as well (Riger & Galligan, 1980).

Lack of a Mentor: Another difficulty that women face is the lack of role models. A promotion to an executive position usually results from the interaction between the individual's abilities and efforts and formal support and encouragement. Therefore, an individual's career advancement requires a joint effort between the organization and the employee (Knutson & Schmidgall, 1999). Having a mentor is one of the elements that reinforces this process.

Access to mentoring is a particularly critical matter (Ragins et al., 1998). A mentor, for a less experienced individual than themselves, is an experienced person who guides, advises, and can be a role model in that individual's personal life, job, and career. Having a mentor has many benefits for an individual's career development (Taşkın & Çetin, 2012). An effective mentor shares their experiences by showing particular attention to the individual, guides in breaking through career barriers, increases motivation, and inspires. It helps to identify long-term goals and paths to reach the goals (Knutson & Schmidgall, 1999). Burke and McKeen (1990) conducted a study on how important the mentoring is for the careers of women. Their study reported that all the successful female managers who took place in the survey had a male mentor that had a big effect on their careers. While all of the women who reached top executive levels indicated that they got help from above as mentoring, only 55% of the men at the identical level had the same help. This reinforces the importance that male mentors play in the success of women (as cited in Simonetti et al., 1999).

Although having a mentor contributes to the career advancement, "gender" is a crucial barrier in developing a mentoring relationship for women (Ragins & Cotton, 1991; Hull & Umansky, 1997). Male managers avoid mentoring female mentees in order to ensure male dominance at the senior management levels and avoid harmful workplace gossip and disturbing implications, and they tend to display negative attitudes toward women who are actively engaged in mentoring relationships. Furthermore, the low number of female executives at the senior management level makes it difficult for female employees to find female mentors (Anafarta et al., 2008). These issues make it relatively more difficult for women to obtain mentoring relationship from senior-level managers.

Inability to Participate in Informal Communication Networks: Many organizations are mostly full of social networks relevant to men's interests. Thanks to these networks, employees can understand what motivates the various levels of the management, get to know which projects the top management level are interested in, and hear which positions are vacant and who are potential candidates for them (Karcıoğlu & Leblebici, 2014). Social support networks provide the circulation of critical organizational knowledge (Cohen et al., 2020). However, women, who are called "tokens" because they are few in the workplace, are not as fortunate as men to access information that may carry them to the senior management level since they are separated from the men who outnumber them (dominants) in behavioral differences (Ergeneli & Akçamete, 2004). In many cases, even if they have the executive chairs women remain outside the informal communication networks where their male colleagues who are in a strong position in the organization and this weakens their power (Snavely, 1993). Additionally, women are less likely to socialize with male partners and male clients through shared experiences such as interests in sports. Because of women's lack in access to these social support networks ("old boys networks"), they are also excluded from the information circulation that occurs within these information networks. This information includes organizational knowledge about important clients, specialization, and informal traditions (Cohen et al., 2020). Since men have a different communication style among themselves from women's and women cannot hear about the information concerning themselves and the organization or they do so late, women's careers may be adversely affected (Taşkın & Çetin, 2012). For this reason, it can be said that "old boy networks" have a critical role in continuation of organizational gender inequality (Hull & Umansky, 1997).

3.3 Social Factors

Social factors including discrimination and stereotypes are important factors affecting the organizational culture. A social structure with discrimination is differentiated by settling into the organization in various forms, and then reproduced by expanding (Ergeneli & Akçamete, 2004). Sexist "occupational segregation" and "stereotypes" that determine which professions are appropriate for women in the present society are effective both in whether women have an opportunity to rise in the management and also in their own decision-making process.

Professional Discrimination: Discrimination against women forms the basis of the glass ceiling. There are various approaches to gender discrimination in the business life. Common point of these approaches is to acknowledge the existence of gender discrimination in the labor force and to reveal the reasons for this. The human capital approach connects choosing different professions of both genders to gender roles and suggests that men focus more on their jobs especially than married women. Another assumption of the aforementioned approach is that women earn low wages when they return to work after birth since they cannot develop their skills while away from work. The human capital approach links gender discrimination in the business life to voluntarily individual choices that women make (Günlük-Şenesen & Pulhan, 2000: 9). According to the class approach, the labor market consists of "primary" markets where one can have relatively high wages and fine working conditions and of "secondary" markets where wages and working conditions are worse, and there are jobs that offer no chance of promotion and lack employment security. This stratification causes employees working in primary markets to perceive themselves differently from those working in secondary markets (Günlük-Şenesen & Pulhan, 2000: 11). Murrell and James (2001) state that men are employed in jobs in the primary market and women are mostly employed in jobs in the secondary market (as cited in Ergeneli & Akçamete, 2004). The approach based on gender-based social roles asserts that gender discrimination in the business life has its root in the differences in the socialization process of girls and boys. According to this approach, since women and men are raised in different manners, they develop distinct characteristics and attitudes, and this is reflected in their choice of profession. For instance, in the socialization process, women are encouraged to do housework and men to be authoritarian and aggressive and concentrate on mechanical and math-related fields. For this reason, women are represented at extremely low rates in jobs that require the implementation of authority (Günlük-Şenesen & Pulhan, 2000: 11). These biased expectations for determining the role that individuals play in the society are called gender roles.

Gender roles help determining which professions and jobs are appropriate for men and women. Hence, prejudices based on gender grow stronger over time. According to the statistical approach, the method employed in the process of evaluating candidates applying for a job causes discrimination. Accordingly, the fact that applicants are evaluated based on the characteristics of the gender group they belong to instead of the qualifications they have given rise to discrimination. Authors in favor of institutional approach attribute these different practices to institutional barriers. These barriers may be in the form of gender, race, and prejudice. Accordingly, the benefits that men gained over time in labor markets confront women as an "institutional power" (Ergeneli & Akçamete, 2004).

Social role theory states that individuals have expectations about appropriate behaviors for both men and women (Cohen et al., 2020). In fact, gender roles are one of the factors that determine the distribution of jobs between genders. Given that female employees are also associated with nurturing social roles such as child-care, biased individuals are likely to view female employees as less committed to their careers. In contrast, given that male employees are more likely to be associated with perceived assertiveness and ambition, male employees are often viewed as more suitable leaders than female employees (Weyer, 2007). In this reason, jobs like engineering, management, and general management are classified as men's, while jobs like nursing, teaching, and secretary are considered to be appropriate for women (Taşkın & Çetin, 2012). The discrimination of jobs as men's and women's adopted in the society prevents women from taking certain positions, particularly management positions, in the business life.

Occupational segregation by gender in the business life has been proven by various studies in many countries. Discrimination can be in the form of the concentration of women and men in different jobs and sectors (horizontal discrimination), as well as the separation of hierarchical levels by gender within the same jobs and sectors (vertical discrimination).

Since a significant part of the jobs in the tourism industry is thought to be "feminine," the female dominance in the labor force is high. However, this dominance exists mostly at lower levels (Tükeltürk-Aydın & Perçin-Şahin, 2008). Although women advance up to a certain level before facing the glass ceiling that keeps them from reaching upper management levels, they represent the minority in the senior management positions (Ganiyu et al., 2018). In this context, it can be said that horizontal and vertical gender discrimination exists also in the tourism industry as in other sectors. While women are mostly employed in positions such as human resources, public relations, housekeeping, and reception that do not offer the opportunity to advance for reaching top management levels, they face difficulties in finding positions as a general manager, finance manager, or food and beverages manager (Tükeltürk-Aydın & Perçin-Şahin, 2008). Vertical professional discrimination implies that female employees working in the same workplace and having similar qualifications as the male colleagues face different attitudes, behavior, and evaluations. This discrimination requires women to work harder and wait longer to get promoted in their jobs.

Stereotypes: Men's stereotypes against women and their desire not to lose their power are another barrier to women in the business life (Öktem, Kubat & Kızıltan, 2018). Negative prejudices are mostly based on the idea that women cannot perform high-level jobs (Örücü et al., 2007). Gender prejudices such as "we will allocate our resources to women, and then they will get married after some time and have children," "women put their families in the first place," and "women do not possess the necessary qualifications to be a fine manager" affect women's careers in many cases (Knutson & Schmidgall, 1999). Since the society expects women to play the roles of a mother and a housewife before that of a businesswoman, their career journey have more difficulties and advances more slowly than men's (Öktem, Kubat & Kızıltan, 2018). In the workplace, every step taken by women is controlled by men, and mistakes made are generally attributed to being a woman. Female managers feel under pressure since they represent other women, and if they fail, other women will be approached with prejudice (Taşkın & Çetin, 2012).

The perception of the woman has a great impact in the emerging picture. Although some studies have revealed that there is no difference between men and women in certain positions with respect to talent, there exists a perception that there is a difference. Successful male and female managers can be perceived differently by the employees. Especially when female managers are successful, their success is accounted for by factors such as "luck" and "the job being easy," and attributed to external causes independent of their personal traits. On the other hand, the achievements of male managers are generally explained by personal traits such as "talent" and "intelligence" (Lucas & Lovaglia, 1998). Also, prejudices that say that management requires "being strong" and women are "weak" and "emotional" and that they lack experience and professional competencies for general management or line jobs, and that their organizational commitment is weak are highly common in the workplace (Knutsan & Shmidgall, 1999). Gender has an effect on aforementioned prejudices.

Based on the explanations above, it can be said that attitudes are one of the important elements that creates the glass ceiling. Even though it is assumed that sexist behaviors in the workplace are generally originated by men, some studies underline the negative role of women who manage to succeed in male-dominant environments in the advancement of their female subordinates (Ellemers et al., 2004). High-ranking women can have a substantial influence on the advancement of other women beneath them (Cohen & Huffman, 2007) but in order to be unique at the point that they have risen to, some female managers would rather not support the promotion of other women (Cohen et al., 2020) and damage their careers (Ellemers et al., 2004). Ellemers and Barreto (2008) note that women who want to maintain their unique position and do not give other women equal chances for career have the queen bee syndrome (as cited in Ellemers, 2014).

Women who are in the authoritarian position as queen bees fight to eliminate the competition from other women in order to maintain their positions. According to them, unsuccessful women should look for the reason in themselves and make individual efforts to eliminate discrimination (Zel, 2002). Such female managers usually make the mistake of taking their efforts to rise as a reference. They think as "for promoting to my actual level, everyone can act in the same way I do." (Örücü et al., 2007). They generally exhibit more masculine behaviors, adopt masculine attitudes, and keep their distance from other female employees physically and psychologically (Derks, Laar & Ellemers, 2016). It can also be said that female managers changing their behaviors in business environment over time and coming close to display male managers' behavior patterns and show similar reactions to other female employees as like male managers prevent women from developing and advancing (Zel, 2002).

The attitudes of families, the society, and those who are in decision-making positions in the business life can contribute to the formation of the glass ceiling. A study conducted by Aycan (2004) indicated that attitudes are effective in women being senior executives in Turkey and sociocultural roles and values and individual, organizational, and family-related factors have a role in the formation of these attitudes.

Many organizations acknowledge the negative effects that the glass ceiling has on finding and keeping talented employees, as well as the long-term costs to the organizations and society that result from failing to eliminate the glass ceiling (Cohen et al., 2020). It is crucial for the welfare of companies to work in harmony without any discrimination. For this reason, the glass ceiling problem that hinder women's career advancement and which have undesired consequences that may damage the organization like as job dissatisfaction, the weakening of the organizational commitment, decrease in the efficiency, and employee leave should be fought (Downes et al., 2014). So, how can we fight with the elements mentioned in detail above?

4 Breaking the Glass Ceiling

Today's business environment is changing very fastly. New and emerging technologies, expanded markets, more global consumer bases, and heightened competition for the best talent mean that many enterprises have to review their business strategies and priorities. While complex socio-economic trends add to challenges enterprises encounter, having staff equipped with the right skills and talent without taking their gender and background into account enables enterprises to better navigate any upheaval. Now, more than ever, using gender diversity is part of their talent management strategy (ILO, 2019).

In many countries, particularly in the Western world, actions are taken to increase the representation rate of the groups not represented or underrepresented, to eliminate inequality and gender discrimination, and to change attitudes. Despite all these efforts, practices reflecting gender discrimination in the sectors where women make important contributions and even constitute the majority could not be eliminated completely (Özyer & Azizoğlu, 2014). In general, regulations and policies regarding gender are inadequate especially in the tourism industry in Turkey. In order to establish strategies and policies to break the glass ceiling, firstly, one has to understand the reasons behind the treatment to women, and the strategies that can be implemented are generally universal. In other words, the strategies that woman can practice to get involved in executive levels are generally not different from those in the corporate world (Knutsan & Schmidgall, 1999; Bazazo et al., 2017).

Several personal and organizational strategies can be used to break the glass ceiling. Suggestions for women, especially those in the tourism industry, for breaking the glass ceiling can be listed as follows.

Personal strategies: Various studies indicates that men are predominantly preferred in senior management positions since women have failed to build trust in making difficult decisions through professional and financial intelligence. Thus, it is essential for women to show that they can take risks, make the right decisions, and see the big picture (Demirel, 2015). At this point, it may be a good strategy to exceed expectations. Accomplishing tough and distinguishable project and working in distinct fields may help women for reaching their goals. Also, women who take up on such tasks may acquire new skills and catch their superiors' attention (Knutson & Schmidgall, 1999).

Many women are as competent as their male colleagues even though they get discouraged and lose their confidence sometimes. Brownell (1994) mentions that it may be effective for women to take a determined stance, to be determined and patient, to show positive behaviors, to develop personal skills and to work hard to get rid of the career barriers that they may encounter on the way to leading organizations.

Communication is also highly essential. Even though it has been proven by studies that women are better at communication, according to Demirel (2015), it may not have the expected effect in the business world since women communicate in a more assertive and aggressive manner. While aggression in men is acceptable to a certain extent, it keeps women from receiving support.

Developing working relationships that will enable women to share their work with their colleagues in emergencies concerning their families can help them to manage their family responsibilities as well as their work. Besides the family support, outsourcing can be another solution for working women in such matters as cleaning, feeding, and caring for their babies. Such methods may reduce the work-family conflict, save women some time to invest more in their jobs, and help them uncover their potential (Sharma & Kaur, 2019).

Organizational Strategies: There are two sides of advancement. In addition to what women's own acts for themselves, companies can develop strategies to support women advancement or can create barriers that can prevent women from advancing (Knutson & Schmidgal, 1999). At this point, supporting women's personal development is one of the common strategies. Therefore, it is crucial to find and employ especially women with leadership potential and to identify women with high potential. Having identified the women with potential, companies can help them develop the management skills necessary for their advancement. Specific actions need to be taken for women to gain experience that prepares them for promotion to strategic business areas (ILO, 2019). Job rotation, executive training, leadership training, backup planning, and mentoring may help with this. Also, these initiatives should be an inclusive part of the organization's strategy and culture for each employee (Bazazo et al., 2017). Providing equal employment opportunities may also be a strategy. Increasing networking opportunities, relocating, or transferring is among some other popular strategies (Knutson & Schmidgall, 1999).

Having a mentor who can be taken as a role model and can guide in the business life shapes the desires of the employees regarding their business lives (Crampton & Mishra, 1999). Therefore, having an effective mentor from within or outside the company is essential for women's careers. Companies may support female employees in this regard and help them overcome challenges about mentoring (Ragins et al., 1998).

We have mentioned that one of the factors that make it difficult for women to advance in their careers is the struggle to balance their work and family lives. In this context, practices that provide organizational support (flexible working arrangements, family-friendly practices such as remote working, childcare services, etc.) and social and organizational initiatives can reduce concerns about the family. Implementing practices that harmonize the work and family life may help women concentrate on their work.

Trainings toward overcoming the stereotypes based on gender discrimination may facilitate a more equitable work environment by helping some organizational leaders to recognize either subconscious or conscious gender-based biases that are used in the workplace. Overcoming stereotypes that cause different treatment for women can help to solve the problem. Training programs, which are organized by companies for fairer treatment, that are appropriate for the nature of the business, and that are aimed to reduce or eliminate the prejudices of men about women's managerial skills (Cohen et al., 2020), may not only help break the glass ceiling but also increase the commitment of female employees to the organization.

In order to ensure equality in organizations, it may be beneficial to provide equal opportunities to all employees in planning their careers, to impose a performance-based promotion system instead of a gender-based one and to conduct the performance evaluation according to objective criteria (Sharma & Kaur, 2019).

Undoubtedly, strategies that will be chosen may change according to the individual, the organization, and priorities. While one may postpone getting married for their career, another may give priority to the family and choose to continue in their career later. No matter what the preference regarding the career is, it involves sacrifice and abandonment. A similar situation is present for companies. Some companies take the investment in female employees into account positively, while others may consider it a costly and unnecessary waste of time. The culture and attitude of the companies and the paths that they follow depending on these differs (Knutson & Schmidgall, 1990).

Conclusion

Discussion about women's position in the modern society is among the most conflicting issues. Some sincerely believe that women manage to obtain an equal position in the society compared to men and that women have the same opportunities as men. However, a vast majority has an entirely different view of women's position in the modern society, and the glass ceiling remains to be a barrier that women often faced in their lives (Bazazo et al., 2017).

The participation and employment of women in labor force is an essential element of sustainable development. The lack of gender balance may prevent the improvement of organizational performance. Companies that focus on the future and the changing global business environment consider gender diversity to be part of the dynamics of a sustainable business practices that help the organizations obtain better results. Most companies of different sizes in the world report that gender diversity helps improve business results (ILO, 2019). However, the qualifications and skills of women are not fully employed in the business world. There is no gender-biased discrimination in the law against women entering into the working life or continue working later. However, there exist certain examples of discrimination such as certain jobs and professions not being accepted as socially appropriate jobs for women, unfair treatment in the distribution of duties, women being dismissed first in an economic crisis, and low wages for women (Ministry of Family, Labor and Social Services, 2020).

Despite certain positive changes, inequalities between women and men in senior management positions in the business world continue, and the business world is still alienated against female managers. It is a long way to ensure gender equality in workplace practices and especially in senior management levels. As Kunyalala Maphisa, the president of the Businesswomen's Association of South Africa, said (2011), "the cracks might be bigger but the ceiling is still not shattered." "Job" is still a man's game and important decisions like who will play the game are made mostly by men.

Bibliography

Albayrak, A. (2019), Gender Discrimination in Tourism Industry. In D. Tüzünkan and V. Altıntaş (Eds.), *Contemporary Human Resources Management in the Tourism Industry*, USA: IGI Global Pub.

Anafarta, N., Sarvan, F., and Yapıcı, N. (2008), "Konaklama İşletmelerinde Kadın Yöneticilerin Cam Tavan Algısı: Antalya Ilinde Bir Araştırma", *Akdeniz İ.İ.B.F. Dergisi*, 15 (8), pp. 111–137.

Aycan, Z. (2004), "Key Success Factors for Women in Management in Turkey", *Applied Psychology: An International Review*, 53 (3), pp. 453–477.

Aydın-Tükeltürk, Ş. and Şahin-Perçin, N. (2008), "Turizm Sektöründe Kadın Çalışanların Karşılaştıkları Kariyer Engelleri ve Cam Tavan

Sendromu: Cam Tavanı Kırmaya Yönelik Stratejiler", *Yönetim Bilimleri Dergisi*, 6 (2), pp. 113–128.

Badjo, L. M. and Dickson, M. W. (2001), "Perceptions of Organizational Culture and Women's Advancement in Organizations: A Cross-culturel Examination", *Sex Roles*, 45 (5), pp. 399–413.

Baroudi, J. J. and Igbaria, M. (1995), "An Examination of Gender Effects on Career Success of Information Systems Employees", *Journal of Management Information Systems*, 11 (3), pp. 181–202.

Baxter, J. and Wright, E. (2000), "The Glass Ceiling Hypothesis A Comparative Study of the United States, Sweden, and Australia", *Gender & Society*, 14 (2), pp. 275–294.

Bazazo, I., Nasseef, M. A., Mukattesh, B., Kastero, D., and Al-Hallaq, M. (2017), "Assessing the Glass Ceiling Effect for Women in Tourism and Hospitality", *Journal of Management and Strategy*, 8 (3), pp. 51–66.

Bergman, B., Hallberg, M., and Lillemor, R. (2002), "Woman in a Male Dominated Industry: Factor Analysis of a Women Workplace Culture Questionnaire based on a Grounded Theory Model", *Sex Roles*, 46 (9/10), pp. 311–322.

Brownell, J. (1994), "Women in Hospitality Management: General Managers's Perceptions of Factors related to Career Development", *International Journal of Hospitality Mnagement*, 13 (2), pp. 101–117.

Catalyst (2020), Women CEO's of the S&P 500, 15 January, available at: http://catalyst.org/research/women-in-sp-500-companies, (10.6.2020).

Clevenger, L. and Singh, N. (2013), "Exploring Barriers that Lead to the Glass Ceiling Effect for Women in the U.S. Hospitality Industry", *Journal of Human Resources in Hospitality & Tourism*, 12, pp. 376–399.

Cohen, P. N. and Huffman, M.L. (2007), "Working for Woman? Female Managers and the Gender Wage Gap", *American Sociological Review*, 72 (5), pp. 681–704.

Cohen, J. R., Dalton, D. W., Holder-Webb, L. L., and McMillan, J. J. (2020), "An Analysis of Glass Ceiling Perceptions in the Accounting Profession", *Journal of Business Ethics*, 164 (1), pp. 17–38.

Crampton, S. M. and Mishra, J. M. (1999), "Women in Management", *Public Personnel Management*, 28 (1), pp. 87–107.

Demirel, A. P. (2015), "Cam Tavana Nokta Atışı", *Harward Business Review Türkiye*, https://hbrturkiye.com/blog/cam-tavana-nokta-atisi, 30.05.2020.

Derks, B., Laar, C., and Ellemers, N. (2016), "The Queen be Phenomenon: Why Women Leaders Distance Themselves from Junior Women", *The Leadership Quarterly*, 27 (3), pp. 456–469.

Downes, M., Hemmasi, M., and Eshghi, G. (2014), "When a Perceived Glass Ceiling Impacts Organizational Commitment and Turnover Intent: The Mediating Role of Distrubutive Justice", *Journal of Diversity Management*, 9 (2), pp. 131-146.

Ellemers, N. (2014), "Women at Work: How Organizational Features Impact Career Development", *Work and Organizations*, 1 (1), pp. 46-54.

Ellemers, N., Van Den Heuvel, H., De Gilder, D., Maass, A., and Bonvini, A. (2004), "The Underrepresentation of Women in Science: Differential Commitment or the Queen Bee Syndrome?", *British Journal of Social Psychology*, 43 (3), pp. 315-338.

Ergeneli, A. (2014), *İnsan Kaynakları Yönetimi*, Ankara: Nobel Yayınevi.

Ergeneli, A. and Akçamete, C. (2004), "Bankacılıkta Cam Tavan: Kadın ve Erkeklerin Kadın Çalışanlar ve Kadınların Üst Yönetime Yükseltilmelerine Yönelik Tutumları", *H.Ü. İktisadi ve İdari Bilimler Fakültesi Dergisi*, 22 (2), pp. 85-100.

Ganiyu, R. A., Oluwafemi, A., Ademola, A. A., and Olatunji, O. I. (2018), "The Glass Ceiling Conundrum: Illusory Belief or Barriers that Impede Women's Career Advancement in the Workplace", *Journal of Evolutionary Studies in Business*, 3 (1), pp. 137-166.

Günlük-Şenesen, G. and Pulhan, E. (2000), *Kadın İstihdamı için Yeni Perspektifler ve Kadın İşgücüne Muhtemel Talep*, Ankara: Başbakanlık ve Kadın Statüsü Genel Müdürlüğü.

Hull, R. P. and Umansky, P. H. (1997), An Examination of Gender Stereotyping as an Explanation for Vertical Job Segregation in Public Accounting", *Accounting, Organizations and Society*, 22 (6), pp. 507-528.

International Labour Office (2019), Women in Business and Management: The Business Case for Change, May, https://www.ilo.org/global/publications/books/WCMS_700953/lang--en/index.htm, (10 June, 2020).

Jakson, J. C. (2001), "Women Middle Managers' Perception of the Glass Ceiling", *Women in Management Review*, 16 (1), pp. 30-41.

Karcıoğlu, F. and Leblebici, Y. (2014), "Kadın Yöneticilerde Kariyer Engelleri: Cam Tavan Sendromu Üzerine Bir Uygulama", *Atatürk Üniversitesi İktisadi İdari Bilimler Fakültesi Dergisi*, 28 (4), pp. 1-20.

Knutson, B. J. and Schmidgall, R. S. (1999), "Dimensions of the Glass Ceiling in the Hospitality Industry", *Cornell Hotel and Restaurant Administration Quarterly*, 40 (6), pp. 64-75.

Levis, J. J. (2019), *The Glass Ceiling and Women's History*, https://www.thoughtco.com/glass-ceiling-for-women-definition-3530823, (1.06.2020).

Lucas, J. and Lovaglia, M. (1998), "Leadership Status, Gender Group Size, and Emotin in Face to Face Groups", *Social Interactions*, 4 (13), pp. 17-21.

Mızrahi, R. and Aracı, H. (2010), "Kadın Yöneticilerin Cam Tavan Sorunu Üzerine Bir Araştırma", *Organizasyon ve Yönetim Bilimleri Dergisi*, 2 (1), pp. 216-223.

Mohammadkhani, F. and Dariush, G. (2016), "The Influence of Leadership Styles on the Women's Glass Ceiling Beliefs", *Journal of Advanced Management Science*, 4 (4), pp. 276-282.

Morrison, A. and Von Glinow, M. (1990), "Women and Minorities in Management", *American Psychologist*, 45 (2), pp. 200-208.

Öktem, Ş., Kubat, G. and Kızıltan, B. (2018), "Otel İşletmelerinde Kadın İşgören Davranışlarını Etkileyen Algılara İlişkin Bir Araştırma", *Anatolia Turizm Araştırmaları Dergisi*, 29 (2), pp. 197-208.

Örücü, E., Kılıç, R. and Kılıç, T. (2007), Cam Tavan Sendromu ve Kadınların Üst Düzey Yönetici Pozisyonuna Yükselmelerindeki Engeller: Balıkesir İli Örneği", *Celal bayar Üniversitesi İ.İ.B.F. Yönetim ve Ekonomi Dergisi*, 14 (2), pp. 117-135.

Özyer K. and Azizoğlu, Ö. (2014), "İş Hayatinda Kadinlarin Önündeki Cam Tavan Engelleri ile Algilanan Örgütsel Adalet Arasindaki Ilişki", *Ekonomik ve Sosyal Araştırmalar Dergisi*, 10 (1), pp. 95-106.

Ragins, B. R. and Cotton, J. L. (1991), "Easier Said than Done: Gender Differences in Perceived Barriers to Gaining a Mentor", *Academy of Management Journal*, 34 (4), pp. 939-951.

Ragins, B. R., Townsend, B., and Mattis, M. (1998), "Gender Gap in the Executive Suite: Ceos and Female Executives Report on the Breaking the Glass Ceiling", *Academy of Management Executive*, 12 (1), pp. 28-42.

Republic of Turkey, Ministry of Family, Labor and Social Services, General Directorate on the Status of Women (2020), *Women in Turkey*, March, available at: https://www.ailevecalisma.gov.tr/media/44013/02-03-2020-tr-de-kadin-donusturuldu.pdf, (10.6.2020).

Riger, S. and Galligan, P. (1980), "Women in Management: An Exploration of Competing Paradigms", *American Psychologist*, 35 (10), pp. 902-910.

Sharma, S. and Kaur, R. (2019), "Glass Ceiling for Women and Work Engagement: The Moderating Effect of Maritial Status", *FIIB Business Review*, 8 (2), pp. 132-146.

Simonetti, J. L., Ariss, S., and Martinez, J. (1999), "Through the Top with Mentoring", *Business Horizons*, 42 (6), pp. 56-62.

Snavely, K. (1993), Managing Conflict over the Perceived Progress of Working Women", *Business Horizons*, March-April, pp. 17-22.

Sumner, M. and Niederman, F. (2003), "The Impact of Gender Differences on Job Satisfaction, Job Turnover, and Career Experiences of Information Systems Professionals", *The Journal of Computer Information Systems*, 44 (2), pp. 29-39.

Şiyve, O. Ç. (2004), "Kadın Erkek Liderlik ve Cam Tavan", *Tügiad Elegans Magazine*, 66, March-April, http://www.elegans.com.tr/arsiv/66/haber018.html, (10 June 2020).

Taşkın, E. and Çetin, A. (2012), "Kadın Yöneticilerin Cam Tavan Algısının Cam Tavanı Aşma Stratejilerine Etkisi", *Dumlupınar Üniversitesi Sosyal Bilimler Enstitisü Dergisi*, 33, pp. 19-34.

Weyer, B. (2007), "Twenty Years Later: Explaining the Persistence of the Glass Ceiling form Women Leaders", *Women in Management Review*, 22 (6), pp. 482-496.

Zel, U. (2002), "İş Arenasında Kadın Yöneticilerin Algılanması ve "Kraliçe Arı" Sendromu", *Amme İdaresi Dergisi*, 35 (2), pp. 39-48.

https://www.turizmgunlugu.com/2019/03/08/istatistiklerle-kadin, (10 June 2020).

Gözdegül BAŞER
The Impact of Tourism on Turkish Economy

1 Introduction

Tourism is one of the largest and dynamic industries in the world. In 2019, tourism made a contribution of US$8.9 trillion to the world's GDP. Globally, travel and tourism has grown by 3.5 % whereas the whole economy grew by 2.5 % in 2019 (Travel & Tourism,2020). Contribution to global GDP in 2019 was 10.3 % by creating 1/10 of all jobs (330 million). It was the third fastest growing sector after information, communication and financial services (Travel & Tourism,2020). In 2019, total international tourist arrivals was 1.4 billion and total international tourism exports was 1.7 trillion USD (6.8 % of total exports, 28.3 % of global services exports) (International Tourism Highlights, 2019). The top 10 destinations received 40 % of worldwide arrivals. In addition, US$948 billion capital investment (4.3 % of total investment) was done for travel and tourism. 2019 was another year of strong growth although slower compared to the exceptional rates of 2017 (+6 %) and 2018 (+6 %). Uncertainty surrounding Brexit, geopolitical and trade tensions, and the global economic slowdown, weighed on growth. 2019 was also the year of major shifts in the sector with the collapse of Thomas Cook and of several low-cost airlines in Europe (World Tourism Barometer, 2020).

Tourism is a key part of a growing services economy, generating income and foreign exchange, creating jobs, stimulating regional development, and supporting local communities. The foreign (incoming) tourism movements toward a country primarily increase the foreign currency inputs and foreign currency supply of this country, which are extremely important in foreign trade. As a result, the balance of payments deficits decreases, the value of the national currency increases in the international market, the economy gains vitality as a result of tourist revenues, and tourist investments are accelerated and the tourist supply capacity is improved in order to respond to the increasing demand (İçöz, 2005). Tourism exports are economically significant, and have a larger impact on the domestic economy relative to other export sectors. As an example, every USD 1 of expenditure by international tourists in OECD countries on average generates an estimated 89 cents of domestic value added, compared with 81 cents for overall exports (OECD Tourism Trends and Policies, 2020). There is a clear correlation between tourism and economic development. The most economically developed countries in the world are also the countries that have the highest share in terms of tourism revenues (Özdemir, 1992).

Tourism Economic Impact can be defined as the sum of an expanded set of direct and secondary effects of Tourism Consumption and other elements of Total Tourism Internal Demand on the national economy (OECD Tourism Trends and Policies, 2020). One of the chief reasons that governments support and promote tourism throughout the world is that it has a positive impact upon economic growth and development. Tourism should generate employment and income, lead to a positive tourism balance of payments, stimulate the supplying sectors of tourism, and lead to a generally increased level of economic activity in the country. Thus, tourism should have an impact on the frequently used quantitative measure of the economic development, gross domestic product (GDP). As a result, a specialized literature has developed to measure the impact of tourism upon GDP to deal with measuring how tourism contributes to economic growth (Ivanov & Webster, 2007). There have been many research to determine the impact of tourism in economy. Dristakis (2004) concludes that both the tourism-led growth and growth-led tourism hypotheses are valid for the Greek economy. Brida et al. (2008) examine the validity of the tourism-led growth hypothesis for the Mexican economy and find evidence of uni-directional causality running from international tourism to economic growth. Nevertheless, Oh (2005) does not find a long-run cointegrating relationship between tourism and economic growth for the Korean economy.

The total economic impact of tourism is the sum of direct, indirect, and induced effects within a region. Any of these impacts may be measured as gross output or sales, income, employment, or value added (Stynes, 1997). The direct effects of tourism arise from expenditure by tourists, which immediately generates income for businesses and households, employment and revenue from taxation. Indirect effects arise as initial income is received by households, government and local businesses and is re-spent on activities necessary to provide the products and services purchased by tourists. This is sometimes referred to as "upstream expenditure." In addition, some of the income received by governments, households, and businesses will be re-spent "downstream" i.e., on consumption goods and services unrelated to the supply of tourism products. At each stage, some tourism expenditure is lost because it is used to purchase imported goods and services and some induced expenditure may be lost through savings. These losses from the system are generally referred to as leakages (Ennew, 2003).

The initial expenditure by tourism can have significant additional effects throughout the rest of the economy, resulting in increased income and expenditure by a range of different groups, many of whom are not directly connected with tourism. This process of spending and re-spending is commonly described as the multiplier effect. The term "multiplier" is used to describe the final change in output in an economy relative to the initial change in tourist expenditure and is central to any measure of the economic impact of tourism. The true impact of tourism is not the actual expenditure by tourists; it is the final impact that this expenditure has on the economy (Ennew, 2003).

Tourists contribute to sales, profits, jobs, tax revenues, and income in an area. The most direct effects occur within the primary tourism sectors like lodging, restaurants, transportation, amusements, and retail trade. Through secondary effects, tourism affects most sectors of the economy. An economic impact analysis of tourism activity normally focuses on changes in sales, income, and employment in a region resulting from tourism activity (Stynes, 1997). In addition, there are several other categories of economic impacts that are not typically covered in economic impact assessments, at least not directly like changes in prices, changes in the quality and quantity of goods and services, and changes in property and other taxes.

According to the World Tourism Organization (International Tourism Highlights, 2019), Turkey ranks as the 6th country in tourist arrivals and as the 15th in tourism receipts in the world. Turkey has been taking a considerable share in the world's tourism and travel industry. Turkey has the highest increase by 22 % compared to the previous year whereas Turkey ranks as the 11th among Tourism earners in 2019 (International Tourism Highlights, 2019).

Tourism has a high potential in Turkey from many aspects. Turkey is a huge open-air museum, a repository of all the civilizations nurtured by the soils of Anatolia. For centuries, Turkey has also been a crossroads of religions. Turkey is a very rich country in terms of natural beauties and cultural heritage. Tourism in Turkey is focused largely on a variety of historical sites, and on seaside resorts along its Aegean and Mediterranean coasts (Kervankiran, 2015). Turkey has 7,200 km of coastline and ranks 3rd among all countries with its 463 blue-flag beaches. Tourism has a high potential to contribute to the Turkish economy as a result of a good planning and well-developed tourism policies. Turkey has determined its strategic plan and aims to reach its goals with a strong belief in its potential for the development of tourism all through the country with many alternative tourism facilities. Turkey aims to get a higher economic contribution from tourism.

2 Tourism Development in Turkey

The development of the tourism industry in Turkey can be divided into two main stages as pre-planned period (1923–1963) and the planned period (1963 to present). Developments related to the tourism sector ahead of the planned period remained extremely limited due to the general economic reality in those years but still some important steps were taken. It is possible to divide the part of the planned period between 1963 and 1983 as the first period and from 1983 to the present, as the second period.

The main goal of tourism policy in the period of 1963–1983 was to contribute to the balance of payments, increase foreign currency revenues, create new jobs, and provide holiday opportunities for Turkish citizens. Many tools were used to achieve these goals, including organizations, legal and financial regulations, and special projects. The establishment of the Ministry of Tourism was one of the most important steps taken. Another important development was the establishment of Tourism Bank (Yağcı, 2007). Turkey had economical development plans for every five years which also included tourism. Since 1963, tourism targets were determined in these plans. Although the targets were not completely reached, tourism developed systematically. Either the number of tourists or the tourism income increased regularly (Kervankıran, 2015).

Turkey adopted tourism as an alternative economic development strategy to support new export-led growth strategies recommended by international agencies (e.g., the International Monetary Fund and World Bank) to create more jobs and to establish a favorable image on the international platform by exemplifying immediate implementation of an outward-oriented economic development policy, which seemed to have been essential just after the 1980 military coup (Tosun, 1998).

In 1982, the government passed the Tourism Encouragement Law, which introduced regulations regarding land use, property rights, and incentives for private sector tourism investments. To tackle the lack of adequate infrastructure and modern facilities, the government initiated large-scale physical infrastructure (roads, waterworks, sewage, etc.) upgrading programs in the late 1980s in the newly emerging tourism sites in the western and southern coastal areas. The idea of planned development was completely abandoned since 1983. 1983 was a period of time when the state withdrew from superstructure investments and started to accelerate privatization activities, and new incentives for the private sector came into force. For example, the Law for the Encouragement of Tourism No. 2634 has brought many incentives to the tourism sector in Turkey. Some of these were investment allowance, low interest loan, building construction exemption, financing fund exemption, incentive premium, tax, picture, fee exemption, foreign personnel employment, foreign exchange allocation, reduction in electricity, gas and water fees, communication facilities, and value added tax deferral (Tutar & Tutar, 2004). Incentives, ranging from tax exemption to granting of public land, were provided to private entrepreneurs in order to develop the international tourism sector (Aykaç Alp, 2009). These attempts envisaged the establishment of a tourism industry based on free market principles (Akkemik, 2012). The tourism revenue of Turkey within the international tourism started to increase gradually thereafter (Yağcı, 2007).

3 Assessing the Impact of Tourism

Assessment of the overall impact of tourism in economic terms will require detailed information relating to tourist expenditures, prices, tax revenues, and expenditures by other sectors of the economy, prices for tourism and non-tourism products, patterns of arrivals etc. (Ennew, 2003). The problem with measuring the impact of tourism spending is quite simply that tourism does not exist as a distinct sector in any system of national accounts. Systems of national accounts are the main mechanism for tracking what is produced and sold within an economy and tracing what happens to expenditure. However, tourism is essentially an activity that is defined by consumers at the point of consumption. In effect anything that tourists buy and any form of expenditure that tourists make is a contribution to the economy that is generated by tourism. Of course, a very large proportion of tourist expenditure goes into identifiable tourism characteristic sectors such as transport, hotels, recreation, etc. However, tourists also spend money in other sectors – clothing, gifts, cosmetics, food, petrol etc. – which are not normally associated with tourism (Ennew, 2003).

An examination of studies of the economic impact of tourism published in academic journals and elsewhere yields four tools that have been applied generally for macroeconomic analyses of tourism as Tourism Satellite Account (TSA), Input-output (I-O) table and model, Social Accounting Matrix (SAM), and Computable General Equilibrium (CGE) models (Statistics and TSA,2020).

a. Tourism Satellite Account (TSA)

World Travel and Tourism Council simulated Tourism Satellite Accounts for a range of countries in order to be able to evaluate the economic contribution of tourism (Ennew, 2003). The Tourism Satellite Account (TSA) has been employed by more than 60 countries to measure the direct effects of one special kind of aggregate demand – Tourism Consumption – on their national economies (Tourism Satellite Account, 2008). It is designed to be a distinctive method of measuring the direct economic contributions of Tourism Consumption to a national economy TSA determines and re-examines the significance of tourism in the general economy (Küçükaltan & Terzioğlu, 2013). The first TSA exercise took place in 2010, with 23 participating countries. Eurostat's aim was to collect readily available and voluntarily submitted TSA data from Member States, EFTA, and candidate countries at regular intervals but not each year. The next exercises took place in 2013 (22 countries) and 2016 (19 countries) (Eurostat, 2019).

The purpose of tourism satellite accounts is threefold: to analyze in detail all aspects of demand for goods and services associated with the activity of visitors, to observe the operational interface with the supply of such goods and services within the economy and to describe how this supply interacts with other economic activities (Eurostat, 2019). TSAs typically concentrate on measuring the direct impact of tourism expenditure, and often they do not directly address the issues of the indirect and induced effects. Traditionally tourism impact analysis relied heavily on simple Keynesian multipliers. These are calculated based on estimates of leakages from a given economy and seek to provide a single figure that relates tourism expenditure to output, income employment, sales, or any other aggregate outcome that is of interest. Such multipliers are relatively straightforward to calculate and provide a quick and simple way of assessing the overall magnitude of a change in tourism expenditure (Ennew, 2003). As Turkey has not participated to TSA, yet, TSA data for Turkey is not available.

b. Input-Output Model (I-O Model)

I-O model is a mathematical model that describes the flows of money between sectors within a region's economy. Flows are predicted by knowing what each industry must buy from every other industry to produce a dollar's worth of output. Using each industry's production function, I-O models also determine the proportions of sales that go to wage and salary income, proprietor's income, and taxes. Multipliers can be estimated from input-output models based on the estimated re-circulation of spending within the region. Exports and imports are determined based upon estimates of the propensity of households and firms within the region to purchase goods and services from local sources (often called RPC's or regional purchase coefficients). The more a region is self-sufficient and purchases goods and services from within the region, the higher the multipliers for the region (Stynes, 1997).

As an example to input-output model, Atan and Arslanturk (2012) state that the tourism and economic growth nexus by means of input-output analysis, covering forward and backward linkage effects, based on the 2002 input-output table, prove the impact of tourism on economic growth where variables used for economic growth are income and production outputs.

c. Social Accounting Matrix (SAM)

A social accounting matrix (SAM) is a comprehensive and economy-wide database recording data about all transactions between economic agents in a specific economy for a specific period of time. A SAM extends the classical input-output framework, including the complete circular flow of income in the economy (European Commission, 2018). A SAM is an extended input-output (I-O) table and demonstrates total transactions in an economy as depicted by the circular flow diagram. An I-O table demonstrates interdependence among production sectors in an economy. In addition to intersectoral links, a SAM shows the links between production sectors and all institutions in the economy (households, enterprises, government, and the rest of the world). Therefore, it is a useful tool to investigate the impact of a change in an exogenous account on all economic sectors and institutions (Akkemik, 2012).

Akkemik (2012) notified that the contribution of international tourism to the Turkish economy using two SAMs for 1996 and 2002, respectively. Two analyses were conducted using the SAM impact model: (i) sectoral comparison of GDP elasticities, and (ii) SAM impact analysis of international tourism on output, value-added, and employment. The results showed that the GDP elasticity of international tourism was relatively low and the impact of foreign tourist expenditures on domestic production, value-added (GDP), and employment in Turkey were modest. The results implied the possibility of leakage of foreign tourist expenditures out of the economy.

d. Computable General Equilibrium (CGE) Models

CGE modelling is based around a mathematical specification of key relationships within the economy (what determines levels of supply, demand, etc.), and is calibrated to real data to ensure that the model provides a good representation of the economy. With a comprehensive model of the economy, which incorporates businesses, governments and consumers, it is possible to analyze the economy-wide impacts of changes in tourism spending, changes in subsidies or taxation, and other policy and market changes (Ennew, 2003).

Akkemik and Şenerdem (2014) established the CGE analysis on tourism sector in Turkey for the first time. They critically analyzed the tourism strategy of Turkey using a multi-sector dynamic CGE model with a 20-year horizon. The model included two tourism sectors, domestic and international where they built a social accounting matrix. They found that any incentive to the tourism sectors will have a positive impact on the economy.

Gül (2015) analyzed the economic effects of a 10 % increase in foreign demand for the tourism industry in Turkey by using a computable general equilibrium model based on a social accounting matrix year of 2002. The tourism industry was specified by using tourism satellite accounts and set as a separate sector in input–output table. Simulation results showed that increase in foreign tourism demand caused an increase in the GDP and employment, households' income, and real private consumption and investment expenditures in the economy and put pressure on domestic prices at different levels.

Another study examined the relationship between tourism revenues and economic growth in Turkey, using VAR analysis using annual data for the period 1980–2016 (Şahin, 2016). Cointegration and causality analysis were used in determination of the relationship between short- and long-term tourism revenues and economic growth in Turkey. The results of the analysis showed that there was a relationship between the variables in short and long term, from tourism income toward economic growth.

4 The Impact of Tourism on Turkish Economy

Tourism industry is one of the most important sources of growth in Turkey, and tourism is an important cause of economic growth for the long term (Çoban & Özcan, 2013). However, the empirical findings indicate that even though there is growth-promoting effects of tourism revenues in the long run, there is no short-run relationship between tourism and economic growth (Yıldırım & Öcal, 2004). Tourism is among the sectors at the center of economic and social balances in terms of foreign currency; it helps to close the deficits in the balance of payments, and it provides job opportunities as well as contributing to the national income. Tourism, which is the biggest foreign currency source after export, is extremely vital as it affects many other sectors. Tourism within the economy directly affects around 50 sectors in the economy such as food industry, transportation, communication, wholesale and retail trade, entertainment, real estate, construction, etc. (Türsab, 2014).

The Impact of Tourism on Turkish Economy 183

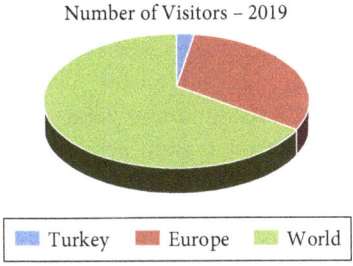

Fig. 9-1: Number of Foreign Visitors in 2019
Source: Turizm İstatistikleri, 2020

There are many studies that point to the relation among growth, foreign exchange and tourism receipts in Turkey (Arslantürk & Atan, 2012), stating tourism as an important determinant of overall long-run economic growth (Savaş et al., 2012); tourism and investment contribute to Turkish economy positively and significantly (Gökovalı, 2010), economic growth contributes to the sectoral development of tourism while tourism contributes to the economic growth (Demiroz & Ongan, 2005). Under the light of the findings, it is seen that the total amount of the input from the other sectors for the output in tourism sector is high and tourism sector is in a structure of nurturing other sectors; hence, it is thought that tourism sector will support the production in other sectors, with a considerable impact on growth (Atan & Arslantürk, 2012).

Turkey received 51.747.199 million visitors in 2019 (Turizm İstatistikleri, 2020) and Europe received 742,3 million foreign visitors whereas the world received 1,461million foreign visitors in 2019 (World Tourism Barometer, 2020, world tourism barometer). It is possible to say that Turkey received almost 6,97 % of the total visitors in Europe and 3,54 % of the total visitors in the world (Fig. 9-1).

According to data known as total backlink coefficient in input purchases, 1 unit increase in demand in the tourism sector causes an increase of 2.12 units of production in other sectors in Turkey. This increase is 1.97 units in Spain, 2.2 units in Italy, 2.1 units in France, and 1.7 units in Greece (Türsab, 2014). Therefore, it is possible to say that tourism in Turkey is contributing to other sectors which are named as a "multiplier effect."

Accommodation facilities with a total investment value of more than 70 billion dollars make important contributions to economic and social life every year via goods and services purchases, salaries and social security payments, and taxes paid to the government. The accommodation sector purchases around $ 9 billion of goods and services every year from the domestic market for its guests. It allocates $ 2–2.5 billion for investments and renovations. Employees are paid around $ 0.2 billion as salary and social security share. Moreover, the hospitality industry makes these contributions using the least imported input as 1.5 % in investments, 5 % in domestic purchases, and 1 % in labor (Türsab, 2014).

In Turkey, 8-9 % of the total workplaces are assumed to operate in the tourism sector (Türsab, 2014). Approximately 2–3 % of the employees in Turkey are employed in tourisim, travel, and hospitality sector enterprises. On the other hand, the wages paid in the tourism sector reach a magnitude of 5 % of the overall (Türsab, 2014). In addition to its rapid job creation capacity, the sector is also the leader in terms of job creation with minimum cost in employment investments per capita. On the other hand, the tourism sector is one of the areas where the increase in employment is directly parallel to the increase in total demand (Türsab, 2014). In general terms, the employment increases during the summer months where it is called as the "peak season" covering April to October, and there is a serious decline in employment during the "low season" of October to April.

More specifically, the impact of Tourism on Turkish economy can be summarized as follows:

a. Tourism Revenue in Total Exports and Balance of Payments

It can be argued that the tourism receipts can be regarded as a stable source of foreign currency in Turkey and the payoff of economic policies directed to develop tourism will be higher in the long run (Kaplan & Çelik, 2008). Tourism is an important industry for Turkey and represents one of the most important sources of foreign currency earnings (Savaş et al., 2012). There is a one-way causality from Tourism revenues to real Exchange rates (Öncel et al., 2016). Tourism can be considered as an invisible export item which has considerable impacts on the balance of payments (Saray & Karagöz, 2010). Tourism is heavily dependent on labor and listed in the section of international services under current accounts of the balance of payments. Therefore, tourism incomes have an export effect, for tourism incomes are of foreign exchange nature (Arslantürk & Atan, 2012). Growth of Tourism industry has been regarded as an important source of balance of payments surpluses and as an additional revenue source for GDP (Savaş et al., 2012). Exports generate the main foreign currency in the economy, and tourism revenue in exports occupies a ratio of around 21.4 % of

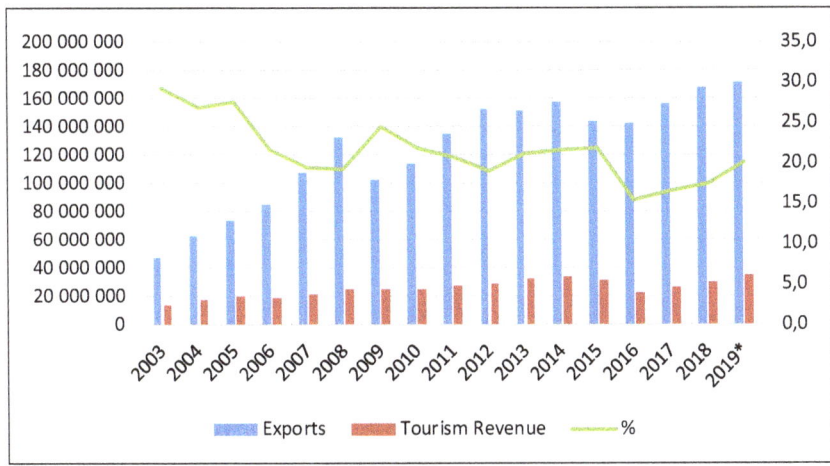

Fig. 9-2: Tourism Revenue in Total Exports
Source: Turkish Statistical Institute (Türkstat) & Türsab

total exports and 2019 figures show that Turkish tourism revenue is around 20.1 % of total exports in Turkey.

Tourism receipts in Turkey do not only involve foreign visitors, Turkish citizens residing abroad contribute to the Turkish economy in terms of tourism receipts. As an example, in 2018, 18,11 % of total tourism receipts belonged to Turkish citizens residing abroad. Although, there is an increase in the number of visitors to Turkey, it is not possible to say that there is a continuous increase in the total tourism receipts and average expenditure per person.

b. Contribution to GDP

Tourism has a significant positive effect on Gross Domestic Product (GDP) both in the short term and in the long term in Turkey (Esen & Özata, 2015). Empirical results show that there is no cointegration between tourism sector and economic growth in the long run; however, there is a positive unidirectional causality running from the tourism sector to the GDP in the short run (Terzi & Pata, 2016). Tourism incomes in Turkey are known to have a significant position within Gross National Product (GNP). It is seen that Turkey's tourism incomes and the ratio of these incomes in GDP are directly proportionated to number of visiting tourists (Işık et al., 2019). Zortuk (2009) found a unidirectional positive causal relation from tourist arrivals to GDP. The share

Tab. 9-1: Tourism Receipts per Person

Years	Foreigners			Turkish Citizens Residing Abroad			Total		
	Number of Foreigners	Tourism Receipts (US$ 1000)	Average Expenditure Per Person (US$)	Number of Citizens	Tourism Receipts (US$ 1000)	Average Expenditure Per Person (US$)	Number of Visitors	Tourism Receipts (US$ 1000)	Average Expenditure Per person (US$)
2008	26.431.124	19.612.296	742	4.548.855	5.418.439	1.191	31.137.774	25.415.067	820
2009	27.347.977	19.063.702	697	4.658.172	5.690.629	1.222	31.759.816	25.064.482	783
2010	28.510.852	19.110.003	670	4.517.091	5.558.366	1.231	32.997.308	24.930.997	755
2011	31.324.528	22.222.454	709	4.826.800	5.638.484	1.168	36.769.039	28.115.692	778
2012	31.342.464	22.410.364	715	5.121.457	6.354.378	1.241	37.715.225	29.007.003	795
2013	33.827.474	25.322.291	749	5.398.752	6.760.180	1.252	39.860.771	32.308.991	824
2014	35.850.286	27.778.026	775	5.564.784	6.289.260	1.130	41.627.246	34.305.903	828
2015	35.592.160	25.438.923	715	6.025.370	5.843.074	970	41.114.069	31.464.777	756
2016	25.265.406	15.991.381	633	6.099.924	5.964.853	978	30.906.680	22.107.440	705
2017	32.079.527	20.222.971	630	6.540.819	5.908.752	903	37.969.824	26.283.656	681
2018	39.488.401	24.028.311	617	6.624.191	5.345.472	801	46.112.592	29.512.926	647
2019									

Source: Turkish Statistical Institute (Turkstat)

of tourism in GDP increased from 0,8 % in 1980 to 7 % in 2007 (Zortuk, 2009). Although international tourism was generally deemed as an important sector in previous studies, the impact of foreign tourist expenditures on domestic production, value-added (GDP), and employment in Turkey can be best described as modest. In addition, compared to international tourism, domestic tourism activities are found to be more important in terms of GDP elasticity, which indicates the need to diversify tourism activities (Akkemik, 2012).

One of the principal targets of Turkey's tourism policy is to use foreign tourism income to eliminate a portion of the deficit in the balance of payments. Because the service industries experiencing greater labor-intensive developments in tourism constitute a growth opportunity for Turkey, which lacks capital, generating greater foreign currency inflow, employment, and national income through service exports within the tourism industry is possible. In addition to an increase in national income and the employment tax base, Turkey has also seen an expansion and intensification of competition, resulting in more efficient use of resources throughout the economy. In this way, tourism has contributed to the Turkish economic growth (Küçükaltan & Terzioğlu, 2013).

Tab. 9-2 shows tourism revenue in GNP (%). During the last 17 years, tourism contributed to GNP in Turkey by 3,6 % on the average.

c. Employment

Tourism is a labor-intensive sector and it is quite important in terms of labor and consequently employment (Bozkurt & Bahar, 2015). Employment in tourism can be divided in two parts as direct employment and indirect employment. The direct employment effect of tourism is seen in services that have a direct or significant close relationship with tourists. Lodging, food and beverage business, travel agencies, tourist guides, entertainment business, transportation, gift shops, etc. are some of the examples. Indirect employment is the employment that occurs in the sectors producing for tourism. These are the construction and repair of touristic facilities, the production of consumer goods such as food and beverage, the production of goods such as equipment and furniture, the employment by local administrations or the government for tourism-related work. Approximately 1.5 million people are estimated to be employed indirectly for tourism in Turkey (Tutar & Tutar, 2004). SGK (Social Security Institution) states that there are around 1.036.142 people insured in sectors related to tourism and hospitality (Tab. 9-3) in Turkey who are directly employed by tourism (Statistical Yearbook,2018).

Tab. 9-2: Tourism Revenue in GNP (%)

Years	Tourism Revenue (000) $	Tourism Revenue in GNP (%)
2003	13.854.866	4,4
2004	17.076.607	4,2
2005	20.322.111	4,1
2006	18.593.951	3,4
2007	20.942.500	3,1
2008	25.415.067	3,3
2009	25.064.482	3,9
2010	24.930.997	3,2
2011	28.115.692	3,4
2012	29.007.003	3,3
2013	32.308.991	3,4
2014	34.305.903	3,7
2015	31.464.777	3,7
2016	22.107.440	2,6
2017	26.283.656	3,1
2018	29.512.926	3,8
2019	34.520.332	4,6

Source: Türsab

Tab. 9-3: Number of Insured Persons by SGK in Tourism and Hospitality

Sector	Number of Insured Persons (2018)
Lodging	226.760
Food and beverage facilities	668.387
Travel agency/tour operator facilities	47.727
Sports, entertainment facilities	51.022
Airport transportation	27.542
Total	**1.021.438**

Source: Statistical Yearbook (2018), SGK

d. Regional Development

As well as having a positive impact on economic growth and development overall, tourism can also have a positive impact on regional development, and may help to even out some of the inequalities between different parts of a given country. Regions which do not have access to other major resources or do not have major urban centers may be able to use tourism to improve regional incomes and reduce out-migration (Ennew, 2003). As an example to regional development, economic impact of religious tourism in Mardin was examined by Egresi and Bayram (2012). The authors state that tourism had a strong economic and social impact on local businesses and the local community. Many new jobs were created; pressure was put on the infrastructure, and the majority of the tourists were satisfied with their experience and many expressed their intention to return in the near future and/or to recommend the religious sites in Mardin to others.

Tourism Strategy of Turkey has an objective as rerouting all tourism investments toward reducing the imbalances of welfare and development imbalances throughout the country and treats them with an approach that safeguards, conserves and improves the natural, historical, cultural, and social environment (Tourism Strategy of Turkey, 2023. Turkish Ministry of Culture and Tourism aims to make arrangements for handling in the first place and strengthening the infrastructure of areas where mass tourism activities grew intense and for extending tourism season throughout the entire year in these regions. Mass tourism that consists of sun-sea-sand activities concentrates on the coastal regions has already reached a saturation level; therefore, a number of arrangements for maximizing the receipts and having tourism facilities for 12 months in different regions in Turkey will be organized like health, sports, and culture tourism alternatives (Tourism Strategy of Turkey, 2023).

e. Investments

Turkey has around 1.7 million bed capacity. 1.172.886 million beds are ministry-licensed and have described standards and stars; the others (542.222) are municipality licensed. There are 4.761 ministry licensed lodging enterprises and 8.081 municipality licensed lodging enterprises (Turizm Sektörünün Yapısı, 2014). Turkish government supplies investment incentives to many sectors including tourism. Analyzing the records for 1998 till 2018, the number of incentives given for Tourism investments is around 6,03 % among all the total number of investments (TTYD). The Turkish government offers incentives such as reduced utility prices and reduced tax rates, while also pursuing policies aimed at eliminating any bureaucratic barriers that may hinder growth in the tourism sector. (Tourism, 2019)

f. Costs of Tourism

Tourism also entails costs to the economy, making demands on the infrastructure and scarce resources of the economy in addition to the environmental damages, such as pollution and uncontrolled development. Moreover, the development of tourism industry in developing countries requires some of scarce resources, such as capital and skilled labor, to be diverted from their alternative uses, leading to the production loss in other sectors. Thus, the overall impact of tourism sector on economy depends primarily upon the nature of the country's economy and alternative forms of development that are applicable (Yıldırım & Öcal, 2004).

Tourism development may impose some significant and direct financial costs on governments. These costs include the costs of advertising and marketing the country as a destination and the establishment and operation of national tourism organizations as well as the costs associated with developing and maintaining relevant infrastructure. Additional costs may be incurred in instances where governments need to provide subsidies and other incentives to attract private sector investment (Wood, 1996). Over time, of course, the costs of developing and subsidizing tourism may be offset by additional government income in the form of taxation on tourism related activities, but the initial, upfront costs often remain highly visible (Ennew, 2003).

There are other short-term economic costs that may result from tourism. If tourism development is heavily reliant on imported goods and services, there is a risk that existing local production may be displaced or its development inhibited. This effect can be particularly significant if "demonstration effects" result in the local population copying tourists and increasing their consumption of imported goods and services rather than domestically produced ones (Ennew, 2003).

Revenues from tourism are also potentially quite variable. Part of this variability is simply a reflection of seasonal patterns in demand which, although predictable, can create inefficiencies in terms of the utilization of resources. More problematic, perhaps are the unpredictable reductions in tourist arrivals and expenditure which arise as a consequence of unanticipated events – terrorist actions are perhaps the most obvious example, but extreme weather conditions can also have similar effects as can natural disasters. Where economies are heavily dependent on tourism, sudden drops in demand can have significant negative impacts on income and employment. Although sudden falls in tourism arrivals may be problematic, the same is true of sudden increases. High levels of tourist arrivals may also cause problems. Large numbers of tourists may overload local infrastructure placing pressure on water, electricity, sewage provision, and on transport links. Such infrastructure problems may be particularly acute in developing countries and may add to tourism development costs because of the need for additional investment (Ennew, 2003).

Finally, it is often argued that many of the economic benefits of tourism are lost because of high levels of leakages in the form of increased imports. To the extent that tourism expenditure is lost overseas, the host economy fails to realize one of the major benefits of tourism. A high level of leakages suggests that, other things being equal, the multiplier for tourism expenditure will be lower and hence the economic benefits of tourism expenditure will be reduced. High levels of leakages may lead to low multipliers, but if the level of expenditure is relatively high in total, then the benefits may still be significantly greater than those which would arise with lower levels of leakages, higher multipliers but lower initial levels of tourism expenditure (Ennew, 2003).

Conclusion

Turkey is a fastly developing country and tourism receipts are quite important to contribute to the development of the economy. International tourism receipts support the economic development as they are a major source of foreign exchange, especially for less-developed countries confronted by foreign exchange constraints (Oh, 2005). International tourism receipts contribute to debt repayment and recover current account deficits as well as contributing to national income levels and employment. It is possible to state that certain industrial policies, including public infrastructure investments and promotion of entrepreneurial initiative via specific financial supports in the tourism sector may contribute to the country's growth performance (Akal, 2010).

Tourism sector has been described as "an industry without chimney." It is a labor intensive industry and creates a high amount of employment opportunities. Tourism has direct, indirect, and induced effects on the economy, and it creates a multiplier effect by influencing many other sectors. The impact of foreign tourists' spending on GDP far exceeds the commonly held figure of around 1%. In fact, with respect to value added and output, foreign tourists' spending is 2–3 times on total output (Tohamy & Swinscoe, 2000). It is possible to state that Tourism contributes to Turkish economy by impacting many other sectors and creating a high level of employment. The need for the monetary effects of tourism in developing countries is much higher than in developed countries (İçöz, 2005). Growth in the Turkish tourism industry has been above the global average in recent years, and the direct positive contribution of the industry to the current account deficit in 2018 was 52%.

On the other hand, it should be noted that it is quite difficult to test the impact of tourism on economy as tourism creates direct, indirect, and induced impacts. It might not be possible to find evidence to support entirely every contention about tourism growth and national development owing to formidable difficulties in obtaining information (Tosun, 2009). The impact of tourism on economy can be tested via tourism satellite account, input-output model, social account matrix, and computable general equilibrium models. The literature review points out that Turkish researchers have mainly concentrated on computable general equilibrium models and the other instruments have not been heavily preferred.

Turkey has a high tourism potential with its natural and historical attractions. Turkey offers a high-quality service with its accommodation availabilities. Tourism in Turkey can be developed in many different alternatives and in many different regions of Turkey. In its geographical location, Turkey has strong rivals like France, Spain, Italy and Greece in the Mediterranean area. Therefore, Turkey must utilize opportunities evaluating advantages against its rivals (Keskin & Cansız, 2010). Among these opportunities, low value of Turkish lira in comparison to European Euro and US Dollars is another advantage to attract tourists from Europe and USA. In short, it can be stated that real exchange rate is an important variable for tourists who prefer Turkey (Sarı & Oğuz, 2018).

Turkish Ministry of Culture and Tourism is prepared by the "Tourism Strategy of Turkey – 2023." Culture and Tourism Ministry's 2023 vision is to get 75 million tourists and $ 65 billion revenue (Tourism Strategy of Turkey, 2023). Although there has been a loss in 2020 due to the Covid 19 pandemic, Turkish authorities still hold the target. In 2020, the tourism economy has been heavily hit by the coronavirus (COVID-19) pandemic, and measures introduced to contain its spread. Turkey received almost 4.3 million visitors in the first 3 months of 2020; however, the months thereafter was a collapse of almost 100 %. This figure is very similar to the situation in the world. OECD announced that depending on the duration of the crisis, revised scenarios indicate that the potential shock could range between a 60 and 80 % decline in the international tourism economy in 2020. Beyond immediate measures to support the tourism sector, countries try to develop recovery measures. These include considerations on lifting travel restrictions, restoring traveler confidence and rethinking the tourism sector for the future (OECD Tourism Policy Responses, 2020).

Finally, there are also economic consequences of most social and environmental impacts that are not usually addressed in an economic impact analysis. These can be positive or negative. As an example, tourism also provides a social impact on societies. There is a social exchange among different cultures. Contributing to the distribution of income from rich countries to poorer countries, from developed to developing and underdeveloped countries, tourism also causes convergence among countries (Bozkurt & Bahar, 2015).

Among negative impacts, traffic, inflation can be listed. Traffic congestion will increase costs of moving around for both households and businesses. Improved amenities that attract tourists may also encourage retirees or other kinds of businesses to locate in the area (Stynes, 1997). An intermediate position would point to the presence of significant economic benefits but also highlight potential costs, which if not monitored, could significantly erode the gains made to employment, income and other economic aggregates (Ennew, 2003).

Finally, tourism is a sensitive sector and is influenced by many factors. Turkey has a high potential for tourism and tourism has been contributing to the Turkish economy in a significant way. Turkey is on its way to develop its tourism potential, go through alternative tourism facilities and reach its targets in a sustainable way.

Bibliography

Akal, M. (2010), Economic Implications of International Tourism on Turkish Economy", *Tourismos: An International Multidisciplinary Journal of Tourism*, 5 (1), pp. 131–152.

Akkemik, K. A. (2012), "Assessing the Importance of International Tourism for the Turkish Economy: A Social Accounting Matrix Analysis", *Tourism Management*, 33 (4), pp. 790–801.

Akkemik, K. A. and Şenerdem, E. D. (2014), CGE Assessment of Tourism Policies in Turkey, Conference Paper, 54th Congress of the European Regional Science Association: "Regional Development & Globalization: Best Practices, 26–29 August 2014, St. Petersburg, Russia, 1–37.

Arslanturk, Y. and Atan, S. (2012), "Dynamic Relation between Economic Growth, Foreign Exchange and Tourism Incomes. An Econometric Perspective on Turkey", *Journal of Business, Economics*, 1 (1), pp. 30–37.

Atan, S. and Arslanturk, Y. (2012), "Tourism and Economic Growth Nexus: An Input Output Analysis in Turkey", *Procedia-Social and Behavioral Sciences*, 62, pp. 952–956.

Aykaç Alp, E. (2009). *Türkiye'de Turizm Sektörünün Tarihsel Gelişimi ve Turizm Talebi İle Hizmet Sektörü Arasındaki İlişkinin Analizi* (No. 0023). Yildiz Technical University, Department of Economics.

Bozkurt, K. and Bahar, O. (2015), "Turizm Gelirlerine Yönelik Bir Yakınsama Analizi", *Uluslararası İktisadi ve İdari İncelemeler Dergisi*, 8 (15), pp. 157–178.

Brida, J. G., Carrera, E., and Risso, W. A. (2008), "Tourism's Impact on Long-Run Mexican Economic Growth", *Economics Bulletin*, 3 (7), pp. 1–10.

Çoban, O. and Özcan, C. C. (2013), "Türkiye'de Turizm Gelirleri-Ekonomik Büyüme Ilişkisi: Nedensellik Analizi (1963–2010)", *Eskişehir Osmangazi Üniversitesi İktisadi ve İdari Bilimler Dergisi*, 8 (1), pp. 243–261.

Demiroz, D. M. and Ongan, S. (2005), "The Contribution of Tourism to the Long-run Turkish Economic Growth", *Ekonomický časopis*, 9, pp. 880–894.

Dristakis, N. (2004), "Tourism as a Long-run Economic Growth Factor: An Empirical Investigation for Greece Using Causality Analysis", *Tourism Economics*, 10, pp. 305–316.

Egresi, I., Kara, F., and Bayram, B. (2012), "Economic Impact of Religious Tourism in Mardin, Turkey", *Journal of Economics and Business Research*, 18 (2), pp. 7–22.

Ennew, C. (2003), "Understanding the Economic Impact of Tourism", *World*, 477, pp. 463–466.

Esen, E. and Özata, E. (2015), "Turizmin Ekonomik Büyümeye Etkisi: Turizme Dayalı Büyüme Hipotezinin Türkiye İçin Geçerliğinin ARDL Modeli ile Analizi", *Anadolu Üniversitesi Sosyal Bilimler Dergisi*, 17 (1), pp. 43–58.

European Commission (2018), https://ec.europa.eu/jrc/en/publication/social-accounting-matrices-basic-aspects-and-main-steps-estimation(26.06.2020).

Eurostat (2019), "Tourism Satellite Accounts (TSA) in Europe (2019 edition)", https://ec.europa.eu/eurostat/documents/7870049/10293066/KS-FT-19-007-EN-N.pdf/f9cdc4cc-882b-5e29-03b1-f2cee82ec59d(26.06.2020).

Gokovali, U. (2010), "Contribution of Tourism to Economic Growth in Turkey", *Anatolia*, 21 (1), pp. 139–153.

Gül, H. (2015), "Effects of Foreign Demand Increase in the Tourism Industry: A CGE Approach to Turkey", *Anatolia*, 26 (4), pp. 598–611.

Işık Maden, S., Bulgan, G., and Yıldırım, S. (2019), "The Effect of Tourism Sector on Economic Growth: An Emprical Study on Turkey", *Journal of Yasar University*, 14, pp. 215–225.

Ivanov, S. and Webster, C. (2007), "Measuring the Impact of Tourism on Economic Growth", *Tourism Economics*, 13 (3), pp. 379–388.

İçöz O. (2005), *Turizm Ekonomisi*, Ankara: Turhan Kitabevi Yayınları.

International Tourism Highlights (2019), https://www.e-unwto.org/doi/pdf/10.18111/9789284421152, (10.10.2020).

Kaplan, M. and Çelik, T. (2008), "The Impact of Tourism on Economic Performance: The Case of Turkey", *The International Journal of Applied Economics and Finance*, 2 (1), pp. 13–18.

Kervankiran, I. (2015), "Contribution of the Five Year Development Plans to Tourism in Turkey", *Electronic Turkish Studies*, 10 (2), pp. 587–610.

Keskin, A. and Cansiz, H. (2010), "Tourism, Turkey and Economic Development", *Atatürk Üniversitesi İktisadi ve İdari Bilimler Dergisi*, 24 (4), pp. 23–46.

Küçükaltan, D. and Terzioğlu, M. K. (2013), "Economic Impact of Tourism Demand: Evidence from Turkey", *Anatolia*, 24 (3), pp. 484–488.

OECD, *OECD Tourism Trends and Policies 2020*.

OECD Tourism Trends and Policies (2020), https://www.oecd.org/cfe/tourism/OECD-Tourism-Trends-Policies%202020-Highlights-ENG.pdf, (10.10.2020).

OECD Tourism Policy Responses (2020), https://read.oecd-ilibrary.org/view/?ref=124_124984-7uf8nm95se&title=Covid-19_Tourism_Policy_Responses, (10.10.2020).

Oh, C. (2005), "The Contribution of Tourism Development to Economic Growth in the Korean Economy", *Tourism Management*, 26 (1), pp. 39–44.

Öncel, A., İnal, A. G. V., and Torusdağ, A. G. M. (2016), "Türkiye'de Reel Döviz Kuru-Turizm Gelirleri İlişkisi: 2003–2015 Dönemi İçin Ampirik Bir Uygulama", *Yüzüncü Yıl Üniversitesi İktisadi ve İdari Bilimler Fakültesi Dergisi*, 2, pp. 125–142.

Özdemir, M. (1992), *Turizmin Türkiye'nin Sosyo-Ekonomik Yapısına Etkileri*, Ankara: Kök Sav.

Saray, M. and Karagöz, K. (2010), "Determinants of Tourist Inflows in Turkey: Evidence from Panel Gravity Model", *Uluslararası Yönetim İktisat ve İşletme Dergisi*, 6 (11), pp. 33–46.Sarı Y. and Oğuz, Y. E. (2018), "Reel Döviz Kurlarının Turizm Talebine Etkisi Üzerine Karşılaştırmalı Nedensellik Analizi", *Electronic Journal of Social Sciences*, 17 (66), pp. 603–620.

Savaş, B., Beşkaya, A., and Şamiloğlu, F. (2012), "Analyzing the Impact of International Tourism on Economic Growth in Turkey", *Uluslararası Yönetim İktisat ve İşletme Dergisi*, 6 (12), pp. 121–136.

Statistics and TSA, (2020), UNWTO, https://www.e-unwto.org/doi/pdf/10.18111/9789284417100, (10.10.2020).

Statistial Yearbook, (2018), SGK (Social Security Institution), www.sgk.gov.tr/wps/portal/sgk/tr/kurumsal/istatistik, (28.06.2020).

Stynes, D. J. (1997), *Economic Impacts of Tourism*, Illinois Bureau of Tourism, Department of Commerce and Community Affairs.

Şahin, B. E. (2016), "Türkiye'de Turizm Gelirleri ve Ekonomik Büyüme Arasındaki İlişki (1980-2016)", *Yönetim ve Ekonomi Araştırmaları Dergisi*, 16 (3), pp. 239-253.

Terzi, H. and Pata, U. K. (2016), "Türkiye'nin İktisadi Büyümesinde Turizm Sektörünün Katkısı", *Erciyes Üniversitesi Iktisadi ve Idari Bilimler Faküeltesi Dergisi*, (48), pp. 45-64.

Tohamy, S. and Swinscoe, A. (2000), *The Economic Impact of Tourism in Egypt*, ECES The Egyptian Center for Economic Studies, Working Paper No. 40.

Tosun, C., Timothy, D. J., and Öztürk, Y. (2003), "Tourism Growth, National Development and Regional Inequality in Turkey", *Journal of Sustainable Tourism*, 11 (2-3), pp. 133-161.

Tosun, C. (1998), *Community Participation in the Tourism Development Process at the Local Level: The Case of Urgup in Turkey*, PhD Thesis, Strathclyde University.

Tosun, C. (2009), *Promoting a Sustainable Tourism: National Study for Turkey. UNEP/MAP/BLUE PLAN: Promoting Sustainable Tourism in the Mediterranean*: Proceedings of the Regional Workshop, pp. 963-1022. Sophia Antipolis, France, 2-3 July, 2008. MAP Technical Reports Series No. 173. UNEP/MAP, Athens.

Tosun, C. and Jenkins, C. L. (1998), "The Evolution of Tourism Planning in Third World Countries: A Critique", *Progress in Tourism and Hospitality Research*, 4, pp. 101-114.

Tourism, (2019), https://www.invest.gov.tr/en/sectors/pages/tourism.aspx(28.06.2020).

Tourism Satellite Account, (2008), https://unstats.un.org/unsd/publication/Seriesf/SeriesF_80rev1e.pdf(10.10.2020).

Tourism Strategy of Turkey, (2023), https://www.ktb.gov.tr/Eklenti/43537,turkeytourismstrategy2023pdf.pdf?0&_tag1=796689BB12A540BE0672E65E48D10C07D6DAE291(28.06.2020).

Travel and Tourism, (2020), Global Economic Impact & Trends, http://wttc.org. (10.10.2020).

Turizm İstatistikleri, TTYD, http://ttyd.org.tr/tr/turizm-istatistikleri(28.06.2020).

Turizm İstatistikleri, (2020), https://yigm.ktb.gov.tr/TR-9851/turizm-istatistikleri.html(10.10.2020).

Turizm Sektörünün Yapısı, (2014), http://www.tuyed.org.tr/wp-content/uploads/2014/02/AKTOB-arastirmasi.pdf(10.10.2020).

Turkish Statistical Institute (Türkstat), https://data.tuik.gov.tr/tr/display-bulletin/?bulletin=turizm-istatistikleri-ivceyrek-ekim-aralik-ve-yillik-2019-33669#(10.10.2020).

Türsab, https://www.tursab.org.tr/istatistikler-icerik/turizm-geliri(Access date: 10.10.2020).

Tutar, F. and Tutar, E. (2004), *Turizm Sektörünün Ekonomiye Katkıları Açısından Türkiye'nin OECD Ülkeleri İçerisindeki Yeri*, Ankara: Seçkin Yayıncılık.

Türsab (2014), "Turizm Sektörünün Yapısı, Büyüklüğü ve Ekonomiye Katkısı", https://www.tursab.org.tr/istatistikler/turizmin-ekonomideki-yeri, (Access date: 09.06.2020).

UNWTO & WTO (2019), "UNWTO, World Tourism Barometer", 4 (4), November.

UNWTO (2019), World Tourism Barometer, 4 (4), November.

UNWTO1, International Tourism Highlights, 2019 Edition.

UNWTO2, The Economic Impact of Tourism, Overview and Examples of Macroeconomic Analysis.

Wood, R. C. (1996), "The Last Feather-Bedded Industry? Government, Politics and the Hospitality Industry during and after the 1992 General Election", *Tourism Management*, 17 (8), pp. 583–592.

World Tourism Barometer, (2020), https://www.unwto.org/world-tourism-barometer-n18-january-2020(10.10.2020).

Yağcı, Ö. (2007), *Turizm Ekonomisi*, Ankara: Detay Yayıncılık.

Yıldırım, J. and Öcal, N. (2004), "Tourism and Economic Growth in Turkey", *Ekonomik Yaklaşım*, 15 (52–53), pp. 131–141.

Zortuk, M. (2009), Economic Impact of Tourism on Turkey's Economy: Evidence from Cointegration Tests", *International Research Journal of Finance and Economics*, 25 (3), pp. 231–239.

Gul YILMAZ

The Importance and Role of Gastronomy Tourism within the Context of Sustainable Destination Management

1 Introduction

When comparing to previous years, the tourism sector is developing and changing rapidly by going on offering many alternatives. Tourism activities are not considered only within the scope "sea, sand, sun" understanding any longer, but it has diversified in various fields, and has gained new meanings by providing service for four seasons.

The options, such as golf tourism, health tourism, tableland tourism, and botanical tourism have emerged by means of different alternatives, and the gastronomic tourism, which has become quite popular particularly in recent years, is also included within these options. The individuals, previously wishing to know and experience the cuisines of destination regions, have begun to travel only for discovering different food cultures, taste different flavors. This has enabled gastronomy tourism, which can adapt to many types of tourism, to become more prominent among other alternative tourism types.

However, gastronomy tourism's being important on its own is not sufficient when tourism activities are considered as a whole. Because everybody taking advantage of tourism must develop awareness in terms of protecting tourism and carry it into the future. This is only achievable by being "sustainable."

Although today's economic and technological developments brought along with globalization seem to improve life standards, they may sometimes damage the nature in certain points. Therefore, there may occur problems in long-term use of natural resources. Solutions to these problems actually lie on "sustainability management" enabling to use of resources for a longer time. For this reason, it is possible to implement sustainability and its management in tourism.

Sustainable tourism is a type of tourism integrated into development and supporting development at the same time. Sustainable destination management refers to ensuring long-term prosperity development of the local people, the cooperation of all tourism stakeholders for the purpose of carrying the sociocultural values, the natural heritage of tourism and all tourism resources into the future, and benefiting from today's resources in the following years.

One of the most important stakeholders within the scope of sustainable destination management is the local people. Gastronomy tourism involves in the process at this point. After recognizing the gastronomy tourism potential of Turkey, studies have begun to be carried out for using the destination as a marketing tool and maintaining the process sustainably. In this study, it was attempted to develop suggestions for gastronomy potentials in our country and to determine what should be the goals for ensuring sustainability.

2 The Concept of Sustainability in General Terms and Sustainable Tourism

Sustainability is defined to ensure continuity without consuming main resources as a result of overuse and overload by maintaining the general flow and operation of a system. In other words, sustainability is a type of development meeting today's needs without damaging the ability of the next generations to fulfill their needs. Resources should be evaluated continuously, and maximum protection should be provided in accordance with this evaluation; environment protection awareness should be especially prioritized (Demircan, 2016: 12–13).

Accordingly, the evolvement of the term of sustainability from thought to concept is a result of worldwide problems such as substantial damages given to the ecosystem by social, technological, and economic developments in the 21st century, rapid population growth, poverty, environmental pollution, not distributing limited resources fairly and people's concerns about a healthy future (Bozlağan, 2008: 1026).

In order to mention successful sustainability, development should also follow it. To put it simply, "the sustainable development" means programming the life and development of today and future without consuming natural resources by providing a balance between human beings and nature, enabling future generations to meet their needs and to provide development. The purpose of sustainable development is to allow future generations to live in the same prosperity level of today's generation without decreasing natural capital stocks (Çetin, 2006: 2–3). The sustainable development focuses on providing both development and protection; thus, it is considered as an approach trying to improve the quality of natural resources and human resources, to provide benefit from these for a long time (Doğan & Gümüş, 2014: 7).

The common point of the concept of sustainability, used in many fields, is to protect future and human resources. When considering this in terms of

Gastronomy Tourism and Sustainable Destination   201

tourism, it is argued that local administrations have marketed natural, sociocultural and historical resources to tourists along with the development of mass tourism; however, they have failed to satisfy the protection of these values, and cultural and natural resources have even been damaged due to unplanned and unconscious use. For this reason, the concept of sustainability in tourism has been raised as a result of damages given to the environment by touristic activities (Altanlar & Akıncı Kesim, 2011: 2–3).

To sum up, the problems such as people damaging the environment to achieve economic gain, their destructing natural resources, unconscious consumption have caused the concept of "sustainability" to gain importance. The problems caused by activities have also brought along different solutions, and the concepts of "Sustainable Development" and "Sustainable Improvement," which have same meanings, have been developed (Özkök & Gümüş, 2009: 54–55).

When considering in this respect, sustainable tourism is not only a type of tourism. Sustainable tourism may be considered as a philosophy covering all policies and applications sensitive to resources for all types of tourism (Can, 2013: 26–27). The concept of sustainable tourism may be stated as an approach aiming to carry a destination into future generations by preventing or minimizing any possible damage, to occur as a result of touristic activities, to social, cultural, natural, historical, and environmental resources and aiming that all types of tourism should be mindful of the environment.

2.1 Types of Destinations

The destination is a name given to places where people prefer to stay for participating in activities by traveling. In other words, the target place (destination) desired to be reached during travel is somewhere determined by an individual as the point where s/he plans to go.

The product mostly emerges in two ways in tourism. The first one is the tourism product originating from historical, natural, and cultural resources of a country or region. The second one is the touristic services allowing tourists to have a holiday by changing their places, in other words, all touristic services constituting a "package tour" (Hacıoğlu & Avcıkurt, 2008: 5). Therefore, it is likely to define the touristic destination based on the above information as a geographical region with local people visited by tourist groups (Özdemir, 2007: 4).

Destinations may be categorized into different groups in accordance with their features and geographic sizes. What is more, types of destinations are addressed according to the elements, leading tourists for travel, such as distance, geographical conditions, climate, and duration of stay. A tourist, who will travel, prefers a destination for touristic purposes such as education, culture, religion, business, sports, and holiday (Güripek, 2013: 52). Tourism destinations are generally classified as follows:

The Destinations Providing Resting Services: This is the type of destination with options, such as thermal, mountain tourism, which are provided for meeting resting expectations of tourists (Kozan, Özdemir & Günlü, 2014: 116).

The Destinations Providing Entertainment Services: This is the tourism service prepared in accordance with the entertainment expectation of tourists. Bodrum may be given as an example of an entertainment destination in our country. In addition, the travels made for gambling are also regarded within entertainment destinations.

This may be exemplified to go to TRNC, Las Vegas for gambling (İçellioğlu, 2014: 40).

The Destinations Providing Historical Services: Tourists from many parts of the world travel in order to see historical monuments, religious buildings, museums, old palaces, etc. Turkey is an important destination center with its historical and cultural values (Uygur & Baykan, 2007: 35).

The Destinations Providing Cultural Services: The destinations meeting expectations of tourists in a wide range, such as festivals and events, handicrafts, language learning, music and art activities, heritage resources, gastronomy, technology, education, and religion are included in this group (Ayaz, Apak & Batı, 2016: 87). What are more, tourists, wishing to meet people from different cultures, to participate in the festivals of where they go, to benefit from the sports facilities of where they go, prefer these destinations.

The Destinations Providing Ethnical Services: This is the type of tourism resulting in the fact that people from different nations or cultures live together. The geographical areas, where such a tourism option is provided, are named after "the touristic destinations offering ethnical service" (Özdemir, 2014: 23).

The Destinations Providing Financial Services: The travels made for economic purposes, such as attending trade fairs, business meetings, conferences, seminars, and congresses may differ from holiday purpose travels. The visitors traveling for business and the places they go are generally considered within this scope (Özdemir, 2014: 25).

Tab. 10-1: Destination Types and Markets

Type of Destination	Consumer	Activities
Urban	Business	Meetings, Encouraging elements, Conferences, Exhibitions,
	Free Time	Education, Religion, Health, Travel, Shopping, Shows
Coast	Business	Meetings, Encouraging elements, Conferences, Exhibitions,
	Free Time	Sea-Sand-Sun, Sex, Sports
Mountain	Business	Meetings, Encouraging elements, Conferences, Exhibitions,
	Free Time	Ski, Mountain Sports, Health
Rural	Business	Meetings, Encouraging elements, Conferences, Exhibitions,
	Free Time	Relaxation, Agriculture, Training Activities, Sports
Authentic Third World	Business	Business Opportunities, Research, Incentives, Adventure, Authentic,
	Free Time	Helpfulness, Special Attention
Unique – Exotic Special	Business	Meetings, Incentives, Moving Away
	Free Time	Special Situation, Honeymoon, Anniversary

Source: Buhalis, 2000: 101

The Destinations Providing Political Services: Tourism destinations are also considered as areas where political and social relationships are managed. Moreover, political tourism refers to the mutual interaction of individuals attending political events and meetings held in different parts of the world and tourism establishments in these regions (Özdemir, 2014: 26).

When addressing today's tourism, the tourism regions, to where people travel for a while from their residential place for various purposes, have some different features. Therefore, the tourism destinations are not only places providing cultural, resting, entertainment, and similar services but also are divided into types such as urban, rural, coastal, mountain, and authentic. The tourism destinations are addressed in accordance with their features related to the abovementioned issues (Tab. 10-1):

2.2 The Factors Affecting Destination Preferences of Tourists

The preferences or characteristics of tourists are not sole effective factors in leading them to determine the touristic places to where they will go, but the cultural activities and socio-economic status of the region to be preferred also gain importance in choosing a destination (Beerli & Martin, 2004: 624).

The factors, such as physical landscape, transportation facilities, accommodation facilities, restaurants of the place to be gone, maybe affective in destination preferences of tourists (Özdemir, 2014: 177). There are many studies in the literature on why tourists prefer or do not prefer a destination. When considering these studies, the primary factors that are effective in destination preferences of tourists may be ranked as follows (Kılıç et al., 2011: 368):

- Touristic potential, touristic establishments, and tourism services
- Environmental safety and quality
- Natural, cultural, and historical resources
- Transportation facilities
- Prices of products
- Image
- Service quality
- Advertising and promotion activities
- Activity facilities

In addition to these, there are internal factors that are effective in the preference of tourists. Of the internal factors which are also called "push factors," psychological and physical factors, social interaction and search/discovery are the leading ones.

The factors affecting the destination preferences of tourists are attempted to be summarized as follows (Fig. 10-1):

2.3 Sustainable Destination Tourism

The sustainability, which is an important concept in tourism as in other sectors, is a type of development enabling to provide the needs of the current period without limiting the rate of future generations to meet their needs. Resource usages should be considered under this understanding, and the required awareness raising should be carried out related to the protection of nature and environment. Therefore, it is likely to say that sustainability is required in the development and marketing of every destination wishing to have a share from tourism (Can, 2013: 28).

Gastronomy Tourism and Sustainable Destination

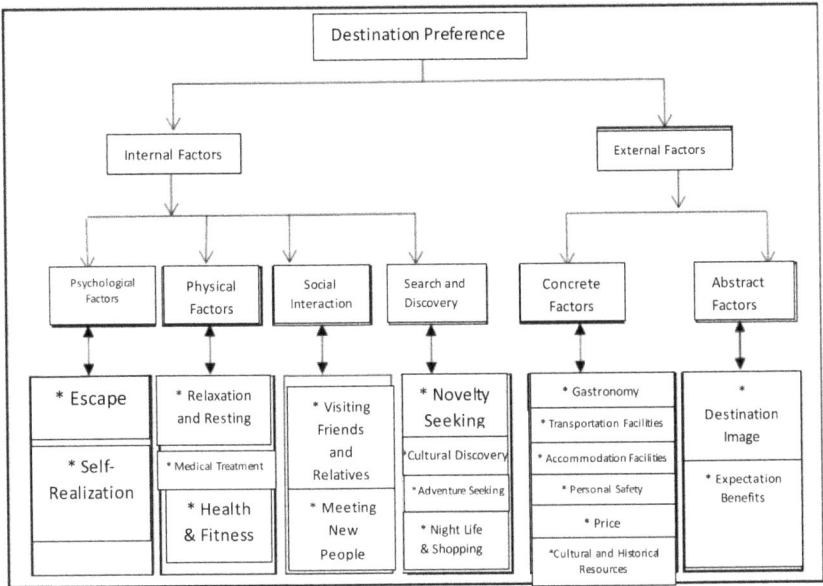

Fig. 10-1: The Factors Affecting the Destination Preferences of Tourists
Source: Hsu, Tsai, ve Wu, 2009: 291

Since tourism has become massive today, most of the destinations are in competition. The desire of a destination to host more tourists by leaving its rivals behind and following a different strategy plays an important role in creating a sustainability perception for destinations (Demircan, 2016: 14).

In order to ensure sustainability in destinations, stakeholders need to cooperate in developing appropriate policy and planning. In this regard, "making a destination sustainable and to maintain this is an element of management" (Doğan & Gümüş, 2014: 8). Therefore, when mentioning about the sustainability of destinations, it is also important how they are managed.

3 Concept of Gastronomy and Its History

Gastronomy, a Greek word, originated from the relationship between food and culture. It was formed by the combination of the words "gaster," meaning stomach in Greek, and "nomos" referring to the law (Kivela & Crotts, 2006: 355). The gastronomy may be defined as a science discipline addressing foods in artistic and scientific respects by beginning from the historical process of foods and covering the studies aiming to adapt these to today's conditions in detail. The gastronomy has a wide range covering the fields such as food presentation, food hygiene, and food order (Özbay, 2017: 31–32).

The gastronomy, of which history dates back to thousands of years, started with the hunting activities of the human being in order to be fed, the first step was taken for cooking with the invention of fire. As of ancient times, all civilizations made foods in accordance with their geographies, climates, cultures, and different cultures have been affected from each other by migrations, wars, marriages, and trade. Thus, civilizations could develop their own cuisine cultures (Özgen, 2013: 2–4).

It is argued that the term of gastronomy was first used by Sicilian Greek Archestratus, and the term of gastronomy was even used in the work, considered as the first guide, in which he mentioned about the foods and wines of the Mediterranean region in the 4th century BC, and another name of this work was "Gastronomia." It can be urged that two leading civilizations in the gastronomy were Chinese and Roman civilizations in earlier periods (Karamustafa & Ülker, 2018: 9–10). French and Italian cuisines were the ones that directed gastronomy with the Renaissance.

It is also likely to express that eating is not only for nutrition but have gained meanings such as social status and prestige in today's gastronomy understanding (Özbay, 2017: 34).

3.1 The Gastronomy Tourism as a Destination Factor

The gastronomy is considered one of the goals of a destination visit and has importance in this respect with the development of the tourism sector and the spread of mass tourism. "The gastronomy tourism" was originated from the reasons, such as tourists' wish to try new flavors, their search for alternative tourism options (Rand & Heat, 2006: 210).

The gastronomy is of importance as a touristic marketing factor for many regions. Local cuisines are known more by means of gastronomy tourism, and the cuisine and food culture are raised in the promotion of a destination, region, or country.

In today's tourism sector in which there is an intense competition, it is an increasing trend that a destination attempts to develop special products in order to differ from other destinations, and this is a significant resource as a marketing tool for local cuisine. The gastronomic richness of a region may influence many tourists since they represent cultural identity, cultural experience, communication, and sharing in destination marketing activities. While potential values such as food, wine, and olive oil, having the characteristics of a region, serve for strengthening the image of a region and its branding, they may be effective in developing business opportunities to contribute to the economy of a region. When considering the successful gastronomy tourism examples, gastronomy is not only an event for food & beverage, but it has encountered visitors as the applications in which food & beverage and knowledge have been combined (Çalışkan, 2013: 43).

The fact that the eating experience can become the most unforgettable part of the tourist's journey provides the opportunity to differentiate the destination and provide an important competitive advantage. In this sense, food & beverage consumption should not be considered meeting physical needs, but it should be recognized that it has symbolical meanings. Because food and beverage activities, which may become indicators of status and prestige in symbolical respect, cause people to have social relationships and entertain (Çalışkan, 2013: 43). To exemplify, while "Beluga caviar" is an indicator of high class and a noble taste, beans has been considered the food of lower- or middle-class food. Accordingly, it is an important indicator defining the status of any person, where, when, with whom, what s/he eats (Çalışkan, 2013: 43). While gastronomy tourism is made as a factor of destination, such details should be attached importance, and image studies of destination should be done carefully.

4 The Importance of Gastronomy Tourism within the Context of Sustainable Destination Management

With increasing competition, providing the sustainability of resources of a region in ecological, economic and sociocultural respects is of utmost importance in terms of both tourism and destination management. A region may become prominent more than its rivals by recognizing the values of a destination region and providing the sustainability of these values by developing new resources. This also applies to the food cultures of destinations. The issue that should be considered is to be able to promote any destination by both protecting its food culture and adding new values (Aksoy & Sezgi, 2015: 87–88).

In today's world, where the understanding of eating is not just a physical need has become widespread, people enjoy landscape, environment, space, and other surrounding elements by eating out of their homes. Therefore, eating is neither only meeting a physical need nor only a leisure time activity. All of them should be addressed altogether. Tourists may evaluate a destination they go not only by its foods but also its characteristics, such as landscape, nature, historical structure. Some tourists travel only for gastronomical purposes and participate in gastronomy tourism activities for both tasting different flavors and observing the stages of cooking (Özdemir, 2014: 101–103). For example, tourists may come from a distance of kilometers for wine tasting, and they may spend both time and money in this respect. For this reason, such tourism activities have become the main travel motivations. Becoming prominent with food culture in a region helps in the increase of destination attraction, marketing of the region, and sustainability of its resources.

Special food and beverage cultures of regions are important in the destination tourism. Because these foods are culture of that region and represent that region.

The gastronomy has a quite significant place within the scope of sustainable tourism and sustainable destination management. The fact that tourists coming to a region consume local food makes a lot of contribution to a region. Accordingly, it is likely that a region may develop, provide competition advantage, local economy may be invigorated, and gain may be made (Aksoy & Sezgi, 2015: 89).

Presenting local foods carefully and in high quality causes also quite significant effects. To exemplify, if a tourist coming to a region likes the local food s/he has eaten, s/he may want to export it. This helps to provide sustainability in the tourism sector and destination management and provides positive effects on the economy of the country. What is more, local cuisines are introduced on television, Internet, social media, and blog pages and are associated with travel in recent years. These promotions result in the fact that these cuisines are known by everybody, and tourists desire to go to these destinations (Özbay, 2017: 41–42).

A number of attempts are drawn attention recently for providing the development of gastronomy tourism and ensure its sustainability. These attempts cover followings: Tours made for introducing and tasting regional foods, fairs, and festivals where regional foods are introduced, tours giving training on regional foods besides enabling to taste them, activities explaining the stages of how to cook these foods. These tours contribute much to a destination that is visited. The importance of destination management is big in organizing touristic activities without any problem, providing products and services more than expectations of visitors by fulfilling demands and requests of them. For this reason, it does not seem likely that destinations may develop by themselves; good destination management is required especially for the successful development of gastronomy tourism (Yüksek, 2014: 25–27).

This provides sustainability of a region and thus, affects its development and maintenance. In addition, the local cuisine of a region contributes to other attraction factors in that region. Accordingly, more tourists come, and positive development is achieved in economic respect.

Conclusion

Although sustainable tourism is an issue based on the environment, it is a managerial process on fulfilling the needs related to vital functions, cultural integrity, ecological processes, and biodiversity by both local people and tourists. Sustainable destination management refers to ensuring long-term prosperity development of the local people, cooperation of all tourism stakeholders for the purpose of carrying the sociocultural values, natural heritage of tourism, and all tourism resources into the future and benefiting from today's resources in the following years.

The gastronomy tourism, which has become popular in recent years, has been included in the destination tourism, and activities such as visiting local wine producers, wine tasting, wine training, vintage trips, visiting local cheese producers and cheese tasting, visiting local markets, visiting olive oil producers have begun to be organized within the scope of gastronomy tourism. The important point in this respect is to guarantee the long-term maintenance of tourism development with the sustainable tourism understanding and to provide the sustainability of application areas. Therefore, the thought of sustainability aiming long-term development and improvement is of quite important in the gastronomy as it is in tourism. It is considered that the thought of sustainability, which may play a significant role in increasing gastronomic experiences and discovering new tastes, may have share in the sustainability of tourism and gastronomy.

Bibliography

Aksoy, M. and Sezgi, G. (2015), "Gastronomi Turizmi ve Güneydoğu Anadolu Bölgesi Gastronomik Unsurları", *Journal of Tourism and Gastronomy Studies*, 3 (3), pp. 79–89.

Altanlar, A. and Akıncı Kesim, G. (2011), "Sürdürülebilir Turizm Planlaması için Yöre Halkı ve Yerli Turistlerin Davranış ve Beklentilerini Anlamaya Yönelik Bir Araştırma: Akçakoca Örneği", *Ankara Üniversitesi Çevrebilimleri Dergisi*, 3 (2), pp. 1–20.

Ayaz, N., Apak, Ö. C., and Batı, T. (2016), "Yöneticilerin Kültür Turizmi Algısı: Safranbolu Destinasyonu Örneği", *Uluslararası Türk Dünyası Turizm Araştırmaları Dergisi*, 1 (2), pp. 84–96.

Beerli, A. and Martin, D. (2004), "Tourist Characteristics and the Perceived Image of Destinations: A Quantitative Analysis- A case study of Lanzorate, Spain", *Tourism Management*, 25 (5), pp. 623-636.

Bozlağan, R. (2008), Sürdürülebilir Gelişme Düşüncesinin Tarihsel Arka Planı, İçinde: H. Y. Ersöz (Ed.), *Sosyal Siyaset Konferansları Kitap 50*, İstanbul, 1011-1028.

Buhalis, D. (2000), "Marketing the Competitive Destination of the Future", *Tourism Management*, 21 (1), pp. 97-116.

Can, E. (2013), "Turizm Destinasyonlarında Sürdürülebilir Turizmin Sürdürülebilir Rekabet Açısından Değerlendirilmesi", *İstanbul Journal of Social Sciences*, 4, pp. 24-40.

Çalışkan, O. (2013), "Destinasyon Rekabetçiliği ve Seyahat Motivasyonu Bakımından Gastronomik Kimlik", *Journal of Tourism and Gastronomy Studies*, 1 (2), pp. 39-51.

Çetin, M. (2006), "Teori ve Uygulamada Bölgesel Sürdürülebilir Kalkınma", *C.Ü. İktisadi ve İdari Bilimler Dergisi*, 7 (1), pp. 1-7.

Demircan, Ş. (2016), Sürdürülebilirliğin Boyutları (Çevresel, Ekonomik, Sosyal ve Kültürel Boyutlar), İçinde: H. Çeken (Ed.), *Sürdürülebilir Turizm Temel Kavramlar ve İlkeler*, Ankara: Detay Yayıncılık.

Doğan, M. and Gümüş, M. (2014), "Sürdürülebilir Destinasyon Yönetimi, Sürdürülebilir Bozcaada: Bir Model Önerisi", *Seyahat ve Otel İşletmeciliği Dergisi/Journal of Travel and Hospitality Management*, 11 (3), pp. 6-25.

Güripek, E. (2013), *Turizm Destinasyonlarının Rekabet Gücünün Arttırılmasında Stratejik Destinasyon Yönetimi: Çeşme Alaçatı Destinasyonu Üzerine Bir Uygulama* (Yayınlanmamış Doktora Tezi), Dokuz Eylül Üniv. Sosyal Bilimler Enstitüsü.

Hacıoğlu, N. and Avcıkurt, C. (2008), *Turistik Ürün Çeşitlendirmesi*, Ankara: Nobel Yayınevi.

Hsu, T. K., Tsai, Y. F. and Wu. H. H. (2009), "The Preference Analysis for Tourist Choice of Destination: A Case Study of Taiwan", *Tourism Management*, 30 (2), pp. 288-297.

İçellioğlu, C. (2014), "Kent Turizmi ve Marka Kentler: Turizm Potansiyeli Açısından İstanbul'un Swot Analizi", *İstanbul Üniversitesi Sosyal Bilmler Dergisi*, 2014/1, pp. 37-55.

Karamustafa, K. and Ülker, M. (2018), *Yiyecek ve İçecek Sektörü*, İçinde: K. Karamustafa (Ed.), *Yiyecek ve İçecek Yönetimi* (pp. 9-46), Ankara: Detay Yayıncılık.

Kılıç, B., Akyurt, H. K., and Sop, S. A. (2011), Çekici Faktörlerin Destinasyon Seçimine Etkisinin Belirlenmesi ve Hüzün Turizmi İlişkisi. İçinde M. A.

Öncü (Ed.), *Türk Turizminde Sürdürülebilir Rekabet Avantajının Elde Edilmesi, 12. Ulusal Turizm Kongresi Bildiriler Kitabı* (pp. 362-370), Akçakoca, Düzce.

Kivela, J. and Crotts, J. (2005), "Gastronmy Tourism: A Meaningful Travel Market Segment", *Journal of Culinary Science and Technology*, 4 (2/3), pp. 39-55.

Kivela, J. and Crotts, J. (2006), "Tourism and Gastronomy: Gastronomy's Influence on How Tourists Experience a Destination", *Journal of Hospitality&Tourism Research*, 30 (3), pp. 354-377.

Kozan,Y., Özdemir, S., and Günlü E. (2014), "Turizm Yazınında " Deniz Turizminin" Olgusal Gelişimi", *Dokuz Eylül Üniv. Denizcilik Fak. Dergisi*, 6 (2), pp. 115-129.

Özbay, G. (2017), *Dünden Bugüne Gastronomi*. İçinde: M. Sarıışık (Ed.), *Tüm Yönleriyle Gastronomi Bilimi* (pp. 1-40), Ankara: Detay Yayıncılık.

Özdemir, G. (2014), *Destinasyon Yönetimi ve Pazarlaması*, Ankara: Detay Yayıncılık.

Özdemir, B. and Aktaş, A. (2007), *Otel İşletmelerinde Mutfak Yönetimi* (2. Baskı), Ankara: Detay Yayıncılık.

Özgen, I. (2013), *Uluslararası Gastronomiye Genel Bakış*. İçinde: M. Sarıışık (Ed.), *Uluslararası Gastroonomi*, Ankara: Detay Yayıncılık

Özkök, F. and Gümüş, F. (2009), "Sürdürülebilir Turizmde Bilginin Önemi", *18 Mart Üniversitesi Biga İktisadi ve İdari Bilimler Fakültesi Yönetim Bilimleri Dergisi*, 7 (1), pp. 51-72.

Rand, G. E. and Heat, E. (2006), "Towards a Framework for Food Tourism as an Element of Destination Marketing", *Current Issues in Tourism*, 9 (3), pp. 206-234.

Uygur, S. M., and Baykan, E. (2007). "Culture Tourism and Effect of Cultural Tourism and Tourism on Cultural Wealth", *Ticaret ve Turizm Egitim Fakultesi Dergisi*, 2, pp. 30-48.

Yüksek, G. (2014), *Turizm Destinasyonları*, Ankara: Detay Yayıncılık.

Oğuz TAŞPINAR
Branding and Image in Gastronomy Tourism

1 Introduction

Effects of tourism on economic, sociocultural and political spheres are influential for the increase in the importance of it. Its economic effects cause tourism to develop with pace and create great competitional environment. Countries and regions try to keep up with continuously changing tourism activities in order to have a saying in this competitional environment. Some of the most prominent of changes are novelties on touristic products, consumer habits and changes in consumer needs. Constant update is needed by destinations to able to follow mentioned changes. Countries are on a race to use their resources most efficiently in order to precede their competitors in tourism market. Without doubt, foremost of these resources is gastronomy. Gastronomy plays an important role to make subject country distinct by providing value with its products. It is demanded that destinations constantly evolve strategy for usage of their gastronomic values in tourism. Thus, bringing a brand image for gastronomy that would ensure superiority to an extent is a priority among strategies to be applied.

Looking into gastronomy products show us it is a necessity for the presentation of a product that its apparent features should be complete and perfect. Yet, apparent features do not mean much since presentation is mostly valued by its discrete features. Although food and beverages addressed as gastronomic products have their similar and even copied versions, their discrete worth cannot be produced easily, which makes them distinguishable against other competitors, for instance, with statement of "Every Cognac Is a Brandy but not every brandy is a Cognac!" France adds a discrete feature to Brandy it produced, which brings distinct position for country on the matter. Differences at gastronomy than competitors can only be presented in this regard, and desired efforts for creating a brand can succeed. During the process, brand image, which is the most important component of a brand, should always be checked and kept updated. In order to be able to reach targeted audience of gastronomy tourism, general understanding of gastronomy image and consumer perception of it is utterly significant. In this context, when approaches about the matter evaluate, it would be seen that efforts usually focus on creating an image for the destination's general features.

2 Branding in Tourism

Today, tourism is a cultural, social and economic structure. As long as tourism develops, participants of tourism show development and changes as well. Every passing day, expectations and demands of visitors change. It is been observed that visitors became pickier and more affected by different messages of countries, cities, destinations and businesses. Their desire to benefit from these concepts is also increased. Continuation of this trend is improved competition in tourism market (Ciğerdelen, 2007: 215). Therefore, branding became the most efficient tool for destination and businesses, which seek to make their services superior or/and different than their competitors.

Similarity of tourism products and services offered by destinations limits the profits of countries and regions. For this reason, destinations and connected businesses in constant competition create difference in the market. Branding is essential for countries and regions to keep existing values and make them continuous. Mentioned values include gastronomy and gastronomy tourism as well. Yet, it is more proper to discuss branding in tourism before discovering the subject on gastronomy.

Tourism became a continuously growing industry on universal dimensions. Tourism has the biggest share worldwide in terms of employment offered. Both direct and indirect employment included, tourism industry offers job vacancies for 200 million people. If it is evaluated in terms of exporting, it is again at the top against industries such as oil, motorized vehicles and electronics. In such great industry, countries, regions and businesses always compete with each other and in order to shine out they use different technics. Most effective of these technics is branding (Caldwell & Freire, 2004: 50).

Branding of tourism products and destinations shows differences completely than other sectors. Branding in tourism provides following benefits:

- Tourism industry involves high participation and different alternatives; therefore, typically it requires a comprehensive research process. As a result, it is seen as a helpful source for decreasing the difficulties of selecting of brand and information research processes.
- It provides competitional support for destinations with its apparent effects on collecting values of potential customers. In other words, brand helps to decrease effects of limited attainability.
- Branding correlates with different sale departments and time, which ensures destination's service guidance and balances diversity of potential customer. In other words, branding establishes continuity against effects of time. Moreover, it makes customers feel confident and happier because

heavily inventing businesses would act accordingly for customer satisfaction in order to protect their investments.
➤ Branding minimizes influences of risk involving factors. Thus, customers think they would face with less risk by choosing a branded product.
➤ Branding pays attention to differentiation, which helps effective branching. This eases entrance to several market segments with more than one brand.
➤ Branding is seen as a focus point to offer an image of the destination's customers causally by providing necessary conditions for motivation and teamwork at the destination's businesses (Clark, 2006: 157).

3 Branding in Gastronomy Tourism

Nowadays, countries, regions and businesses work hard to become desired destinations for tourists by bearing great efforts and costs. Because of our study's topic, emphasis will be given to gastronomy and branding as an attraction factor. Gastronomy is known as an element that improve image of destinations. Yet, not all regions are equal in that regard. While some regional gastronomy continues to develop, others might be non-existent. Furthermore, there are countries and regions in tourism industry that are not aware of their gastronomy values. Gastronomically rich destinations can have stronger position for competition in tourism market. Important point is that not only existence of a destination's gastronomy but also approaches toward gastronomy needs to be addressed.

Brand is an effort for creation of identity in consumers' and visitors' minds about touristic products. Branding in gastronomy; a name, food, drink-beverage, logo that enables gastronomy to be both identified and distinguished can also be explained as the expectation of an unforgettable meal or beverage experience that evokes gastronomy, and the fulfilment of the task of strengthening and reinforcing the recall of pleasant memories from gastronomy tourism. First part of the definition given tells traditional role of a brand for detecting and distinguishing. Different from traditional branding, second part explains how branding in gastronomy represents, directly or indirectly, the real reason for a traveling – an unforgettable experience – which is only possible for the mentioned gastronomy values. Also, while product brands want to state expectation of satisfaction for consumption of the product, gastronomy brands usually have functional distinctiveness addressing either performance of a food beverages or a special service (festivals etc.).

Within tourism, it is demanded in order to be able to compete and successful, a country region or destination needs to offer high-quality service. At the same time, expectation and memory are significant elements of quality experience (Ritchie & Ritchie, 1998: 17–18). All actions for gastronomy branding should fulfill function of entertainment in terms of future satisfaction and expected offers and expectations. After a touristic experience, brand can play an efficient role for strengthening traveling memories with gastronomy values Therefore, the main reason behind the branding of gastronomy; To create a preferred image that will attract visitors, to differentiate the gastronomy of a country from its competitors and to bring more income and to make the regions a better place to live with the ever-increasing economic impact of tourism.despite gastronomy, branding is a new concept, academicians and researchers have a general understanding on applicability of concept on gastronomy like other products and services. Countries, regions and cities already started to evaluate gastronomy values and efforts for branding. Yet, branding is gastronomy and is rather new, and academic studies in this field are limited. Efforts for gastronomy branding are involved in touristic operations of countries, regions and destinations. In other words, gastronomic values of a region are among touristic products of same region. For this reason, food-beverages that are components of gastronomy branding operations are also being used in branding of a destination. As a result, branding of a destination should be understood in order to understand branding in gastronomy. Destination branding operations within literature should be applied on gastronomy tourism with necessity adjustments. Destination branding, which needs to be explained to cover branding in gastronomy, is pictured as such by some researchers:

Ritchie and Ritchie (1998) pay attention to past states and future predictions of destination branding in their research. Despite the rapid development of the marketing of the destination, the branding function cannot be noticed at the stage of increasing the destination awareness and determining the basic qualities in choosing a touristic destination. It is supported that branding plays an important role within tourism administration for destination identity and usage of this identity for attraction of tourists. The result is it is considered branding, which is one of the basic functions of marketing, and is indispensable for destination administration.

Cai (2002) mentions destination image is a quite important subject but also states there are limited number of studies for destination branding, which has more meaningful input than destination image. In study, features of destination brand, destination image and destination brand identity are covered. Cai sees brand identity as a precondition of brand image and essential element of and effective brand. New process of branding is pitched in study, which was interested in organization of destination identity elements for a better image. An operation called *Partnered Branding* bases on idea that organization of sub-destinations, which do not have the capacity to offer their brand just by themselves. Conclusion of study is that elements of sub-destination identity have serious effect on success of partnered branding.

Iliachenko (2003) emphasizes a branding image can be created by using distinct values of destinations such as historical, archaeological, cultural and natural-geographical. Destination image can only be served by such values. In addition, importance of gastronomy and image and their essential place in destination values are suggested in this thesis study.

Morgan, Pritchard and Piggott (2003) mention the partner's role in destination branding in their research at New Zealand. In research, it is focused on how selection of a destination is an important aspect representing lifestyle and necessity for these destinations to be elegant, attractive and meaningful for tourists who spend their hard-earned money this way. Gastronomic activities such as tasting and applying unique dishes in different festivals, etc., are affordable activities that can be bought by visitors. Tourism revenues can be increased by attracting more tourists with several gastronomic values. Yet, support of existent values with an image is among priority subjects for gastronomy tourism and destinations.

Caldwell and Freire (2004) suggest that destination branding is new development and researches in this field are also new. A research made in European countries, including Turkey, and American countries, shows different destination types, differences of branding process and particularities of branding in countries, regions and cities. Research uses Brand Box Model. In this model, physical and service products are developed and application of brand to countries, regions and cities guided subject research. The purpose of study is the evaluation of tourists' decisions regarding how they select countries, regions and cities within subject countries. There are some benefits for researchers from this line of study. These benefits occur as a result of complexity of destinations' functionality (sun, sky, culture, nature etc.); therefore, focus needs to be given toward emotional or physical functions. On the other hand, regions and cities need to focus on other factors than emotional and physical functions due to being smaller and distinct. Gastronomy is a distinct value of a region or city; thus, it needs a functional approach in terms of branding. Actions such as discovery of gastronomic values, transformation of gastronomic values to unique structure and formation of a touristic product are primary field for branding.

Foley and Fahy (2004) express that branding has a unifying role for strength of tourism activities. Supportive role of image in brand positioning is exemplified by Ireland case in their study. It is emphasized that the first positioning strategy for Ireland's tourism brand is the hospitality and unique natural beauty of the local people.. In addition to behaviors of local population, tourist demands and perceptions are also covered in this research piece, which proves its significance as a strategic guide for destinations. Food and beverages used in gastronomy tourism are usually produced and served by local people; therefore, it supports them in several ways as well.

Tosun and Bilim (2004) argue that there is a change in tourism marketing strategies worldwide; country-focused approaches are on decline and region-based approaches should be equipped. To prove their point, they took city of Hatay as an example. Study tries to explain city marketing efforts done under tourism field with a local application. At the same time, it is emphasized that branding is an important factor for marketing of cities in tourism field. When it is thought gastronomy has local roots, it makes sense that gastronomy is effective for branding and marketing on regional level.

Konecnik (2004) states in a study that the evaluation of destination image has significance. Moreover, role of identity in branding is focused on and image evaluation, as a primary spontaneous process, is mentioned. Gastronomy is consisted by accumulation of food and beverages of a region. Identity of these products forms gastronomic identity, in this context. Marketing of a destination would be much easier and efficient with a strong gastronomic identity.

Hankinson (2005) explains branding of places as a strategic marketing effort shows increase in last fifteen years. It is stated that branding for places can be applied on smaller spaces, shopping malls and activity fields. Branding of place involves in destination marketing and has great effects in tourism marketing. Gastronomy is not only consisted of food and beverages. Places offer these products such as historical points, and businesses are also parts of gastronomy. Places prepare a food or beverage that can show some differences in certain cases in the eyes of visitors. Long life of business or partnership with a world knows chef attract attention of guests. Importance gained by branding of mentioned places improves region's stand in gastronomy tourism as well.

Kavaratzis and Ashworth (2005) tell destination branding done according to the product marketing. It is expressed administrators should see destination as a product and prepare branding strategy in this regard. When gastronomy is seen as a part of a destination, it should be taken as product for marketing in branding activities. In other words, gastronomy is a product and is pivotal for destination marketing and branding.

4 Brand Image in Gastronomy Tourism

Nowadays, countries and regions make serious commitments and bearing huge costs to become favorite destination in competition for visitors globally. Yet, this does not occur at same levels. While there are places with rich natural resources thanks to their geographical positions, there are also poorer regions in terms of natural resources too. In addition, there are destinations with rich cuisine advantages but somehow unable to function it properly. Although existence of various gastronomic products is worth mentioning, discovery of gastronomic products and their usage is equally important. Strategic significance of gastronomy and gastronomy tourism lies in brand usage and brand image creation. Recognition level and brand image of gastronomy support competitiveness of destinations in world tourism market.

Brand image, which is encountered frequently in marketing, is appearance, identity of a business, tried to be reflected on society (Öter & Özdoğan, 2005: 129). In this context, gastronomy branding image, perception of an identity, formed by regionally unique cuisine or gastronomy activities, by targeted masses. How a region's gastronomy is perceived by target masses, especially potential and existing visitors, has great significance on development and enrichment of destination and increasing revenues. Destination brand image, which separated into five subdivisions, also includes gastronomy brand image. Mentioned subdivisions can be listed as such: economic, physical environment, activity and operations, brand attitude, features of people. Here, brand attitude addresses consisted by perception of customers and satisfaction related with gastronomy brand (Temeloğlu, 2007: 21).

Branding image achieved at gastronomy is one of the most influential factors for being chosen. Having a brand image at gastronomy plays an important role for subdivision of tourism marketing. Even though visitors do not attend only gastronomy tourism, they feel its effects, which can influence other tourism efforts either negatively or positively. Therefore, destination with a gastronomy image helps creation of an image for preferred regions. Existence of strong gastronomy and brand image for a region or country leads to positive effect on tourism market. A destination with a strong gastronomy brand image:

➣ Would be regarded as a luxury product in comparison to others; thus, would be priced higher.
➣ Would be distinguishable in tourism market.
➣ Would create multiplying effect which would result increase on other products' value.
➣ Would gain trust of potential visitors, increase chance of being selected.

➤ Would decrease costs and increase efficiency of buzz marketing.
➤ Would have a long life in tourism market due to being distinct.

For all these reasons, gastronomy should not be overlooked in tourism activities for destinations. Perception of visitors should be detected and weak-strong points should be analyzed properly. In this regard, first, countries should know their gastronomy values and reveal them. Then, subject products should be made distinguishable than their counters and offered to customers with a brand image. It is worth mentioning that countries should not enter market with only one product. Targeted masses tried to be reach by creating several products. Owned gastronomic values should be considered as products and be included in tourism strategies. Because tourism industry is quite competitive, it became a necessity to develop effective plans and strategies for touristic regions. These regions often focus on improving their images rather than competitive positions in market (Baloğlu & Mangaloğlu, 2001: 1). Image of gastronomy in tourism contains gastronomy activities of a region, thoughts on food-beverages, beliefs and impressions. Gastronomy image is established by accumulation of several elements. From visitors' experiences of food-beverages to information gathered via advertisement-promotions, everything seen, read and experienced helps creation of an image.

There are limited amount of studies about gastronomy which is a decisive factor for visitors for choosing a destination. Studies regarding gastronomy can be observed on countries' tourism strategies. Importance given by Spanish state towards gastronomy can be proved with plan of *City Strategy on Gastronomy as a Tool for Tourism and Employment Development*. In addition, gastronomy is frequently encountered in tourism strategies of Asian, European and African countries. Although this is the case, there is no study regarding gastronomy is worth mentioning. It is seen that destination image is mostly likely been covered under image within tourism. Studies about gastronomy consist approximately one percent of studies within tourism field. Situation shows that importance of gastronomy for touristic destinations is not known thoroughly.

5 Role of Gastronomy Image for Customers' Selection of Destination

Potential visitors usually do not have enough information about a destination they did not travel in past. Information they gathered also comes from media and social cities. Yet, eating and drinking is the most popular among information gathered about destinations. Individuals visit famous restaurants, try special food-beverages of a region in their travels for instance. Visitors formulate gastronomy image of alternative destination with the information. Thus, gastronomy image gains importance for selection of destination. Within destination administration, gastronomic values should be determined, dominant features should be pointed out. Information gathered about destinations by visitors before the actual experience often supported by the mental image about the destination they had. Most of times, it is decided by the image that where would a tourist go for touristic activities. There are many tasks for creation and improvement of destination image. Since gastronomy is an important part of destination, its image is effective on tourists' decision for traveling.

6 Process for Formation of Primary Image for Gastronomic Products

Image formation is the most effective element for visitors before arrival to a touristic place; thus, it should be known that how image is formed to know how to affect behavior (Öter & Özdoğan, 2005: 130). In this context, it is needed while marketing to prepare for formation of primary image for gastronomic products. Visitors' desire for traveling a destination is the initial moment for image formation. Primary image formation starts with detection of visitors' demands and continues with selection of destinations which are capable of answering these expectations (Lubbe, 1998: 22). Gastronomy comes to picture for fulfillment of eating drinking needs and provides support for formation of primary image. Gastronomy amalgamates with other specialties of destination and becomes a product and forms the image that can be chosen by visitors. Since an unsatisfied need would create problem, destination marketers should offer a distinct, different and original ways for a solution (Hacıoğlu, 2005: 23). Nowadays, many European countries use products connected to gastronomy and its differentials for solving such problem. It is known traveling motivation can be explained with Maslow's hierarchy of needs. In other words, psychology, security, love and belonging, feeling appreciated and self-fulfillment influence decisions of visitors but we can also support these concepts with information and aesthetics (Lubbe, 1998: 22). These needs and motivations push elements are shown in Fig. 11-1. Primary image formation is connected to "pushing" and "pulling" elements.

Traveling motivations are grouped in two categories. These are pushing and pulling elements (Heung, 2001: 260). Pulling elements are labeled as attractiveness of a region and its additional factors. Elements supporting attractiveness of destination evaluated in three primary categories: static

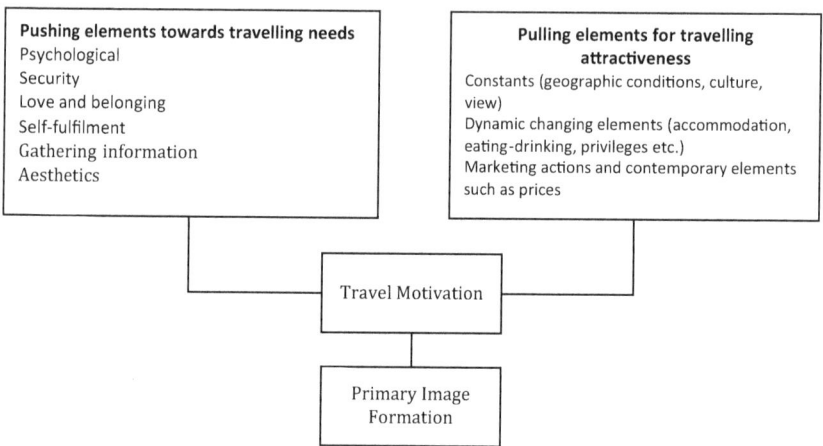

Fig. 11-1: Travel Motivation
Source: Lubbe, 1998: 25

elements, dynamic elements and spontaneous decision elements. Static elements cover natural surroundings, breath taking view, climate, transportation possibilities and historical-cultural values. In terms of static elements, gastronomy deals with unique food-beverages of region, cuisine based on region's agricultural-livestock products and historical evolution of special cuisine. Dynamic factors are accommodation, possibilities for eating-drinking, personal care, service, entertainment and sports, distance to market, political position and trends in tourism. In this context, gastronomy festivals, culinary workshops are dynamic factors for gastronomy. Spontaneous decision elements involve marketing of destination under gastronomy umbrella, prices of culinary products and regional gastronomy festivals and national culinary identity.

There are many features of a destination. Gastronomy is among the essential ones. Primary gastronomy image forms with visitors' desire for gastronomy tourism. It is thought that pushing and pulling elements usually act in correlation but there are cases when some elements act independently. These are mostly related with magnitude of the image. Organic image, created image and mixed image thought to be formed earlier than primary image. Primary image can be involved in mentioned image dimensions since a destination becomes touristic once visitors show interest (Lubbe, 1998: 22–24).

7 Dimensions of Gastronomy Image

Image is multi-dimensional and consists both symbolic and discrete qualities (Avcıkurt, 2005: 24). Yet, literature analyzes image in three different dimensions (Gunn, 1972; Fakaye & Crompton, 1991: 10–16; Lubbe, 1998: 21; Chen & Kerstetner, 1999: 256–266). These are organic image, created image and mixed image which is formed by both mentioned images. Basic image is formed by essential values of a destination; created image is formed by sale increasing actions and mixed image is mostly formed by visitors' experiences and information after their travel.

Organic image is priorly shaped image of a country, region in terms of gastronomy (Lubbe, 1998: 22). Contemporary news, articles, television broadcasts, other communication sources and public mass communication means establish organic image (Chen & Kerstetner, 1999: 256). Organic images evolve in long term, are permanent and cannot be changed in short notice. In other words, it has long roosts. Organic image is not formed with intention or purpose; it emerges by itself with time. Some gastronomies have really strong organic images continuously published in media channels (Like French gastronomy). Elements creating organic image are not related with tourism most of the time. In a way, it is formed sources other than touristic businesses and national tourism organizations. To illustrate; it cannot be said Turkey and Turkish gastronomy has a nice organic image (Kuşluvan, 2011).

Created image, on the other hand, is formed by marketing of gastronomy of a country (Lubbe, 1998: 23). Outcome is that it is an image shaped within consumers' mind with the intentional actions and marketing efforts of touristic businesses and national tourism organizations on gastronomy. In addition, mixed image is formed by experiences of an individual caused by gastronomy of visited country. Gastronomy image formed in potential customers' mind might not represent reality but important result is actually the image created. Perceptions are more significant than facts because people act on subjective basis, perceptions usually can be wrong, prejudiced and obsolete. Yet, in the eyes of customers perception replace reality (Kuşluvan, 2011).

8 Effective Factors on Formation of Gastronomy Image

There are many analyses and evaluations on image. It helps to understand tourists easily. Gastronomy image tried to be explained with destination image in above chapters since gastronomy has all components of a destination in its body. Yet, there also differences between factors affecting gastronomy and

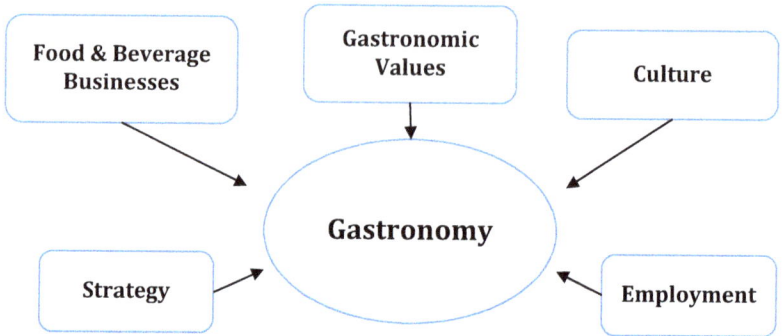

Fig. 11-2: Effective Factors on Formation of Gastronomy Image

destination images. Models and formulas are developed to cover what affects destination image, in studies focusing branding and image. Due to its novelty, there is no existing model on gastronomy and gastronomy tourism in academic terms. Therefore, it is thought it would be useful to picture factors affecting gastronomy image as in Fig. 11-2:

- **Food and Beverage Businesses:**

In order to offer, serve gastronomy products some facilities are needed. These facilities, usually known as eating-drinking businesses, should be designed according to the needs of gastronomy tourism. Some necessities should be encountered in order to have an image in gastronomy tourism. Best application for such businesses for acquirement of an image might be **Michelin Guide**. At the beginning of 1900s, boss of Michelin tiers Andre Michelin creates a marketing strategy who did not expect to able to affect 100 years later. Andre, with his brother Edouard, decides to prepare a guide to help his customers. This guide shows customers, later also to drivers, locations they can eat, fix their car and buy fuel. Initially they manage to reach 3000 vehicles with this guide, later they publish same guide in Belgium, 1904 and tips the domino piece would be able to influence 100 years future in food industry.

While Michelin Guide was only giving locations, with time it started to point location due increasing number of businesses. Point system involves stars for grading, three stars being top grade. Candidate restaurants are inspected by anonymous taste inspectors of Michelin who decides which restaurant deserves a star. Star is given to a restaurant only for a year. The number of stars can be increased or decreased with continuous inspections.

Criteria of Michelin changed for the last time in 1939. In other words, five chapters criteria which consists product quality, mastery in preparation and cooking, creativity, equivalency of price and consistency on quality standards are in application for 78 years. Furthermore, harmony of menu with ambiance, quality of service, hygiene, being a novelty, chef also being a business administrator, work experience of chef in other Michelin starred restaurants are known as criteria needs attention.

In 2014, there are 349 one starred, 81 two starred and 50 three starred, 480 in total, food-beverages businesses. In London, which is one of the major cities of Europe, there is only one business with three stars. There are no Michelin starred restaurants in Turkey since Michelin Guide does not cover Turkey. Outcome is that Michelin businesses have a prestigious image in terms of food-beverages. Existence of such restaurants has positive effects on gastronomy image of a country. Yet, it does not mean businesses; destinations without a Michelin star cannot have an image. Every business can have an effective image by advancing its existing qualities.

- **Gastronomic values:**

Gastronomic values formed by distinct cuisines of a country are treasures cannot be used by others. Reason for is that gastronomic values are products which gets brand with their own identity (Okumuş vd., 2007: 253). In other words, local gastronomic values offer a unique experience to visitors, special to a very specific region. World Tourism Organization detected in research made in 2000 that gastronomic values of destination are most effective way for expressing its culture. Support given on brand image of a region by destinations' gastronomic values cab be listed as such (Du Rand & Heath, 2006: 210):

➢ Gastronomic values add value to agricultural products.
➢ Gastronomic values create topic cause attraction.
➢ Transforms culinary culture to gastronomy activities.
➢ Merge food beverages with greater activities.
➢ Gastronomic values create brand image and develop it.

Countries, destinations use distinct food and beverages they possess in gastronomy field. Of course, not every food and drink can be considered as gastronomic products, there are some features needed for such qualification. Products can be selected by visitors, can be used as attraction points are able to be labelled as gastronomic values. Same products can be utilized by different countries. Uniqueness of process such production, preparation and serving can put a product under gastronomy roof. In this regard, countries with worldwide know cuisines have an advantaged position gastronomic image because long rooted cuisines attract tourist's attention. French cuisine can be offered as a successful case of gastronomy image, for example. Destinations should recognize gastronomic values they have and utilize them. Moreover, gastronomy product with an image not only valued in its home country but also can spread its positive image in different regions by marketing. This should be considered about other gastronomic activities apart from food-beverages as well. Festivals, feasts and exhibits are useful on destinations' gastronomy image. October Beer Festival, Toscana Olive Oil Festival, French Wine Festival continue for weeks and show gastronomy image of a country to world.

In order to have a gastronomical image both internally and externally, products marked with geographical features, in other words unique products of a region labelled by public institutions, are quite important. Registered products of Turkey are rather less, especially when diversity and distinctiveness of Turkish cuisine is considered. There are 124 products have taken special geographical mark from Turkish Patent Institute. Most products with geographical marking are from city of Erzurum. Erzurum is followed by Kayseri, Manisa, Mersin, Mardin, Afyon, Bursa and İzmir. There are currently 185 application waiting for a resolution from Turkish Patent Institute to be able to label as geographical marking.

Following questions need to be answered to discover existent resources and develop gastronomic values (Wolf, 2006: 38):

➢ What are the unique gastronomic values of the destination?
➢ Are there any food and beverages produced in the region special to the place?
➢ Are there any agricultural products special to the region?
➢ Are there any region-specific preparation cooking and serving technics?
➢ How many sources are existent for businesses such as restaurant, café, producer, winery, and butchery?

- **Strategy:**

One of the most important factors in image formation in gastronomy tourism is strategy studies. Each activity must work meticulously to create an effective image. However, effective implementation and control is also essential for success. The first step in the gastronomic tourism strategy process is to make a strategy plan. Defining active participants in the strategy plan and making task distributions is the next step. It should also be known to which group the strategic plan will be prepared for creating a successful gastronomic image. The preparation of a region's gastronomic strategy plan begins with creating a draft with the strategic analysis of that region, including vision and mission studies.

Strategic planning is the study of determining long-term goals and objectives and choosing effective strategies to reach these goals and targets. The gastronomy strategy plan should include the following five stages:

1. Research of the gastronomic region: To determine the gastronomic attractiveness of the country, the local places, the competitors, the strengths and weaknesses of the region in terms of gastronomy, and the main trends.
2. Vision and objectives: Whatever the region wants to look like in terms of gastronomic tourism, it is to make appropriate vision and target activities. These activities should include the prerequisites required for reaching the targets, and long and short-term targets.
3. Strategy development: By comparing the advantages of the region in the gastronomy dimension, determining the strategy and determining the necessary resources for the successful implementation of this strategy.
4. Action plan: A plan that defines the responsibilities and details of the application, including the cost and time limit.
5. Implementation and control: Implementing the plan and providing information to the public and partners in the form of annual and monthly reports.

Preparing a gastronomic strategy plan is not a process that a person or institution can carry out alone. Since it is an active, continuous and a combination of many elements, everyone has responsibility from the local people to an organization with international connections. Today, countries that do not create the image of gastronomy with important cuisines can only compete with price-oriented, while holding in a market and maintaining their existence is realized by the consumer's preferences, and this is largely provided by the brand image.

- **Employment:**

It is known that gastronomy is not only about food and drink. Therefore, it is not right to mention only those who cook while talking about the employment of gastronomy. In the name of gastronomy, there are many people who have studies in every field. These can be scientists as well as normal citizens who are committed to gastronomy. The presence of such people contributes to the development and appreciation of the gastronomy of the country in which it is located, and to have an image in the world gastronomy. Considering the relationship between gastronomy and other disciplines, it is possible to see gastronomy in every field. However, if it is necessary to mention only the people who are interested in gastronomy, it is possible to make some classifications in this sense.

Gourmet: The gourmet, generally known as gluttony and clumsy, is stated as a tasting person who knows the taste of food and drink and can evaluate a meal or drink unlike other people (Dilsiz, 2010: 5). It is also referred to as gourmet in the gastronomic dimension, either in theory or in practical practice, and in the person who is experienced and expert (Öney, 2013: 161–162). He is a person who is knowledgeable about good food, foody, is fussy to eat, enthusiastic about food, lives only for pleasure, and is very meticulous and passionate about food and drink. Another definition is gourmet taste, which is the name given to people with sensitive palate, who can distinguish the flavors of different types of dishes, wine and coffee (Hatipoğlu, 2010: 7).

The gourmet is the person who knows and appreciates the details about food and drink, the aroma and harmony in the ingredients of a food, the details in its cooking, the spirit that participates in cooking. Gourmet is a peculiar feature of smart, sensitive, adult, sensitive and tasteful people. This title is mostly used for those who have a high food culture, who know the food and drink manners, and who know not only food but also the kitchen works (Öney, 2013: 162).

Gastronomy: who has experience both in theory and practice in food and beverage. It is the person who carries out studies on the field of science, which examines the issue of food and drink, necessary to sustain their lives (Gürsoy, 2013: 20). It is known that many definitions exist as a gastronome. As can be understood from these definitions, gastronomy is the name given to people who know how to cook, prepare and present according to some sources, but who do not need to know how to cook, but who need to know food and drink well.

The first use of the term gastronomy is based on the period after the publication of his book *Gastronomy in Paris*. After this definition, different terms are also derived in the literature. Gastronomy was seen as one of the important titles in the 1800s. Charles Monselet, who is a journalist, novelist, poet and dramatist who has researches on food and drink, has been nicknamed King of Gastronomers by his contemporaries (en.wikipedia.org).

Degustator: The name given to people whose job is to drink and drink tastes (www.iskitabi.com). They are the people who evaluate the beverage in various fields. However, this situation is known to be a degustator not only for beverages but also for butter, olive oil and chocolate products (Larousse Gastronomique, 2005: 229).

Epicure: It is also referred to as the Greek philosopher Epicure, which is named after pleasure and eating and drinking (Yıldırım, 2012: 270). Ancient Greece is the homeland of Epicurus, who believes in the ideology of delicious food and beautiful life and devotes his life to popularizing this ideology. As a result, Epicure is the definition used for people who are fond of eating, and who love the comfort of eating and drinking (Gürsoy, 2013: 18).

Gastronomy Author; Information about products and materials within the scope of gastronomy is known as those who publish publications about the art of gastronomy and criticize food and beverage businesses. Gastronomy writers share their comments about flavors in gastronomy guides, mostly seasonal publications, where they describe their preferred hotels and restaurants (Larousse Gastronomique, 2005: 395)

The basic concepts of gastronomy in terms of employment have been tried to be mentioned above. The presence of the mentioned people helps us to see how many people care about that country in terms of gastronomy and the image gained by gastronomy. The worldwide evaluation of gourmets, gastronomers and authors in terms of tourism contributes to the country's being a preferred destination in gastronomic tourism. For this reason, highlighting and promoting the values of such professional groups is very important for the sustainability of gastronomic tourism. Considering that this is happening across the country, the awareness of the cuisine culture and values of each geography has appeared and it is understood that it has an image. As a result, employment in gastronomic science should not be understood only as cooks and service personnel. It should not be forgotten that there are many people who are influential in creating image of gastronomy tourism.

- **Culture:**

Gastronomy culture is undoubtedly one of the most basic elements in the formation of the image. And Turkish gastronomy in this respect is one of the richest gastronomies in the world, perhaps the first. Because it was founded on the ruins of an empire that covered the entire Arab world, partly the Caucasus, and part of the Rumeli Territory along with the entire Balkans. And today in the territory of the Republic of Turkey, it is possible to see the traces of people from every corner of this vast geography. in the lands of the Ottoman Empire once directed, if there are today considered to be a different state of Turkey's social heritage over how it easily understandable. Today, the branding of Turkish gastronomy causes some confusion. In this sense, some researchers who do and analyze studies speak of an Ottoman Gastronomy, while some researchers speak directly of a Turkish gastronomy. In fact, as mentioned, today's Turkish gastronomy was founded on the ruins of the Ottoman Empire. And it is very normal that this magnificent Empire took over the culture of gastronomy. But our republic we live in, that geography takes over this heritage, which has led today to reach a synthesis and an interesting take on gastronomy of Turkey (Ateş, 2005: 56).

Gastronomy should not only be considered as eating and drinking. The rules that determine what can be eaten and how to eat and the compelling effect of these rules on us make food a cultural element different from its effect on meeting the physiological need. In this period of intercultural interaction, the importance of local gastronomy and even the increase in the examples that occur during the image of gastronomy makes the subject even more important. At the same time, the dominant character of globalization increases interaction. In this regard, it is essential to have a ground where we can talk about Turkish gastronomy. Reading and writing texts displaying various views and ideological approaches to the image formation of this gastronomy is now a common phenomenon.

Gastronomy is one of the main topics of cultural studies. As it is known, culture is everything that people add to nature. Gastronomy itself is a concrete object, but the events and network of relationships around it form the culture of the food. From geography to agriculture, nutrition to medicine, from sociology to psychology, from literature to aesthetics, from history to religion, folklore to art, there are many interests, fields of knowledge and science that are not related to gastronomy. This broad relationship represents the richness of the gastronomic culture. For example, when we walk, seeing the oil lamp in one of the oven brings me to mind that it was oil lamp that night. We think that important days will be visited afterwards, it is customary not to go to these visits empty handed and a gift should be given to our elders

The ingredients of the gastronomy in a country are made of products produced from the geography on it. Accordingly, many cultural events take place on this road from agriculture to nutrition, from health to food economy. Common habits in societies are ethnological and social anthropological, institutionalized continuities, sociological, nutritional individual preferences and preferences are related to psychological culture.

Literature and art also have a close relationship with the culture of gastronomy. In written sources, the strings that begin with the date of Yesevi, the thorn apple of Yunus, the apple of Mevlana and extend to the gastronomic epics of today's minstrels are examples of the connection between literature and gastronomy. The effort to ensure the visuality of the food in the process from the preparation of the food to the presentation, as Feyzi Halıcı expresses, food is a result of fine arts. In this sense, the cultural values of the countries also appear in gastronomy. Turkey has a long history of cultural rooted will play an effective role in the formation of the gastronomic image. Turkey should be in line with the culture of gastronomy and tourism will contribute to the social structure.

Conclusion

In this study, the factors affecting the image of Gastronomy were tried to be explained. The best example of this situation is France. The image of France in the world in terms of gastronomy is quite high. If we consider the factors affecting the image as the gastronomy of France, today, every country wants to have a Michelin star restaurant in order to create an image in the world as food and beverage businesses. As of 2012, there are 594 Michelin star restaurants in France.

France established the Tourism Engineering Agency in 1993 to establish the image of gastronomy and stated what needs to be done and continues its research. This agency conducts gastronomic product research at home and offers suggestions to relevant institutions for marketing these products abroad. In the light of these researches and suggestions, gastronomy and image issues have always maintained their place in tourism policies.

Authorities in the field of gastronomy in France first engaged in the registration activities of gastronomic products owned by France. The best example of this is; brandy cognac is called Cognac only if it is made from grapes grown in the Cognac region of France. The qualification of others as brandy is finalized by law. It is possible to see this situation in many more foods and beverages.

The fact that France announces that part of the culture it owns is gastronomy shows how much importance it attaches to the image. The fact that the gastronomy book was first written by Antonie Careme, then the gastronomy encyclopedia Larousse Gastronomique was published in France and the names of the food and beverage businesses - restaurant, brasserie - show the world that gastronomy has a long history and is a part of culture..

The fact that the cooks who are trained in France can find jobs all over the world and are among the most valuable employees reveals how great the image of France's gastronomy is. The fact that the gastronomy schools in our country cooperate with the schools in France, receive trainer assistance or partner in internship, indicates the importance and image of French chefs.

As a result, countries with a good gastronomic image will benefit from gastronomic tourism in every sense. Of course, it will not be easy to come to this situation. However, the amount of effort, work, research and importance will be received

Bibliography

Ateş, T. (2005), *Mutfak Kültürü, Yemek ve Kültür*, Çiya Yayınları, 1. Sayı, pp. 47-61.

Avcıkurt, C. (2004), Ülke İmajı ve Turizm İliskisi-Türkiye Örneği, Turistik Yerlerin (Destinasyonların) Pazarlanması, *Haftasonu Turizm Konferansı Ix*, 17-19 Ekim 2004, Nevsehir: 1-17.

Avcıkurt, C. (2005), *Turizmde Tanıtma ve Satış Geliştirme*, İstanbul: Değişim Yayınları.

Baloglu, S. (2001), "Image Variations of Turkey by Familiarity Index: Informational and Experiential Dimensions", *Tourism Management*, 22 (2), pp. 127-133.

Baloglu, S. and Mangaloglu, M. (2001), "Tourism Destination Images of Turkey, Egypt, Greece, and Italy as Perceived by US-based Tour Operators and Travel Agencies", *Tourism Management*, 22 (1), pp. 1-9.

Cai, L. A. (2002), "Cooperative Branding For Rural Destinations", *Annals of Tourism Research*, 29 (3), pp. 720-742.

Caldwell, N. and Freire, R. F. (2004), "The Differences Between Branding A Country, A Region And A City: Applying The Brand Box Model", *Journal of Brand Management*, 12 (1), pp. 50-61.

Chen, P.-J. and Kerstetter, D. L. (1999), "International Students's Image of Rural Pennsylvania As A Travel Destination", *Journal of Travel Research*, 37 (3), pp. 256-266.

Cigerdelen, T. (2007), Turizmde Markalaşma, *I. Ulusal Türkiye Turizm Kongresi*, 07-08 Eylül, Sakârya: 213-226.

Clark, M. (2006), "A Case Study in the Acceptance of a New Discipline", *Studies in Higher Education*, 57 (2), pp. 133-148.

Dilsiz, B. (2010), *Türkiye'de Gastronomi ve Turizm- İstanbul Örneği* (Yayınlanmamış Yüksek Lisans Tezi), İstanbul Üniversitesi SBE, İstanbul.

Du Rand, D. E. H. and Heath, E. (2006). "Towards a Framework for Food Tourism as an Element of Destination Marketing", *Current Issues in Tourism*, 9, pp. 206-234.

Durant, R., Champion, H., Wolfson, M., Omli, M., Mccoy, T., D'agostino, R. B. Jr, Wagoner, K., and Mitra, A. (2007), "Date Fighting Experiences Among College Students Are They Associated with other Health Risk Behaviors?", *Journal of American College Health*, 55 (5), pp. 291-296.

Fakaye, P. C. and Crompton John L. (1991), "Image Differences between Prospective, First Time, and Repeat Visitors to the Lower Rio Grande Valley", *Journal of Travel Research*, 30, pp. 10-16.

Foley, A. and Fahy, J. (2004), "Incongruity Between Expression and Experience: The Role of Imagery in Supporting the Positioning of a Tourism Destination Brand", *Brand Management*, 11 (3), pp. 209-217.

Gunn, C. (1972), *Vacationscape: Designing Tourist Regions*, Austin: Bureau of Business Research, University of Texas.

Gürsoy, D. (2013), *Yiyelim, İçelim, Tarihini Bilelim-Dünden Bugüne Gastronomi*, İstanbul: Oğlak Yayıncılık.

Hacıoglu, N. (2005), *Turizm Pazarlaması* (5. Baskı), Ankara: Nobel Yayın Dağıtım.

Hankinson, G. (2005), "Destination Brand Images; A Business Tourism Perspective", *Journal of Services Marketing*, 19 (1), pp. 24-32.

Hatipoğlu, A. (2010), *İnançların Gastronomi Üzerine Etkileri: Bodrum'daki Beş Yıldızlı Otellerin Mutfak Yöneticilerinin Görüşlerinin Belirlenmesine Yönelik Bir Araştırma* (Yayınlanmamış Yüksek Lisans Tezi), Sakârya Üniversitesi SBE, Sakârya.

Heung, V. C. S., Qu, H., and Chu, R. (2001), "The Relationship between Vacation Factors and Socio-demographic and Traveling Charecteristics: The Case of Japanese Leisure Travellers", *Tourism Management*, 22, pp. 259-269.

Iliachenko, E. (2003), *Culture, History and Nature as Tourist Destination Constructs: The Case of Östra Norrbotten, Sweden*. Working Paper, Ies, Luleå University of Technology Division of Industrial Marketing and E-Commerce.

Kavaratzis, M. and Ashworth, G. J. (2005), "City Branding: An Affective Assertion of Identity or a Transitory Marketing Trick?", *Tijdschrift Voor Economische En Sociale Geografie*, 96 (5), pp. 506-514.

Konecnik, M. (2004), "Evaluating Slovenia's Image as a Tourism Destination: A Self-Analysis Process Towards Building a Destination Brand", *Brand Management*, 11 (4), pp. 307-316.

Kuşluvan, S. and İlhan, İ. (2011), Nevşehir'de Turizm Gelişiminin Temel Sorunları ve Bazı Öneriler, *I. Uluslararası Nevşehir Tarih ve Kültür Sempozyumu*, Cilt 7, 81-93. Nevşehir: Nevşehir Üniversitesi. (PDF) Gastronomi ve Turizm İlişkisi Üzerine Bir Değerlendirme.

Larousse Gastronomique (2005), İstanbul: Oğlak Yayınları.

Lubbe, B. (1998), "Primary Image as a Dimension of Destination Image: An Emprical Assessment", *Journal of Travel and Tourism Marketing*, 7 (4), pp. 21-30.

Morgan, N. J., Pritchard, A., and Piggott, R. (2003), "Destination Branding and the Role of the Stakeholders: The Case of New Zealand", *Journal of Vacational Marketing*, 9 (3), pp. 285-299.

Okumus, B., Okumus, F., and McKercher, B. (2007). "Incorporating Local and International Cuisines in the Marketing of Tourism Destinations: The Cases of Hong Kong and Turkey", *Tourism management*, 28, pp. 253-261.

Öney, H. (2013). Gastronomi Turizmi. İçinde S. Bahçe (Ed.), *Alternatif Turizm* (pp. 158-188), Eskişehir: Anadolu Üniversitesi Yayınları.

Öter, Z. and Özdogan, O. N. (2005), "Kültür Amaçlı Seyahat Eden Turistlerde Destinasyon İmajı: Selçuk-Efes Örneği", *Anatolia: Turizm Araştırmaları Dergisi*, 16 (2), pp. 127-138.

Ritchie, B. J. R. and Ritchie, R. J. B. (1998), The Branding of Tourism Destinations – Past Achievements & Future Challenges, *Annual Congress of the International Association of Scientific Experts in Tourism*, Marrakesh, Morocco, September, 1998, 1-31.

Temeloglu, E. (2007), "Hizmet: İşletmelerinde Markalaşma", *Kalkınmada Anahtar Verimlilik*, 19 (224), pp. 20-21.

Tosun, C. and Bilim, Y. (2004), Hatay'ın Turistik Bir Şehir Olarak Pazarlanması, *Balıkesir Ulusal Turizm Kongresi*, 15-16 Nisan 2004, 269-288.

Wolf, E. (2006), *Culinary Tourism: The Hidden Harvest*, Dubuque, IA: Kendall/Hunt Publishing Company.

Yıldırım, A. (2012). "Epikürizm ve Bâkî'nin Bir Gazeli Üzerine", *Turkish Studies - International Periodical for the Languages, Literature and History of Turkish or Turkic*, 7 (3), pp. 2701-2709.

Suna Muğan ERTUĞRAL, H. Neyir TEKELİ and Sezgi
GEDİK ARSLAN

Effect of Environment in Increasing Sustainable Competitiveness at Tourism Destination

1 Introduction

Changes in political structure after the World War II, together with emergence of new nation-states and fast structuring in winners and losers of the war, have brought development efforts to the forefront. Economic development process is considered a comprehensive factor with its social, political and environmental dimensions. Efforts on how to develop states include policies of governments and scientific studies. Development policies are established in line with such efforts based on utilization of countries' natural resources and geographical opportunities.

Development, which means economic advancement and qualitative improvement of societies, is of importance for diversification and expansion of economic sectors and social structure. In particular, expansion of global tourism industry in the post-war era during when development efforts were speeded up has been an extremely important progress. These advancements have also increased competition among the regions in time.

Tourism industry has developed from the past to the present and has been primarily preferred by developing countries for attaining development within a sustainable development perspective.

Tourism sector displays a progress at an equal rate with the rapid growth observed in the world economy. Share of tourism revenues in the world economy has been increasing recently. Tourism has positive economic effects, including those on balance of payments, revenue, domestic import, workforce multiplier, employment increase and creating new job opportunities, economic development and generating foreign exchange earnings and its regulatory effect on export and income distribution. Positive economic effects of tourism also increase competition among tourism regions. Ever-increasing competition factor in tourism sector comes with various competition strategies. Sustainable environmental factors are among these strategies.

Environmental factors take an important place in developing tourism and competitiveness in a tourism region. Particularly, tourism must be environmentally sustainable. However, together with industrial and technological developments, air, water and soil are polluted, natural resources are wasted and self-renewing character of the nature is harmed. Therefore, protection of environment and approaches to environmental sustainability have gained importance in recent years. In traditional approaches, comparative advantages of production factors provide competitive advantage while so do environmental factors in tourism. It has been observed that some regions prioritizing environmental factors have attained higher development in terms of tourism in recent periods. The fact that environment is center of attraction as a competition factor is of primary importance.

2 Determining Sustainable Competitiveness of Tourism Destinations

Disappearance of country borders due to globalization, developments in communication and information technologies and strengthening of international organizations and companies have increased competition. Competition today is not only among companies but also at the national and even international level. This has caused countries to make alterations in their competition strategies.

Competitiveness can be defined as the power of generating higher revenue and employment of an industry/sector in a country compared to the corresponding industry/sector in another country. A country should be at a level that enables it to compete with the products of another country in terms of certain points including price, quality and design. Competitiveness helps to put forth relative positions of countries or current sectors. While some businesses describe competitiveness as productivity, amount of added value per unit of production and the rate of increase in this amount, some others define it as convincing buyers to prefer their products among all alternatives (Koç & Özbozkurt, 2014: 86; Gürpınar & Sandıkçı, 2008: 106).

Competition is much more intense especially in sectors, like tourism, in which there are international activities. Tourism competition is a destination's sustaining its market share and power as well as protecting and continuously improving it in time (d'Hautesserre, 2000: 23). In order to gain a competitive advantage in tourism industry, a destination should provide all its attractions and tourist experiences that are superior to alternative destinations open to potential visitors (Dwyer & Kim, 2003: 369).

Countries resort to different methods to gain advantage of competitiveness on a sectoral basis in the international market. Diamond model developed by Porter in the 20th century (1990) sets the criteria for countries to be more competitive in certain sectors than other countries. According to Porter's Diamond model, competitiveness of a country is based on 4 core interrelated factors which are factor conditions, firm strategy, structure and rivalry, demand conditions and related and supporting industries. The model's variable determinants in addition to the core factors are the role of government and chances, which are asserted to affect the others by the model (Porter, 1998: 171).

Another model for tourism destination competitiveness is Conceptual Model of Destination Competitiveness developed by Ritchie and Crouch building on Porter's Diamond model. The model bases competitiveness on two dimensions: "comparative advantage" and "competitive advantage." The model also suggests that global macro-environment (global economy, terrorism, cultural and demographic tendencies etc.) and competitive micro-environment affect tourism system which influences destination (Crouch, 2007: 2). Ritchie and Crouch's model comprises 5 core components, namely Supporting Factors and Resources, Core Resources and Attractors, Destination Management, Destination Policy, Planning and Development and Qualifying and Amplifying Determinants, and 36 sub-components related to them (Ritchie & Crouch, 2003: 63).

Kim's model of competitiveness is another study aimed at determining factors that constitute competitive power. Kim (2000) defines competitiveness in tourism as the ability of tourism market structure and conditions, tourism resources, tourism human resources and tourism infrastructure in a country to generate added value and enhance national prosperity (Sert & Şahbaz, 2017: 80). There are four dimensions in Kim's model that determine competitiveness. Accordingly, the resources are categorized into primary resources as environment, subject, resources; secondary resources as tourism policy, tourism planning, tourism investments, tourism duties and prices, tourism management; tertiary resources as tourism infrastructure, hosting system, attractions, promotion and tourism labor force and quaternary resources as tourism demand, tourism employment, tourism performance and tourism export (Kim, 2000: 39).

In 2003, Dwyer and Kim developed "Integrated Model of Destination Competitiveness" by combining core elements of Porter's model of business and national competitiveness with core components of Ritchie and Crouch's model of destination competitiveness. Main elements of this model are as follows (Dwyer &Kim, 2003: 377–379):

- Resources category is divided into two types: Endowed (inherited) and Created. Endowed Resources, in turn, can be classified as natural (mountains, lakes, beaches, rivers, climate etc.) and heritage or cultural (cuisine, handicrafts, language, customs, belief systems etc.). Created resources include tourism infrastructure, special events, and the range of available activities, entertainment and shopping.
- Situational conditions are forces in the wider external environment that impact upon destination competitiveness (economic, social, cultural, demographic, environmental, political, legal, governmental, regulatory, technological vb.)
- A distinction is made between public and private sector activities with regard to destination management factors. For instance, public sector may carry out activities such as developing national tourism strategies and private sector may apply private sector industrial training program.
- Demand conditions comprise three main elements of tourism demand awareness, perception and preferences.

The model of competitiveness of tourism destinations designed by Bahar and Kozak (2005) advances the idea that destination competitiveness is under the influence of a wide range of complex factors, which interact with each other including individual, regional, territorial, quantitative (human capital and role of education, IT and technological development, supply conditions of tourism, demand conditions of tourism, cost, investment, incentives and financial regulations) and qualitative (sustainable tourism and environment, service quality and customer satisfaction, productivity and efficient use of resources, touristic product diversification, image and innovation, tourism marketing strategy and market share, state and bureaucracy, tourism competition strategy) factors. For a destination to gain competitive power, individual, regional-territorial, quantitative and qualitative competitiveness factors must act at the same time, as a whole and in harmony (Bahar, 2004: 66).

Although models developed to explain destination competitiveness are different from each other, they all underline the relationship between tourism and environmental sustainability. Protection and sustainability of environmental values provide competitive advantage in terms of improvement of tourism.

3 Tourism Industry and Its Relationship with Environment

Relationships between tourism and environment show that environment forms creative elements of tourism while tourism is a destructive element for environment. Besides, tourism is a means of protecting environment. This demonstrates presence of a contradiction in the relationship between tourism and environment; on the other hand, it brings tourism's function of protecting environment, arising from supplemental nature of the same case and well organization of tourism, into force (Olalı & Timur, 1988: 365).

We see environmental impact of tourism as unfavorable situations as pressure on natural resources and harms given to ecosystems. Depletion of natural resources and environmental degradation associated with tourism activities cause huge problems in many regions that are rich in terms of tourism. Moreover, it is widely accepted that not only uncontrolled tourism expansion causes environmental degradation but also environmental degradation poses a serious threat to tourism activities. The fact that most of the tourists prefer relatively high consumption patterns when they arrive at the destination may be a significant challenge regarding natural resources and local ecosystems especially in developing countries and regions (Neto, 2003: 4). Hence, tourism planning is important for sustainable development and sustainable tourism.

The system where rapid population growth and consumption are in the forefront, particularly in the period of development efforts, has caused damage to natural resources. Disruption of ecological balance within this period is observed to be a consequence of not paying regard to relationship between environment and development. This has developed in a structure that depletes natural resources and damages environmental values notably due to unlimited consumption of resources and maximum profit target brought about by a production approach based on such consumption. The fact that production has developed in a structure, which is based on natural resources, has led to unconscious consumption of resources. Therefore, destructive damages caused to natural resources and environmental values have created disruptive impact on ecological balance. Some notable examples of widespread environmental problems have been unplanned urbanization, global warming and climate changes, depletion and pollution of water sources, extinction or mutation of species, emergence of contagious diseases, polluting effect of wastes, which cannot be disposed by nature on soil, water and atmosphere and degradation of natural habitat. Upon emergence of these threats in the late 1960s, it has become compulsory to handle these issues that threaten environment with a new point of view.

In her research titled "Silent Spring" conducted in 1962, Rachel Carson highlighted fatal impacts of various chemicals used as pesticides on natural life. Silent Spring, which is among the first studies with regard to sustainable development, had a vital role in raising today's environmental awareness

(Çetiner, 2012: 18). Later on, a vast number of countries of different socio-economic structures and development levels came together for the environment for the first time at UN Conference on the Human Environment in Stockholm in 1972 (Stockholm Conference) and adopted UN Declaration on the Human Environment. Sustainable development concept was first defined in Brundtland Report prepared by the World Commission on Environment and Development in 1987 as "a development that meets the needs of the present without compromising the ability of next generations to meet their own needs" (http://www.mfa.gov.tr/). The report of Brundtland Commission has provided a focal point in global fight against environmental destruction and poverty (McChesney, 1991: 13).

Sustainable development is a holistic approach that aims to establish balance between the requirements of human life and sustainability of natural resources and thus to make a coherent planning from the present to the future with environmental and social dimensions (TÜSİAD, 2012: 10). This planning process has resulted in prominence of fields of activity that protect the environment and provide environmentally friendly achievements.

Activities aimed at protection of worldwide environmental values have been continued in parallel with environmental awareness of the human being. However, the motivation for production and earning has speeded up environmental problems. United Nations Conference on Environment and Development held in Rio de Janeiro in 1992 and also called Earth Summit pointed out fight against environmental problems at global scale and the phenomenon of sustainability.

Rio Summit held in 1992 also led to preparation and opening for signature of the United Nations Framework Convention on Climate Change. The main objective of the convention is to keep the greenhouse gas concentrations in the atmosphere at a level to prevent hazardous human-induced impact on climate system. The said convention, in general, establishes principles, action strategies and obligations with regard to protection of global climate system and reducing greenhouse gas emissions, in other words, it constitutes the roof of climate regime and determines the rules of the game. Declarations that serve as a guide for protection of environment and finding solutions and mutually signed agreements are of importance.

Sustainable development in tourism adopts the principle of meeting the needs of the present tourists and host regions as well as looking out for protection and improvement of future opportunities. In this way, continuity of cultural integrity, compulsory ecological processes, biodiversity and life support systems is ensured and management of all resources is brought to the forefront so as to satisfy economic, social and aesthetic requirements. Sustainable tourism products are those treated in harmony with beneficial local environment, society and cultures instead of being harmed by touristic development. Despite strong ties of people with the environment, unplanned and uncontrolled growth of cities due to industrialization has caused destruction of nature, and this has led to lower livability of cities for people. Ultimately, it has increased demands of people for traveling to places with natural beauty with a wish to feel the presence of natural environment (Kaya et al., 2011: 256; Özdemir, 2006: 35).

Human, who lies at the heart of touristic activities, is a creature, which cannot be considered independent from social or natural environment (Boudreau & Newman, 1993: 141). Some of the declarations and conferences adopted/held for the purpose of raising an overall environmental awareness for tourism industry and putting sustainable activities into force and with the support of institutions, including mainly UNWTO (United Nations World Tourism Organization), UNFCCC (United Nations Framework Convention on Climate Change), UNEP (United Nations Environment Programme), WMO (World Meteorological Organization), WEF (World Economic Forum) and IPCC (Intergovernmental Panel on Climate Change) are as follows (UNWTO, 2008: 3):

- 1980, Mania Declaration on World Tourism,
- 1982, Acapulco Document on the Rights to Holiday,
- 1985, Sofia, Tourism Bill of Rights and Tourism Code,
- 1989, The Hague Declaration on Tourism,
- 1995, Lanzarote Charter on Sustainable Tourism,
- 1998, Lanzarote Conference on Sustainable Tourism,
- 1999, Global Codes of Ethics for Tourism,
- 2003, First International Conference on Climate Change and Tourism in Djerba,
- 2008, UN Climate Conference in Poznan,
- 2009, UN Climate Change Summit in Copenhagen.

Interaction of people with local residents at touristic destinations they visit and presence of natural beauties in the region are considered resources for tourists in terms of tourism. From this point of view, it is clearly understood that environment is much more important for tourism compared to many other industries and that environment directly affects touristic products in the region. In this regard, Turkey, with its significant natural beauties and biodiversity on national and international scale, has natural resources that are attributed value globally. Nevertheless, the said environmental values are damaged and depleted insensibly as a result of wrong planning/planlessness and unsustainable practices (Dal & Baysan, 2007: 70).

4 Sustainable Development and Sustainable Tourism

It is observed that the concept of sustainability is often used also in the fields of business and economy. However, it is of use to state that in terms of sustainability in tourism, environmental resources are highlighted. Another key concept that has significant effect on the rise of sustainable tourism concept is sustainable development. Sustainable development is described as many of dominant issues related to industrialization, trade and urbanization as a symbol of "return to the nature" or natural process (Yavuz & Zığındere, 2000: 326). As importance of sustainable development for the future of the world is recognized, every sector has endeavored to produce distinctive solutions and sustainability has come with a separate assessment and different practices for every sector (Kahraman & Türkay, 2006: 95).

In another perspective, sustainable development is a process of creating a series of opportunities, by protecting flexibility of economic, social and environmental systems, which enables individuals and society to realize their own desires and potential. Phenomenon of sustainable development, which involves economic and ecological principles to direct economic growth and development, is better understood when environmental deteriorations arising from overconsumption of natural resources capital are considered (Çetiner, 2012: 2).

Associating main elements among tourism, environment and sustainable development is based upon three foundations (Kahraman & Türkay, 2006: 95):

- Effects of tourism on environment,
- Strong ties between tourism and natural, cultural environment, and the fact that the environment is a major attraction for the tourists,
- The fact that environmental factors affect tourism.

Considering these three factors, it can be asserted that sustainability, with regard to tourism industry, finds a field of application through its interaction with environmental values.

Sustainable development represents the attraction of continuity of economic development which does not put an unreasonable strain on environmental, sociocultural or economic bearing capacity of the world (Weaver, 2005: 10). Accordingly, sustainability has gained meaning for all developing areas of economy which develop and contribute as they develop. For this reason, phenomenon of sustainability has started to be widely used. Significant progress is made with regard to speed-up of activities that are fed on environmental values and that feed environmental values as they develop and shaping of economic life together with such development.

An evolution has been experienced from a standardized mass tourism characterized by holidays which is produced and consumed similarly, routinely and collectively, without taking norms, cultures and environmental factors of touristic destinations into account toward a new sustainable tourism in the long term. Current mass tourism has turned into an unsustainable one. Within this period, qualitative aspects of tourism have been ignored while quantitative growth is emphasized and necessity of considering cultural, social and environmental facilities of tourism destinations in terms of sustainability has been underlined (Mak, 2003: 181).

Research and literature studies on sustainable tourism which started in the beginning of 1990s have been continuing within the scope of sustainable development and sustainable tourism. In the beginning of 1990s, the term sustainable tourism was handled by academicians and practitioners in a way to cause changes in city and holiday tourism understanding (Weaver, 2005: 25–28). Scenarios of doomsday/ecological disaster in the late 1980s and the fact that sustainable tourism became a trend in travel and tourism for many people have contributed to evolvement of the subject. The phenomenon of sustainable tourism takes protection of quality of natural resources such as environment, landscape, air, seawater, freshwater, flora and fauna, intrinsic values owned and structures that are deemed worth protection and cultural resources into consideration (Middleton & Hawkins, 1998: 10).

Protection and development of resources lie at the heart of sustainable development. Utilizing resources continuously and by protecting them, particularly, support of renewable resources to development without exceeding their limits to renew themselves lay the foundation for the philosophy of development that protects environment. It is clear that it is difficult for environment-oriented tourism sector to develop with a degraded environment and therefore it is difficult to benefit from economic functions of tourism. The main problem here is related to how a healthy development can be achieved in tourism. The solution to this problem lies in protecting and improving the resources tourism makes use of, while ensuring development of tourism (Çakılcıoğlu, 2013: 27). Hence, protection of environment and environmental and cultural values is important. Specifically, development of tourism by protecting environmental values and environment is of great importance for the sustainability of tourism.

Natural resources owned and bestowed upon us by the earth provide fuels, metal ores, food, and water, which are our basic needs for living, and landscapes and natural environments, which are the essential resources for most entertainment and tourism industries. These resources have been used throughout the evolution of humankind; however, they are not infinite. Many of them become scarce or deteriorate because the population needs them for economic development and demands more than ever (Middleton & Hawkins, 1998: 21).

A balanced and coherent relationship between the natural environment and acts of humans can only be established by complying with principles of sustainable development. Sustainable tourism refers to ensuring continuity of factors and assets that make up tourism without losing their qualities and quantities. Increasing tourism activities come with ecological, geological and aesthetical problems and unless necessary precautions are taken, neutralize sustainability opportunities with natural processes. In order to establish and maintain a well-balanced tourism-environment relationship, the targets of a country regarding tourism and resource protection should be consistent and consistency with other related targets should be ensured (Gündüz, 2004: 58–60). Competitive advantage gained by tourism approach which is developed based on environmental values lays the basis for principles of sustainable tourism. Protection of environment and environmentally friendly development of tourism is the advantage of the tourism region.

Factors that provide comparative advantage for the tourism region can be divided into two as qualitative features and price differences. Factors like quality, diversity, distinctness, uniqueness and authenticity, which represent the attractiveness of touristic resources, constitute the qualitative dimension. Environmental problems and solutions to these problems may also be handled as part of qualitative dimension (Ayaş, 2007: 62).

5 Environmental Competitiveness in Sustainable Tourism

Competitiveness is a relative, multi-dimensional and dynamic concept. Therefore, defining and measuring it bears a number of difficulties. In traditional approaches, comparative advantages in production factors provide competitive advantage and mathematical criteria, including market share, market growth rate and earned revenue are considered sufficient for evaluation of competitiveness (Barney, 2001: 643).

Factors that provide advantage for the tourism region can be divided into two as qualitative features and price differences. Factors like quality, diversity, distinctness, uniqueness, and authenticity, which represent the attractiveness of touristic resources constitute the qualitative dimension. Environmental problems and solutions to these problems may also be handled as part of qualitative dimension. Competitiveness of a tourism region is defined as the advantage of attracting tourist population traveling based on environmental values and rendering activities and services that emerge during the visit. In parallel to increase in tourism revenues, competition in sharing the expenditures allocated by people from their budget to tourism activity has become fierce, and "Environmentally Sustainable Tourism Development" has increased the importance of tourism strategies, which are known as methods of improving competitive positions and attractiveness of the regions, in tourism markets. When creating a global competition strategy in markets where diversity of touristic products like ecotourism, coastal tourism, green tourism, historical, cultural and antique tourism, nature and landscape tourism, health, sports and festival tourism is increasing, adoption of a strategy compliant with touristic resources of the region is a significant prerequisite for competing (Jap, 2001: 10).

Tourism regions race for attracting conscious, highbrow tourist population, which respect environmental values. Developed tourism regions play to various strengths of the region such as distinctness, diversity, attractiveness and originality to the detriment of the competing tourism regions in order to direct the budget people allocate to tourism expenditures to their regions. Environmentally friendly tourism development strategy selects tourist population which attaches importance to environmental values and is ready to bear the cost of protection of these values as the target group. Providing distinctive activities and services for relatively low prices compared to competitors in order to attract the aforementioned population to the region is the core of the sustainable competition strategy in tourism.

Conclusion

Traveling need for health-related, cultural, religious and touring purposes has gradually risen since the early ages. Tourism sector which has gained economic and sociocultural importance especially after the World War II has displayed a profound development with diversification of alternative tourism activities. When we look at the development trend of tourism as of periods all around the world, it is observed that holiday tourism understanding has evolved towards alternative tourism types and a tendency to take environment factor into account has arisen. Particularly, new pursuits have been started upon feeling the pressure caused by fast-moving mass tourism due to time and place concentrations. In addition, significant changes have been experienced in tourism consumption patterns in line with economic, social and technological changes.

A wide perspective is crucial for determining factors that affect competitiveness. Economic structure and market structure should be evaluated in terms of competitiveness. Those who want to have advantage in this competitive environment try to do so by achieving excellence in one of management, cost, diversification and focusing strategies. As well as trying to gain competitive advantage by attaining perfection through innovative activities on the said strategies, business world endeavors to maintain sustainability of this advantage. The source of keeping competitive advantage sustainable is the ability of the management to develop innovative activities. Innovative activities are important for both competitive advantage and economic recovery. Thanks to innovation, businesses not only have cost advantage due to reducing costs but also gain competitive advantage because of diversification of products and services. This requires consideration of supply and demand conditions. For this reason, an evaluation of the region in terms of touristic supply and demand and regional economy should be taken into account. Environmental factors are important especially as a touristic attraction. Besides, improvement of tourist infrastructure and superstructure is important.

In this respect, competitive advantage depends on quality, price and economic structure and centralized decisions. Developing demand and supply is crucial for providing competitive advantage. Thus, developing regional supply and demand provides competitive advantage. These factors are of importance for ensuring sustainable competitiveness of a tourism region. It has gained competitive power in development of tourism in a particular region. For a better understanding of tourism competition, factors that lead to competitive advantages should be considered together. Current resources of a destination provide an advantage according to that region, on the other hand, if these resources are used in an effective and efficient manner, competitive advantage can be gained. A tourism region rich in natural resources sometimes cannot compete with another destination which is not. Unless sustainability of tourism resources of a destination is ensured, it is not feasible to talk about long-term competition. It is essential for analysis of tourism sector with regard to competitive advantages including climate, nature, beaches and wildlife. Specifically, natural, cultural and social attractions are the major sources of supply for tourism. It is not possible to reproduce them. Factors such as state policies, competence of work force, management quality and tourism infrastructure indicate the secondary sources of supply for tourism. Thus, these competitive factors play a vital role in development of tourism. This study evaluates economic structure, quality and price advantage especially for tourism sector. In order to achieve enhancement in tourism industry, current environment, competitive and sustainable policy, it will be probable to have a larger share in international tourism revenues in the upcoming years.

Bibliography

Ayaş, N. (2007), "Çevresel Sürdürülebilir Turizm Gelişmesi", *Gazi Üniversitesi İktisadi ve İdari Bilimler Fakültesi Dergisi*, 9 (1), pp. 59-69.

Bahar, O. (2004), *Türkiye'de Turizm Sektörünün Rekabet Gücü Analizi Üzerine Bir Alan Araştırması: Muğla Örneği* (Doktora Tezi), Muğla Üniversitesi Sosyal Bilimler Enstitüsü İktisat Anabilim Dalı, Muğla.

Bahar, O. and Kozak, M. (2005), "Türkiye Turizminin Akdeniz Ülkeleri ile Rekabet Gücü Açısından Karşılaştırılması", *Anatolia: Turizm Araştırmaları Dergisi*, 16 (2), ss. 139-152.

Barney, J. B. (2001), "Resource-Based Theories of Competitive Advantage: A Ten-Year Retrospective on the Resource-Based View", *Journal of Management*, 27 (6), pp. 643-650.

Boudreau, F. A. and Newman, W. M. (1993), *Understanding Social Life: An Introduction to Sociology*. Minnesota: West Publishing Company.

Çakılcıoğlu, M. (2013), "Turizm Odaklı Sürdürülebilir Kalkınma için Bir Yöntem Önerisi", *Mimar Sinan Güzel Sanatlar Üniversitesi Mimarlık Fakültesi Dergisi*, 9 (16), pp. 27-42.

Çetiner, S. (2012), "Sessiz Bahar Ne Diyor(du)?", *Tarlasera*, Temmuz 23012, pp. 18-21.

Crouch, G. I. (2007), *Modelling Destination Competitiveness: A Survey and Analysis of the Impact of Competitiveness Attributes*, Australia: CRC for Sustainable Tourism Pty Ltd.

D'hautesserre, A.-M. (2000), "Lessons in Managed Destination Competitiveness: The Case of Foxwoods Casino Resort", *Tourism Management*, 21 (1), pp. 23-32.

Dal, N. and Baysan, S. (2007), "Kuşadası'nda Kıyı Kullanımı ve Turizmin Mekânsal Etkileri Konusunda Yerel Halkın Tutumları", *Ege Coğrafya Dergisi*, 16 (1-2), pp. 69-85.

Dwyer, L. and Kim, C. (2003), "Destination Competitiveness: Determinants and Indicators", *Current Issues in Tourism*, 6 (5), pp. 369-414.

Gündüz, F. (2004), "Çevre ve Turizmin Sürdürülebilirliği", *Planlama Dergisi*, 1, pp. 58-66.

Gürpınar, K. and Sandıkçı, M. (2008), "Uluslararası Rekabetçilik Analizinde Michael E. Porter'ın Elmas Modeli Yaklaşımı: Türkiye'deki Bazı Endüstrilerdeki Uygulanabilirliğinin ve Sonuçlarının Araştırılması", *Sosyal Ekonomik Araştırmalar Dergisi*, 8 (15), pp. 105-125.

Jap, S. D. (2001), "Perspectives of Joint Competitive Advantages in Buyer-Supplier Relationships", *International Journal of Research in Marketing*, 18 (1-2), pp. 19-35.

Kahraman, N. and Türkay, O. (2006), *Turizm ve Çevre*, İstanbul: Detay Yayıncılık.

Kaya, L. G., Fürüzan A., and Yılmaz B. (2011), "Muğla-Dalyan Turizminin Özel Çevre Koruma Bölgesi Üzerine Etkileri", *İnönü Üniversitesi Sanat ve Tasarım Dergisi*, 1 (3), pp. 255-266.

Kim, C. (2000), *A Model Development for Measuring Global Competitiveness of the Tourism Industry in the Asia-Pacific Region*, Seoul: Korea Institute for International Economic Policy.

Koç, M. and Özbozkurt, O. B. (2014), "Ulusların Rekabet Üstünlüğü ve Elmas Modeli Üzerine Bir Değerlendirme", *İşletme ve İktisat Çalışmaları Dergisi*, 2 (3), pp. 85-91.

Mak, J. (2003), *Tourism and the Economy: Understanding the Economics of Tourism*. Honolulu: University of Hawaii Press.

McChesney, I. G. (1991), *The Brundtland Report and Sustainable Development in New Zealand*, Canterbury: Centre for Resource Management, Lincoln University. Information Paper No 25.

Middleton, V. T. C. and Hawkins, R. (1998), *Sustainable Tourism A Marketing Perspective*, Butterworth-Heinemann, Oxford.

Neto, F. (2003), *A New Approach to Sustainable Tourism Development: Moving Beyond Environmental Protection*, DESA Discussion Papers No. 29, pp. 1-10.

Olalı, H. and Tımur, A. (1988), *Turizm Ekonomisi*, İzmir: Ofis Ticaret Matbaacılık Sti.

Özdemir, Ü. (2006), "Amasra'da Turizm ve Çevresel Etkileri", *Doğu Coğrafya Dergisi*, 11 (15), pp. 33-52.

Porter, M. E. (1998), *Competitive Advantage of National with a new Introduction*, New York: The Free Press.

Ritchie, J. R. B. and Crouch, G. I. (2003), *The Competitive Destination: A Sustainable Tourism Perspective*, Wallingford, UK: CABI Publishing.

Sert A. N. and Şahbaz, R. P. (2017), "Turist Bakış Açısıyla Destinasyon Rekabet Gücünün Belirlenmesinde Sosyodemografik Özelliklerin Etkisine

Yönelik Bir Araştırma", *Journal of Tourism and Gastronomy Studies*, 5 (3), pp. 74–92.

T.C. Çevre ve Orman Bakanlığı, Rio Sözleşmeleri, http://www.ncsa-turkey. cevreorman.gov.tr/, (30.05.2020).

T.C. Dışişleri Bakanlığı, Sürdürülebilir Kalkınma, http://www.mfa.gov.tr/ surdurulebilir-kalkinma.tr.mfa., Accessed 30.05.2020.

TÜSİAD (2012), Sürdürülebilir Turizm, Sis Matbaacılık Prom. Tanıtım Hizmetleri Tic. Ltd., Şti., Yay. No: TÜSİAD-T/2012-09/531, https://tusiad. org/tr/yayinlar/raporlar/ item/6030-surdurulebilir-turizm, Accessed 30.05.2020.

Weaver, D. (2005), *Sustainable Tourism: Theory and Practice*, London: Elsevier Butterworth-Heinemann.

World Tourism Organization (UNWTO) (2008), *Adaptation to Climate Change in the Tourism Sector,* https://www.unwto.org/archive/global/ news/2011-08-16/climate-change-adaptation-and-mitigation-tourism-sector, Accessed 30.05.2020.

Yavuz, E. and Zığındere, Y. Ö. (2000), Sürdürülebilir Kalkınmanın Turizme Etkisi, *Balıkesir Üniversitesi Sosyal Bilimler Enstitüsü Dergisi*, 3 (4), pp. 321–336.

Hidayet KIŞLALI and Mustafa KÖSE

Recreational Second Homes and Sustainable Tourism

1 Introduction

Second home ownership is not a recent phenomenon. On the contrary, having an estate on the countryside was an insignia of being wealthy in the past. However, the second home ownership by middle classes in many countries such as Canada, Germany and various Scandinavian countries can be considered as a contemporary phenomenon (Jaakson, 1986). Even though second homes are not a novel area of exploration, significant increase in second home ownership follows the rapid expansion of roads and car ownership (Hall & Müller, 2004). Second homes are built into locations where it is far enough to offer a detachment from main residences, but close enough to be reached in a short time by car (Timothy, 2004).

While city residences are mostly considered as accommodations with a utilitarian function for the employment, second homes are built for primarily recreational purposes (Jaakson, 1986). "Ranging from rural cottages to farmhouses, villas, hunting or fishing lodges or prefabricated cabins, second homes are a global phenomenon and part of leisure history in many developed countries" (Pitkänen, 2008: 169). Definition of second home may vary according to the approaches scholars choose to study them. The "second home" in this research refers to the homes that were purposefully built or converted to recreational residences on the coastal or mountain areas.

Although the existence and development of the second homes are quite old in Turkey as in the world, the second home ownership has increased since the 1950s as results of factors such as urbanization, industrialization, development of transportation facilities, the increase of leisure time and fashion (Emekli, 2014; Somuncu et al., 2019). The construction of second homes for tourism purposes started to concentrate in areas suitable for coastal tourism in the 1950s (Alkan, 2014; Somuncu et al., 2019). The importance of second home settlements for domestic tourism increased in the 1960s and their construction had growingly continued in coastal areas until 1990s (Kılıçaslan, 2006). This rapid increase in second homes in Turkey is a fact that requires further explorations.

Construction of second homes for tourism purposes in mountains of Turkey followed the developments of second homes in coastal areas (Kılıçaslan, 2006; Alkan, 2014; Somuncu et al., 2019). Highlands faced functional changes and tourism and recreational activities started to develop in 1980s (Somuncu et al., 2010). Highlands have become prominent places to participate in recreational activities and to enjoy natural beauties and good weather for second-home dwellers (Grötzbach, 1984). Even though, it is stated that in addition to agriculture and growing livestock, recreation has been a vital element of many highland settlements in Turkey (Grötzbach, 1984; Koca, 2011), there were unprecedented changes in the functions highlands serve in the daily life of Turks in the last five decades (Grötzbach, 1984).

When second homeowners are considered as tourists, impacts of second homes seem to be similar to tourism impacts. However, because of the longer stays they may cause bigger and more significant changes. Besides, unlike other types of tourism, there must be an accommodation to purchase (Müller, Hall & Keen, 2004). Hence, second homes could be *a curse or blessing* (Müller & Hoogendoorn, 2013). For example, it can be a curse in terms of transportation and emissions Næss et al. (2019), and a blessing when left-behind properties in depopulated regions attract new visitors (Larsson & Müller, 2019). In this research, we will explore the development of second home tourism in Turkey. In the first part, we will discuss the sustainable tourism as a concept with contested meanings. Then, we will focus on second homes developments around the world with a sustainable tourism perspective. Finally, we will attempt to highlight sustainability aspects of second home developments in Turkey. To be able to elaborate on contemporary second home developments, we will shed lights onto developments in some recreational highlands in Adana province.

2 Sustainable Tourism

Sustainability seems to be evolving into a *Swiss Army knife* for scholars and policy makers. These days, concept of "sustainability" can be found in quite unprecedented spaces such as "green prisons" (Moran & Jewkes, 2014). In their critical paper, Torkington et al. (2020) examined tourism policy documents from seven European countries and stated that the term "sustainability" was hijacked and used without a clear definition but implications of infinitely sustaining the current production and consumption of tourism products. Lack of an agreed definition is considered as a serious problem in sustainable tourism research (Alfaro Navarro, Andrés Martínez, & Mondéjar Jiménez, 2020). This, however, does not mean that there is a lack of definition. On the contrary, the term is surrounded by excess amount of definitions (Butler, 1999). While proliferation of the definitions creates flexibility and offers chances to study the phenomenon from various perspectives, it might bring new challenges, too. The danger of becoming a useful tool to *green-wash* highly non-green activities is the *Achilles' heel* of sustainability. The word "sustainable" is defined as "able to be maintained or continued" or "causing little or no damage to the environment and therefore able to continue for a long time" (Cambridge Dictionary, No date) in dictionaries. However, discussions around sustainability and sustainable tourism require further explorations.

There are three highly well established but loosely defined and interchangeably used terms in sustainable tourism studies. These are sustainability, sustainable tourism and sustainable development (Liu, 2003).

the word 'sustainability' at its most basic semantic level simply means the ability to 'sustain', or maintain something indefinitely at a steady rate. This therefore conflicts with the idea of infinite, or exponential, growth. Yet the term has become increasingly contested. On one hand it is now understood by many as being rooted in the discourse of ecology (i.e., avoiding negative environmental impacts and depletion of resources and maintaining an ecological balance), whilst, on the other, it is used (especially in its adjectival form) as synonymous with 'sustained'—as in 'sustained growth', for example, which denotes uninterrupted growth at a continuous rate. Thus, the concept of 'sustainability' has two possible framings; one oriented towards finiteness and the other towards infinity. (Torkington, Stanford, & Guiver, 2020, p. 1046)

One of the main criticism of the studies on sustainable tourism is that while sustainable tourism is considered as an offspring of sustainable development there is a lack of clearly defined links between the two (see Sharpley, 2000). "Arguably, sustainable development originated through the convergence of economic development theory and environmentalism, resulting in the concept of sustainable development" (Hardy & Beeton, 2001: 169). Even though, Bruntland Report (1987) is referred as a document with great importance in sustainable tourism studies, rather surprisingly, the tourism is not mentioned in the report (Wall, 1997).

Since the inception of the *Journal of Sustainable Tourism* in 1993, tremendous amount of studies were published on sustainable tourism. While these studies, noticeably, broadened our understanding of various forms of tourism, they did not create an agreed definition of sustainable tourism. Once, "alternative tourism" was presented as real sustainable tourism within a small, local and autonomous form. However, this conceptualized failed to diffuse. It is understood that both alternative and mass tourism practices can be sustainable or unsustainable based on the contexts (Weaver, Tang, & Zhao, 2020). Tourism not only takes place in established well-planned areas but also in the areas with little to non-planning (World Tourism Organisation, 2004). "Developing sustainable forms of tourism in some areas simply sweeps the problems of tourism under the carpet of other destinations" (Sharpley, 2000: 9). Therefore, rather than a distinct type of tourism itself, sustainable tourism has been evolving into an endeavor to make all forms of tourism more sustainable.

In his thought-provoking study, Müller (1994) likened sustainable tourism development to a pentagon with the angles of "economic health, subjective well-being of the locals, unspoilt nature, protection of resources, healthy culture, optimum satisfaction of guest requirements"(p. 132–133). In his conceptualization, Müller suggests a perfect balance of these angles for sustainable tourism developments. However, there are critics of understanding sustainable tourisms development as a balanced approach. Unlike Müller, they propose that when there are conflicting goals in tourism development, sustainable tourism may advocate compromises rather than balancing acts (Bramwell, Higham, Lane, & Miller, 2017). "Sustainable tourism is extremely complex, because it requires a holistic approach, that includes business, markets, ideologies, environments, societies, cultures and the politics of control – plus the vagaries of fast changing fashion trends" (Lane, 2018: 2).

3 Second Home Tourism

Second home is a multifaceted phenomenon. Scholars approach and analyze the phenomenon from their discipline-oriented angles. In this chapter, we consider it as an important tourism activity. While discussing the second homes in Canada, Jaakson (1986) asserts that visits to second homes are important domestic tourism activities. In this assertion, rather than discussing second homes in the domestic-international domain, he strongly criticizes the idea of not considering visits to second homes as tourism activities. When the reach of the global tourism is taken into account, a domestic-international discussion seems to be futile in these days. Whether it is domestic or international "[v]isits to second homes are a unique form of tourism" (Strapp, 1988, p. 505). The uniqueness comes from characteristics of visitors. In other forms of tourism, a tourist is easily distinguishable, but a second home visitor might change from tourist to resident over time (Strapp, 1988).

Second homes are important components in tourism and housing research (Hall, 2014) and they have increasingly gained academic attention (Larsson & Müller, 2019). Second homes were a research worthy topic even in 1930s.

Scholars such as Ljungdahl (1938) and Sund (1948) studied the developments of second homes on the outskirts of main cities (as cited in Hall & Müller, 2004). However, as many researchers stated that there is no internationally accepted simple and clear definition of "second home" (Müller et al., 2004; Paris, 2011; Hall, 2014; Back & Marjavaara, 2017). "The term of second-home acts as an umbrella expression for a variety of cognate terms including cabin, cottage, crib, holiday home, hut, leisure home, recreational home, summer home, summer house, vacation home, and weekend home" (Hall, 2014: 2). This indicates that there are various terms that refer to certain idea of usage (Hall & Müller, 2004). Besides, the identification and measurement of second homes is difficult due to the dynamic character of them (Paris, 2009). As Paris (2011) states the term "second home" had various meanings depend on cultural contexts. Thus, it can be said that the term "second home" is contested (Müller, 2011). The first home is mostly located very close to source of income; however, second home is generally located in peripheral areas, rural areas, mountain pastures and coastal areas which serve recreational functions (Hall & Müller, 2004; Gallent, 2014). Considering this, the main distinction between dwelling units is that while first homes are frequently located in productive spaces second homes are constructed in consumption spaces (Gallent, 2014). In one of the most influential collections on second homes, Coppock (1977) defines second homes as properties where owners or renters stay occasionally with a purpose of leisure (as cited in Li & Fan, 2020).

There is considerable fluidity between definitions and functions of second homes (Paris, 2009). Rather than constructed as proper "homes," properties for recreational use can also be either a small house or an apartment (Hall & Müller, 2004). Therefore, the term "second home" refers to how a property is mainly used rather than the amenities and physical forms of the buildings (Hall & Müller, 2004; Paris, 2009).

Second home ownership is a phenomenon that has long been a part of tourism activities, especially in Western European and Scandinavian societies (Quinn, 2004). Second homes are very common in Nordic countries, Australia, France, Germany, the United Kingdom, the United States, South Africa and Canada (Visser, 2003; Hall & Müller, 2004; Gallent, 2014; Müller & Hoogendoorn, 2013; Back & Marjavaara, 2017). Some countries, especially Scandinavian countries have very long histories of second home ownership (Tress, 2002). Today, second home ownership is part of daily life in these countries (Hall & Müller, 2004). It is stated that second home developments in Finland started in 18th century as a seasonal routine of wealthy classes, but today it evolved and became a widespread phenomenon (Pitkänen, 2008). Similarly, developments of second homes in Norway can be traced back in 18th century, but significant increases can be observed only after 1950s, thanks to the availability of spare time, easier transport and increased income (Kaltenborn, Andersen, Nellemann, Bjerke, & Thrane, 2008).

Even though it was not a novelty for affluent members of the societies, demand for second homes from middle classes remained low until the postwar economic developments. With the greater economic wealth of the 1960s, demand for second homes increased at unprecedented levels across the United Kingdom, especially in attractive countryside (Gallent & Tewdwr-Jones, 2001). Additionally, in developed countries around the world, especially in Europe and North America, significant number of people own two or more homes (Paris, 2006; Gallent, 2014). For example, as a country with around 60 % forested land (Kaltenborn et al., 2008), in Sweden more than 50% of the population has access to second homes (Back & Marjavaara, 2017). In Switzerland Alpine regions number of second homes exceeded the number of primary residencies (Gerber & Bandi Tanner, 2018).

Additionally, there has been rapid growth of second home ownership due to increasing number of affluent people and the growing mobility for large section of the population, globally (Paris, 2009). Income growth, increased accessibility to remote areas, greater affluence, and increased leisure time are main factors in the rapid growth of second home ownership (Casado-Diaz, 2004). Unprecedented expansion of second homes seems to be a by-product of urbanization. When people moved to the urban areas, they left their primary residences in rural areas. Thanks to the some government initiatives, these left-behind properties were converted to second homes (Larsson & Müller, 2019). Besides, growing number of pensioners in the West with disposable time and excess income is another important force behind the purchases of second homes in various places (Müller, D. K. & Hall, 2004).

Researches on second homes have significantly increased over the last few decades (Clout, 1971; Chaplin, 1999; Gallent & Tewdwr-Jones, 2001; Tress, 2002; Casado-Diaz, 2004; Hall & Müller, 2004; Halseth, 2004; Keen & Hall, 2004; Müller, 2004; Müller et al., 2004; Williams et al., 2004; Gallent, 2007; Paris; 2011; Müller & Hoogendoorn, 2013; Hall, 2015; Back & Marjavaara, 2017; Larsson & Müller, 2019). Second homes have becoming more popular and second home ownership has resulted international and domestic travel for relatively high-income households. In the mountain areas and highland settlements; by contrast, second home ownership is becoming more common for relatively poor households in the homeland of their ancestors. Countryside previously used for farming purposes has become consumption spaces as a result of intensive second home developments (Müller & Hoogendoorn, 2013; Hall, 2014). The increasing second home ownership has an impact on property prices. Besides, it increases demand for rural products. Therefore, second homes have become part of rural economies and they are becoming more popular both for domestic and overseas visitors (Hall & Müller, 2004; Müller & Hoogendoorn, 2013; Hall, 2014).

4 Second Homes in Turkey

This research describes the importance of summer properties within Turkey's rural-recreational countryside and identifies the development of second homes. The second home phenomenon in Turkey is mostly associated with recreational properties and summer vacation homes (Kılıçaslan, 2006; Zoğal & Emekli, 2018). There is a widespread tradition of owning a "summer house" in Turkey (Okuyucu & Somuncu, 2019), varying from well-constructed dwellings to make-shift tents, mostly used in summer months, increasingly at the weekends and holidays (Emekli, 2014). They are usually located in mountain highlands, coastal areas and peripheries of mega cities. Early studies determine that the increasing number of second homes are located on the coastal area of Aegean Sea and Mediterranean region (Kılıçaslan, 2006) and outskirt of growing cities especially in periphery of Istanbul, that dated back to the 1950s (Emekli, 2014; Yazgan, 2018). In 1960s, tourism development in Turkey has been intrinsically likened to the construction sector and the further urbanization of the districts located in coastal areas of Mediterranean and Aegean Sea (Emekli, 2014). Second-home tourism had become a movement signifying the advent of a recreation-oriented seasonal migration to holiday resorts during the 1950s and 1960s. However; tourism development has also spread towards rural areas since 1990s, especially mountain areas which previously used for animal husbandry and farming (Somuncu et al., 2019). With growth of second home ownership, the city dwellers have started to divide their time between their urban permanent residences and their rural second homes. Therefore, it can be said that the flows of seasonal tourists and second home mobility are neither rural nor urban phenomena (Slätmo, Vestergård, Lidmo & Turunen, 2019).

Academic research (Halseth, 2004; Müller & Hoogendoorn, 2013; Müller, 2014) alike highlight that widespread recreational second home ownership is generally associated with tourism development and use of seasonal summer properties as vacation homes. In Turkey, second homes shaped by socio-economic status and cultural values of their owners were overwhelmingly located in close proximity to major urban areas in 1950s (Yazgan, 2018; Somuncu, 2019). With

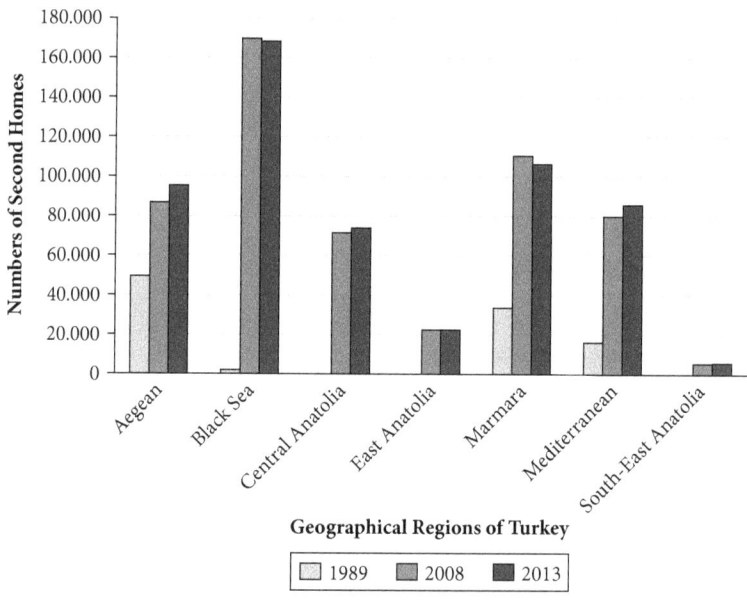

Fig. 13-1: Regional Distribution of Second Homes by Years. Source: General Directorate of Population and Citizenship Affairs of Turkey

the incentives provided to coastal tourism after 1970s, it increased the interest to the coasts and caused vacation homes to be concentrated on the coastal areas (Emekli, 2014). In various parts of Turkey, second home ownership had become an established tourism practice by 1990s and it has substantially increased after 2000s (See, Fig. 13-1). According to the General Directorate of Population and Citizenship Affairs, there were 102,400 summer homes in 1989. In 2013, the number of summer homes reached to 559,934. There was more than fivefold increase in 24 years. These figures suggest that the demand for second homes is growing and second home ownership will dramatically increase in new highland locations, in Turkey. The recent studies show that there are considerable flows of urbanites to temporary highland settlements during the summer vacations and they are growing over the last 30 years (Somuncu et al., 2019).

Fig. 13-1 demonstrates the growth of second homes in Turkey from 1989 to 2013. Marmara, Mediterranean, Aegean and Black Sea regions have experienced phenomenal growth in the last three decades. The other regions have also witnessed significant increases in the second home ownership, but compared to coastal regions their growth is modest.

Increasing second home ownership is associated with higher disposable incomes and hotter summers in Southern Turkey. As stated by Paris (2011), affluence and mobility are main drivers of second home ownership. Increases in car ownership have direct effect on personal mobility. Additionally, Somuncu et al. (2019) mentions that the spread of services to rural and mountainous areas has generated substantial development of second homes around highlands and the repopulation of formerly left-behind temporary settlements.

Much of the growth of second home ownership in Turkey before 1980s was by Turkish families, distinctively from more recent development of summer home in mountain ranges aimed largely at local-based purchasers. This growth is associated in Turkey with a desire to retain links with rural origins by households who had relocated to cities during rapid urbanization since early 20th century.

Second homes are developing in mostly attractive areas such as mountain pastures and coastal areas and they have a huge impact on nature, especially in rural areas (Müller, 2005; Hiltunen, 2007). The establishment of second homes for vacation and recreational use has had significant social, economic and physical effects on the countryside of Turkey, especially in coastal and mountain areas (Ceylan & Somuncu, 2018; Somuncu et al., 2019). The literature of second home ownership demonstrate the substantial diversity in types of summer houses, according to their size, materials, location, and use of different recreational activities. Studies (Kılıçaslan, 2006; Emekli, 2014; Yazgan, 2018; Zoğal & Emekli, 2018; Okuyucu & Somuncu, 2019; Somuncu et al., 2019) on Turkey emphasize the differences of second home ownership in different geographical regions and locations. Economic growth through second home developments in countryside is important. It has become more important, especially, in Black Sea, Marmara, Aegean and Mediterranean. Second home locations in these regions have recently been transforming into recreational activity centers and summer vacation resorts (See, Fig. 13-2).

'Summer houses' are typically located by sea, lakes or peripheral areas in the coastal areas of Marmara, Aegean and Mediterranean regions while; 'country homes in Black Sea and Mediterranean region tent to be located in higher and cooler countryside around cities often on land inherited from family. In the coastal regions, recreational activities connected to the sea are important in the establishment of second homes. However, natural beauty, fresh air, natural environments, rich rural landscapes and mountain related recreational activities are also important factors to second home ownership in highland settlements. The sub-region of Adana, located in the Mediterranean Geographical Region, is one of the places with dense number of second homes. While vacation homes

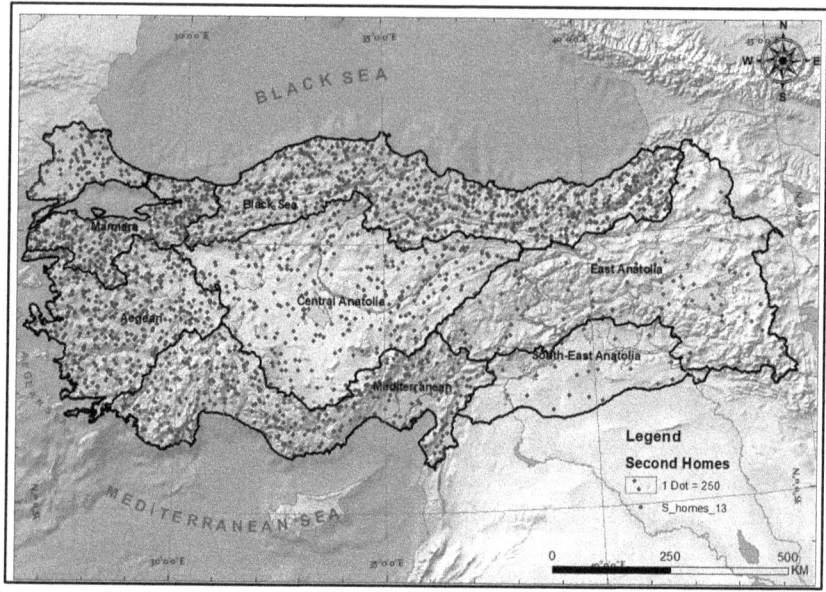

Fig. 13-2: Second Home Patterns According to Geographical Regions of Turkey in 2013.

are generally constructed in the towns located on the Mediterranean coast, summer cottages for recreational purposes are also built in temporary highland settlements in mountainous areas of Adana Province. The most popular and well-known second-home area in Adana sub-region is the skirts of Taurus Mountain where temporary highland settlements area located. In a Turkish context, they are well established as a second home location.

Kozan District has the largest countryside within the province and there are various permanent and temporary highland settlements. The recreational highland settlements which are located on the northern side of Kozan have long functioned as summer holiday destinations. Göller Highland is one of the most popular recreational areas in Kozan. Since 2000s, new second homes have been built in the area and they are purpose-built properties (See, Photo 13-1). The natural features, landscape, pure air of temporary settlement and its closeness to Kozan and Adana are clearly key factors for development of second homes in the region. According to the local government's forecast, total population of settlement reaches 35.000 in summer months and it tends to increase over

Photo 13-1: Development of Second Homes in Göller Highland of Kozan District.

the years. Therefore, developments of second homes have risen significantly in recent years.

Conclusions: "Sustaining" Second Home Development

As we discussed in previous parts, sustainable tourism should be considered as an endeavor to make all forms of tourism more sustainable. Whether it is domestic or international, second home tourism is a form of tourism activity with significant effects on nature and societies. Therefore, development of second homes should be explored with a holistic approach.

In this chapter, we explored second homes with sustainable tourism perspective. We reviewed the definition of second homes and current trends and issues in second home development. We shared places where recreational second homes are built around the world. Then, we explored the second home developments in Turkey. Second homes are mostly constructed along the coastal and mountainous areas as is so often the case elsewhere. Enormous second-home concentrations found in Mediterranean, Aegean, Marmara Sea regions. For nearly 50 years, second homes have been part of the developments in coastal areas. Second home tourism development in Turkey is mainly linked to the growth of mobility, increased leisure time and income. Indeed, during the 2000s, second home ownership in the countryside of Turkey underwent significant changes. The growth of second home construction has created considerable pressures on land and the environment of rural areas of Turkey.

The potential impacts of second home developments on environment, local communities, visitors, and other stakeholders need to be critically studied. In this chapter, we highlighted that there have been important changes in second home tourism in recent years. We tried to shed lights onto developments of second homes. Even though, we tried to build a bridge between sustainable tourism and second homes, we may not call ourselves as absolute "bridge builders." While navigating around the various definitions of *sustainable tourism*, we attempted to highlight second homes. Considering the scope of the chapter, we preferred to save in-dept examination of a selected area for future studies.

As indicated by WTO (2004), environmental, economic, and sociocultural dimensions of sustainable tourism should be considered in future studies. As highlighted by Dolnicar (2020: 8) "to be able to proactively design for environmentally friendly tourism, it is necessary to first develop insight, a theory of the reasons for the occurrence of the environmentally unfriendly behavior." For instance, attitudes of residents about the second home developments in Southern Norway are primarily affected by their attitudes toward environment (Kalterborn et al., 2008).Tourism development and ecological environment might be in harmony when the development is designed considering the characteristics of the environment in terms of scale, material and landscape (Bramwell, 1991). When second home developments in Turkey are considered, complexity of the second home phenomenon becomes apparent. From purpose-built recreational summer houses to converted properties there are many types of second homes. For instance, on the Göller Highlands, second homes vary in size and shape but they seem to be highly similar in terms of (un)sustainability.

Bibliography

Alfaro Navarro, J., Andrés Martínez, M., and Mondéjar Jiménez, J. (2020), "An Approach to Measuring Sustainable Tourism at the Local Level in Europe", *Current Issues in Tourism*, 23 (4), pp. 423–437. doi:10.1080/13683500.2019.1579174.

Alkan, M. Ö. (2014), Osmanlı'da Sayfiyenin İcadı. In T. Bora (Ed.), *Sayfiye Hafiflik Hayali* (pp. 15-44), İstanbul: İletişimYayınları.

Back, A. and Marjavaara, R. (2017), "Mapping an Invisible Population: The Uneven Geography of Second-home Tourism", *Tourism Geographies*, 19 (4), pp. 595-611. doi:10.1080/14616688.2017.1331260.

Bramwell, B. (1991), "Sustainability and Rural Tourism Policy in Britain", *Tourism Recreation Research*, 16 (2), pp. 49-51. doi:10.1080/02508281.1991.11014626.

Bramwell, B., Higham, J., Lane, B., and Miller, G. (2017), "Twenty-Five Years of Sustainable Tourism and the Journal of Sustainable Tourism: Looking Back and Moving Forward", *Journal of Sustainable Tourism*, 25 (1), pp. 1-9. doi:10.1080/09669582.2017.1251689.

Bruntland, G. H. (1987), *Report of the world commission on environment and development: Our common future*. Oslo: Retrieved from https://sustainabledevelopment.un.org/content/documents/5987our-common-future.pdf, (29.05.2020)

Butler, R. W. (1999), "Sustainable Tourism: A state-of-the-art Review", *Tourism Geographies*, 1 (1), pp. 7-25. doi:10.1080/14616689908721291.

Cambridge Dictionary (No Date), *Sustainable*, Retrieved from https://dictionary.cambridge.org/dictionary/english/sustainable, (01.06.2020).

Casado-Diaz, M. A. (2004), Second homes in Spain. In C. M. Hall, & D. K. Müller (Eds.), *Tourism, Mobility and Second Homes between Elite Landscape and Common Ground* (pp. 215-232), Clevedon: Channel View Publications.

Ceylan, S. and Somuncu, M. (2018), İkinci Konut Turizmi ile Tarım Toprağının Elden Çıkışı: Pelitköy'de Kırsal Yoksulluk, *TÜCAUM 30. Yıl Uluslararası Coğrafya Sempozyumu, 3-6 Ekim 2018, Bildiriler Kitabı*, Ankara, 1156-1170.

Chaplin, D. (1999), "Consuming Work/Productive Leisure: The Consumption Patterns of Second Home Environments", *Leisure Studies*, 18 (1), pp. 41-55. doi: 10.1080/026143699375041.

Clout, H. D. (1971), "Second Homes in the Auvergne", *Geographical Review*, 61, pp. 530-533.

Dolnicar, S. (2020), "Designing for More Environmentally Friendly Tourism", *Annals of Tourism Research*, 84, pp. 1-10. doi:10.1016/j.annals.2020.102933.

Emekli, G. (2014), "İkinci Konut Kavrami Açisindan Turizm Coğrafyasinin Önemi ve Türkiye'de İkinci Konutları", *Ege Coğrafya Dergisi*, 23, pp. 25-42.

Gallent, N. (2007), "Second Homes, Community and a Hierarchy of Dwelling", *Area*, 39, pp. 97-106. doi:10. 1111/j.1475-4762.2007.00721.x.

Gallent, N. (2014), "The Social Value of Second Homes in Rural Communities", *Housing, Theory and Society*, 31 (2), pp. 174–191. doi:10.1080/14036096.2013.830986.

Gallent, N. and Tewdwr-Jones, M. (2001), "Second Homes and the UK Planning System", *Planning Practice and Research*, 16, pp. 59–69. doi:10.1080/02697450120049579.

Gerber, J. and Bandi Tanner, M. (2018), The Role of Alpine Development Regimes in the Development of Second Homes: Preliminary Lessons from Switzerland", *Land Use Policy*, 77, pp. 859–870. doi:10.1016/j.landusepol.2017.09.017.

Grötzbach, E. (1984), "Spatial Structure and Development Prospects of Tourism in the Black Sea Region of Turkey", *Ege Coğrafya Dergisi*, 2, pp. 198–207.

Hall, C. M. (2014), "Second Home Tourism: An International Review", *Tourism Review International*, 18 (3), pp. 115–135. doi:10.3727/154427214X14101901317039.

Hall, C. M. (2015), "Second Homes Planning, Policy And Governance", *Journal of Policy Research in Tourism, Leisure & Events*, 7, pp. 1–14. doi:10.1080/19407963.2014.964251.

Hall, C. M. and Müller, D. K. (2004), Introduction: Second Homes, Curse or Blessing? Revisited. In C. M. Hall and D. K. Müller (Eds.), *Tourism, Mobility, and Second Homes: Between Elite Landscape and Common Ground* (pp. 3–14), Clavendon: Channel View Publications.

Halseth, G. (2004), The 'Cottage' Privilege: Increasingly Elite Landscapes of Second Homes in Canada. In C. M. Hall and D. K. Müller (Eds.), *Tourism, Mobility and Second Homes: Between Elite Landscape and Common Ground*, Clevedon: Channel View Publications.

Hardy, A. L. and Beeton, R. J. S. (2001), "Sustainable Tourism or Maintainable Tourism: Managing Resources for More than Average Outcomes", *Journal of Sustainable Tourism*, 9 (3), pp. 168–192. doi:10.1080/09669580108667397.

Hiltunen, M. J. (2007), "Environmental Impacts of Rural Second Home Tourism – Case Lake District in Finland", *Scandinavian Journal of Hospitality and Tourism*, 7 (3), pp. 243–265. doi: 10.1080/15022250701312335.

Jaakson, R. (1986), "Second-Home Domestic Tourism", *Annals of Tourism Research*, 13 (3), pp. 367–391. doi:10.1016/0160-7383(86)90026-5.

Kaltenborn, B. P., Andersen, O., Nellemann, C., Bjerke, T., and Thrane, C. (2008), "Resident Attitudes towards Mountain Second-home Tourism Development in Norway: The Effects of Environmental

Attitudes", *Journal of Sustainable Tourism*, 16 (6), pp. 664–680. doi:10.1080/09669580802159685.

Keen, D. and Hall, C. M. (2004), Second Homes in New Zealand. In C. M. Hall and D. K. Müller (Eds.), *Tourism, Mobility and Second Homes: Between Elite Landscape and Common Ground*, Clevedon: Channel View Publications.

Kılıçaslan, Ç. (2006), "İkinci Konutların Deniz Kıyısına Etkisi", *Süleyman Demirel Üniversitesi Orman Fakültesi Dergisi*, 1, pp. 147–156.

Koca, N. (2011), *Doğal ve Sosyo-Ekonomik Özellikleri Açısından Osmaniye Yaylaları*, Ankara: Pegem Yayınları.

Lane, B. (2018), "Will Sustainable Tourism Research be Sustainable in the Future? An Opinion Piece", *Tourism Management Perspectives*, 25, pp. 161–164. doi:10.1016/j.tmp.2017.12.001.

Larsson, L. and Müller, D. K. (2019), "Coping with Second Home Tourism: Responses and Strategies of Private and Public Service Providers in Western Sweden", *Current Issues in Tourism*, 22 (16), pp. 1958–1974. doi:10.1080/13683500.2017.1411339.

Li, T. and Fan, C. C. (2020), "Occupancy, Usage and Spatial Location of Second Homes in Urban China", *Cities*, 96, pp. 1–10. doi:10.1016/j.cities.2019.102414.

Liu, Z. (2003), "Sustainable Tourism Development: A Critique", *Journal of Sustainable Tourism*, 11 (6), pp. 459–475.

Moran, D. and Jewkes, Y. (2014), "Green" Prisons: Rethinking the "Sustainability" of the Carceral Estate", *Geographica Helvetica*, 69, pp. 345–353. doi:10.5194/gh-69-345-2014.

Müller, D. K. (2004), Second Homes in Sweden: Patterns and Issues. In C. M. Hall and D. K. Müller (Eds.), *Tourism, Mobility, and Second Homes: Between Elite Landscape and Common Ground* (pp. 244–258), Clevedon: Channel View Publications.

Müller, D. K. (2005), Second Home Tourism in the Swedish Mountain Range. In C. M. Hall and S. W. Boyd (Eds.), *Nature-based Tourism in Peripheral Areas: Development or Disaster?* (pp. 133–148), Clevedon: Channel View Publications.

Müller, D. K. (2011), "Second Homes in Rural areas: Reflections on a Troubled History", *Norsk Geografisk Tidsskrift –Norwegian Journal of Geography*, 65 (3), pp. 137–143, doi: 10.1080/00291951.2011.597872.

Müller, D. K. (2014), Progress in Second-home Tourism Research. In A. A. Lew, C. M. Hall, & A. M. Williams (Eds.), *The Wiley Blackwell Companion to Tourism* (2nd ed., pp. 389–400). Chichester: Wiley-Blackwell.

Müller, D. K. and Hall, C. M. (2003), "Second Homes and Regional Population Distribution: On Administrative Practices and Failures in Sweden", *Espace, Populations, Societes*, 21 (2), pp. 251-261. doi:10.3406/espos.2003.2079.

Müller, D. K. and Hall, C. M. (2004), The Future of Second Home Tourism. In C. M. Hall and D. K. Müller (Eds.), *Tourism, Mobility, and Second Homes: Between Elite Landscape and Common Ground* (pp. 273-278), Clevedon: Channel View Publications.

Müller, D. K. and Hoogendoorn, G. (2013), "Second Homes: Curse or Blessing? A Review 36 Years Later", *Scandinavian Journal of Hospitality and Tourism*, 13 (4), pp. 353-369. doi:10.1080/15022250.2013.860306.

Müller, D. K., Hall, C. M., and Keen, D. (2004), Second Home Tourism Impact, Planning And Management. In C. M. Hall and D. K. Müller (Eds.), *Tourism, Mobility and Second Homes: Between Elite Landscape and Common Ground* (pp. 15-32), Clevedon: Channel View Publications.

Müller, H. (1994), "The Thorny Path to Sustainable Tourism Development", *Journal of Sustainable Tourism*, 2 (3), pp. 131-136. doi:10.1080/09669589409510690.

Næss, P., Xue, J., Stefansdottir, H., Steffansen, R., and Richardson, T. (2019), "Second Home Mobility, Climate Impacts and Travel Modes: Can Sustainability Obstacles be Overcome?" *Journal of Transport Geography*, 79, pp. 1-12. doi:10.1016/j.jtrangeo.2019.102468.

Nüfus ve Vatandaşlık İşleri Genel Müdürlüğü (2019), *Niteliklerine Göre Bina Türü İstatistiği*, https://www.nvi.gov.tr/PublishingImages/hizmetlerimiz/istatistikler/ulusal-adres-veri-tabani_istatistikleri/Niteliklerine%20G%C3%B6re%20Bina%20T%C3%BCr%C3%BC%20%C4%B0statisti%C4%9Fi.pdf., (04.04.2019).

Okuyucu, A. and Somuncu, M. (2019), Türkiye'de İkinci Konut Veri Tabanı Oluşturulmasına Yönelik Bir Öneri: Manavgat, Antalya Örneği, İçinde: M. Bulut and Z. Karacagil (Eds.), *Current Debates on Social Sciences 2: Cultural Studies* (pp. 5-19), Ankara.

Paris, C. (2006), Multiple Homes, Dwelling, Hyper-mobility & Emergent Transnational Second Home Ownership, Paper presented at the ENHR conference "*Housing in an expanding Europe: Theory, Policy, Participation and Implementation*" Ljubljana, Slovenia 2-5 July 2006.

Paris, C. (2009), "Re-positioning Second Homes within Housing Studies: Household Investment, Gentrification, Multiple Residence, Mobility and Hyper-consumption", *Housing, Theory and Society*, 26 (4), 292-310. doi: 10.1080/14036090802300392.

Paris, C. (2011), *Affluence, Mobility and Second Home Ownership*, Abingdon: Routledge.

Paris, C. (2011), *Affluence, Mobility and Second Home Ownership*, Abingdon: Routledge.

Paris, C. (2014), "Critical Commentary: Second Homes", *Annals of Leisure Research*, 17 (1), pp. 4–9. doi:10.1080/11745398.2014.890511.

Pitkänen, K. (2008), "Second-home Landscape: The Meaning(s) of Landscape for Second-home Tourism in Finnish Lakeland", *Tourism Geographies*, 10 (2), pp. 169–192. doi:10.1080/14616680802000014.

Quinn, B. (2004), Dwelling through Multiple Places: A Case Study of Second Home Ownership in Ireland. In C. M. Hall and D. K. Müller (Eds.), *Tourism, Mobility and Second Homes between Elite Landscape and Common Ground* (pp. 113–132), Clevedon: Channel View Publication.

Sharpley, R. (2000), "Tourism and Sustainable Development: Exploring the Theoretical Divide", *Journal of Sustainable Tourism*, 8 (1), pp. 1–19. doi:10.1080/09669580008667346.

Slätmo, E., Ormstrup Vestergård, L., Lidmo, J., and Turunen, E. (2019), *Urban–rural Flows from Seasonal Tourism and Second Homes : Planning Challenges and Strategies in the Nordics* (Nordregio Report), Stockholm: Nordregio, doi:10.6027/R2019:13.1403-2503.

Somuncu, M., Akpınar, N., Kurum, E., Çabuk Kaya, N., and Özelçi Eceral, T. (2010), "Gümüşhane İli Yaylalarındaki Arazi Kullanımı ve İşlev Değişiminin Değerlendirilmesi: Kazıkbeli ve Alistire Yaylaları Örneği", *Ankara Üniversitesi Çevrebilimleri Dergisi*, 2 (2), pp. 107–127.

Somuncu, M., Okuyucu, A., and Altundal Öncü, M. (2019), "Second Home Tourism in the Eastern Black Sea Region of Turkey: Development Issue and Mobility Pattern", *Ankara University Journal of Environmental Sciences*, 7 (2), pp. 63–82.

Strapp, J. D. (1988), "The Resort Cycle and Second Homes", *Annals of Tourism Research*,15 (4), pp. 504–516. doi:10.1016/0160-7383(88)90046-1.

Timothy, D. J. (2004), Recreational Second Homes in the United States: Development Issues and Contemporary Patterns. In C. M. Hall and D. K. Müller (Eds.), *Tourism, Mobility, and Second Homes: Between Elite Landscape and Common Ground* (pp. 133–148), Clavendon: Channel View Publications.

Tress, G. (2002), "Development of Second-Home Tourism in Denmark", *Scandinavian Journal of Hospitality and Tourism*, 2 (2), pp. 109–122. doi: 10.1080/15022250216289.

Torkington, K., Stanford, D., and Guiver, J. (2020), "Discourse(s) of Growth and Sustainability in National Tourism Policy Documents", *Journal of Sustainable Tourism*, 28 (7), pp. 1041–1062. doi:10.1080/09669582.2020.1720695.

Visser, G. (2003), "Visible, Yet Unknown: Reflections on Second-Home Development in South Africa", *Urban Forum*, 14 (4), pp. 379–407. https://link.springer.com/content/pdf/10.1007/s12132-003-0020-y.pdf, (01.06.2020).

Wall, G. (1997), "FORUM: Is ecotourism sustainable?" *Environmental Management*, 21 (4), pp. 483–491. doi:10.1007/s002679900044.

Weaver, D., Tang, C., and Zhao, Y. (2020), "Facilitating Sustainable Tourism by Endogenization: China as Exemplar", *Annals of Tourism Research*, 81, pp. 1–13. doi:10.1016/j.annals.2020.102890.

Williams, A. M., King, R., and Warnes, T. (2004), British Second Homes in Southern Europe: Shifting Nodes in the Scapes and Flows of Migration and Tourism. In C. M. Hall and D. K. Müller (Eds.), *Tourism, Mobility, and Second Homes: Between Elite Landscape and Common Ground* (pp. 97–112), Clavendon: Channel View Publications.

World Tourism Organisation (2004), *Indicators of Sustainable Development for Tourism Destinations – A Guide Book*. Madrid. Retrieved from http://www.adriaticgreenet.org/icareforeurope/wp-content/uploads/2013/11/Indicators-of-Sustainable-Development-for-Tourism-Destinations-A-Guide-Book-by-UNWTO.pdf, (28.06.2020).

Yazgan, A. (2018), "Change and Development of the Second House in Istanbul Coastal Areas", *MEGARON*, 13, pp. 422–430.

Zoğal, V. and Emekli, G. (2018), "Urla'da (İzmir) İkinci konutların Değerlendirilmesine Yönelik Nitel Bir Araştırma", *Turizm Akademik Dergisi*, 5 (1), pp. 189–204.

Gonca MANAP DAVRAS

Website Usage in Event Tourism

1 Introduction

The concept of activity is defined as events that have remarkable characteristics in a certain place and time (Getz, 2007). Although people have different reasons for participating in the events, the fact that the event is authentic and unique has cultural characteristics (Thompson & Matheson, 2008) or provides participants with the opportunity to have "once in a lifetime" experience (Neioretti et al., 2001) is an important asset. The activities that are formed as a result of the participation in the activities led to the emergence of the concept of event tourism (Getz, 2008). Event tourism contributes to the destination in an economic, cultural and social area, increases the brand value of the region, the number of tourists and tourism income, and enriches sociocultural structure (Cudny, 2013). In other words, it plays an important role in economic development by increasing the image of the destination.

The continuity of the activities, reaching large masses, quality and success depend on the correct and effective marketing process. The marketing process requires good planning. The most important part of this process is the use of digital tools. With the development of the Internet, virtual platforms such as social networking platforms, blogs and web pages have been used extensively in marketing of international event organizations. Therefore, websites are one of the most used tools. The website of the organization performs direct sales, provides additional resources to consumers, contributes to the image of the event and provides basic information about the event. The interactive structure of the website and its features such as providing continuous information are effective in developing long-term relationships between the organization and consumers (Bauer et al., 2002: 155).

Web marketing provides opportunities such as creating a brand by increasing the image of the event, reducing costs, selling online, fulfilling customer requests in a short time, conducting market research and delivering information about the event to a wide audience (Hoyle, 2002). In order to take advantage of all these advantages, it is necessary to create websites that show the prestige and quality of the organization. Therefore, some basic criteria should be considered while preparing the website. Incorrect or incomplete configuration of the page can turn the positive image of the event into a negative one. This chapter aims to present the advantages of the websites in the marketing process of the activities and reveal the features that these sites should have.

2 Background

Event marketing, which strengthens the communication between the brand and the consumer, is a marketing activity in which the experience based interaction comes to the fore. Event marketing provides advantages such as increasing brand visibility, reinforcing brand addiction, accelerating the decision to purchase, and the opportunity to experience marketing. While increasing the brand value, it aims to create lasting memories and emotional bond by providing a different experience to the target audience. For this purpose, digital communication is used extensively.

According to the data of "Digital in 2019" published by digital marketing agency, we are Social and Hootsuite, Worldwide Internet usage rate has reached 56 %, the number of general users are 4.38 billion whereas the rate of social media users has reached 45 %, and the number of users are 3.48 billion. The Internet provides easy access to a lot of information, increases the availability of search engines in a short time, has a dynamic structure and has gained a different dimension, especially with the advent of Web 2.0 technology, also called the second-generation Internet. The Internet, where one-way communication has been replaced by two-way communication, has made radical changes in event marketing. Traditional marketing communication tools have been replaced by digital communication tools that can appeal to visual and auditory senses and can be used in digital communication technologies.

In web-based marketing, blog sites, forum sites, tools on social media networks are used extensively. The websites are the most used ones. The website is a marketing tool that transfers information or provides services in different ways in accordance with the purposes of individuals or businesses. In the events, websites (Onat & Gülay, 2015: 55) can divide consumers into segments and sub-segments. They also provide strategic Internet platform about how to reach the consumer and how to build lasting relationships with them.

Websites should provide accurate and up-to-date information sources to their target audiences, store them and use them during product and service development phase. For this purpose, applications (IP number or cookie) should be used to obtain more information such as whether visitors visit their websites again, how long they stay on the site, which pages they are interested in, and the number of clicks (counter usage) (Sarı & Kozak, 2005: 265)

While corporate websites are not as interactive as they are compared to social media tools, it is not possible to be active on the Internet without a well-organized website that meets the expectations of its visitors (Onat, 2014: 137). Well-prepared and well-managed websites that are representative of corporate image and brand value in the virtual world provide marketing-oriented services to a wider audience at a lower cost, without space and time limits (Law & Hsu, 2005: 493). Creating a website provides important advantages in terms of promotion and effective marketing as well as cost (Çubukçu, 2010: 43). Providing comprehensive information about the services and reaching the potential activity participants should be the main goal (Aksu & Tarcan, 2002: 94). The loyalty of the site users, who get accurate information about the event, increases (Bauer et al., 2002: 159; O'Cass & Carlson, 2012). Informing thousands and sometimes millions especially in events targeting international participation is very important for the success of the event. For example, more than 3 million people attended the Danube Island Festival (Donauinsel Fest), which was organized for the first time in 1983 and lasted for three days, and the event website played an important role in informing the participants.

Websites have three different functions: pre-sales, online and after-sales (pre, on-line, and after sales) within the scope of marketing. The pre-sales phase involves a company's efforts to attract customers through advertising, public relations, new product or service announcements, and other related activities. The online phase covers the purchasing activities of the customers in electronic environment. The after-sales phase is customer service, problem solving etc. It covers the issues and is aimed at meeting the demands of the customers and ensuring their satisfaction (Liu & Arnett, 2000: 24).

Besides the many advantages of the websites, they also have income generating features. Websites create income for organizers through event tickets, gift vouchers, sales of event-related products or services. They also contribute to revenue growth by publishing sponsor shares and advertising banners that support events. For example, the cost of the 2016 Summer Olympics held in Brazil in 2016 was $ 12 million, while 40 % of the financial resources for games were covered by local sponsors and 12 % from international sponsors from ticket sales.

The connection between the websites of the event organizations and the users is provided through the media. When moving from the initial use to the regular use of the website, the important thing is how to present an attractive and rich content site to the visitors. Thus, the active use of the person and revisiting the site can be provided (Haas, 2002: 639). In order to engage with users, it is necessary to understand user preferences and how users interact with the web (Geissler, 2001: 488). The main purpose of the visitor on the event web pages is to get information about the event and to participate. For this reason, it is important to provide detailed, clear, accurate information about the event and the design of the websites.

The basis of being successful in event marketing activities is to prepare websites with professional design. If the website does not look professional, the quality and success of the product or service offered cannot show itself and cause negative perception (Beri & Singh, 2013). Website design should start with defining the needs, expectations and problems of users (Cox & Dale, 2002; Akhter & Bashir, 2008). It should continue by following the feedbacks and determining the appropriate level of website service quality (O'Cass & Carlson, 2012: 419; Bayram & Yaylı, 2009).

3 Websites in Event Marketing

The design of the website is a significant factor. It should create satisfaction and loyalty by meeting the needs of users and enable consumers to visit the site again (Cox & Dale, 2002). The design of the website should serve the purpose of the event while meeting the needs of the users. The design of a website is important and has a detailed and complex structure. Website design should have the ability to present the information needed for the event organization it represents to the users in the best way. At the same time, the event organization must meet its goals.

Websites are divided into two as Static and Dynamic according to the program language.

- **Static website**: Static websites are websites that are prepared entirely in HTML (interface), which do not receive their data (text, images, videos, etc.) from any database. Static websites are difficult to develop and update. Users who constantly visit the site will see the same information every time they log in because their content is fixed. Static websites are generally more economical than dynamic websites that are set up to provide unchanged little information. The site load time is shorter and the page size is smaller.
- **Dynamic website**: Dynamic websites are prepared on the website background using web programming languages and are easy to update. Dynamic websites can create content according to the user, these contents can be renewed. It can be opened more slowly than static websites, it is more complicated to develop, and its security risk is higher.

The websites used only for information purposes by government agencies and public institutions initially diversified with the expansion of the World Wide Web. Today, there are so many different types of websites that it is sometimes difficult to classify a particular site to a particular category. Various websites can be categorized as follows:

- **Personal website**: These are the sites where people share private information (personal information, photographs, contact information, etc.). It is possible to obtain an environment from social media, earn money, transfer information, and share via personal website. On this site, the person has the opportunity to freely share his own thoughts and to convey his ideas as you wish. Usually, it is created by the owner in order to declare himself, to find friends, like-minded people, people with similar views, etc.
- **Portal sites**: These are the websites that update data on different topics daily and compile the data from many sources. On the portal site home page, events such as news, chat, shopping, music, entertainment, travel, tourism, economy, finance, which have visual and content related to the subject, and which interest the Internet user, appear. Portal sites that facilitate the work of people at the point of access to information help to obtain information in the desired area from a single point.
- **E-Commerce sites**: They are sites that allow you to shop for any product or service over the Internet. Sites that save time and money for customers also provide businesses with competitive advantage in terms of accessibility, ease of sale and entering new markets. These are company owned sites that promote and sell products listed in many categories. Not to be confused with a business website, with an ecommerce website you can add products or services into your cart, and pay for them through the site.
- **Forum sites**: The forum site is the website where people can share and create a forum for discussion on various topics, and share their opinions and work with other people. Forum sites can be in a single area on topics such as education, health, politics, or they can be in multiple categories or even general categories. Although there are strict rules set by the admin of the site, there are many sites that are not actively used today.
- **Entertainment or gaming sites**: These are websites where people enter and play games. Sites that can appeal to all age groups and reach large audiences are used for many different purposes. Most of these websites do aim to make money like business and e-commerce websites do, but usually through the advertisements that show up on the page rather than through selling specific products or services.

- **News websites:** The news site is a powerful information resource dedicated to news from any area. Its main purpose is to deliver the latest information to the user as soon as possible, so the news on the site should be constantly updated. The proper administration and advertising can bring a tangible profit to its owner.
- **Social sites:** These are the sites that inform and gather information from the society on general issues. They are sites where people or their groups create their profiles, share similar interests, common goals and often communicate via message boxes, chat rooms, and comment options.
- **Community sites:** These are the sites that inform and gather information from the society on general issues. They are created to give people a specific space to discuss topics that interest them, give or get a piece of expert advice, seek and hire freelancers, etc. If necessary, the moderator interferes with discussions on the site directed by the moderators. The site, which responds quickly to user problems, also creates a sense of community among visitors.
- **Company sites:** The business website is an official site of a company, its trustworthy online representation. These sites contain pages such as company profile information, product catalog, product sales pages, contact pages. Creating such a website is the optimal solution for all companies that want to become leaders in their field of business.

Since the program language and event activity of the websites generally operating in the web environment are not the same, a different design strategy is required for each website (Scanlon et al., 1998; Spool, 2000; Yen, Hu & Wang, 2007: 161).

Program language, the diversity of event area and the feature of the event also play an important role in the design of the websites. Events include cultural events (New Orleans Jazz Festival, Book Fairs, Art events, etc.), sports events (FIFA World Cup, World Athletics Championships, Australian Open Tennis Tournament etc.), fairs (International car fairs, Shopping fairs, Equipment fairs etc.), festivals (San Fermin Festival, International Film Festival, Coachella Music Festival), business events, social events, award ceremonies (Oscar award ceremony, Grammy Awards, Golden Globe award ceremony, etc.). In addition, it is possible to group the events according to their size as local, major, hallmark and mega events. Local events are the ones with the least impact in terms of geographical coverage (Isparta's Rose Festival, Alaçatı Herb Festival, Şile's Şilebezi Festival, etc.). Major events are held in countries and widely advertised in the international media (Formula One Grand Prix, Istanbul Australian Open Tennis Tournament, Phantom of the Opera Cats musical etc.). Hallmark events are planned to increase the attractiveness of a particular tourism destination (Rio de Janerio Fetivali, Scotland Edinburgh Festival, Octoberfest hallmark festival in Munich, etc.). Mega events are the most comprehensive events (Olympic Games, Paralympic Games, World Athletics Championships etc.) where there are at least one million participants and they require large budgets (Getz, 2008).

It is not possible to compare the websites with the same criteria due to the different characteristics, goals and sizes of the events. However, company websites are evaluated both on functional (online support, information, purchasing, customer service, etc.) and on symbolic features (design, virtual community unity, personalization, etc.) (Karaosmanoğlu et al., 2016: 161). For this reason, in this section, website quality of events is presented under seven main headings: accessibility, content, update, interactivity, security, usability and site design.

3.1 Accessibility

Accessibility means that the user can connect to the website at any time and in a short time. The main factor that provides access to information within a website is links. The links should be accessible and the transitions between the pages should be quick to provide convenience for the user. Failure to reach the website, delay in loading the website causes the users to be distracted and they leave the site. Research reveals that one of the major problems of websites is accessibility (Kaynama & Black, 2000; Cox & Dale, 2001; Perdue, 2001). For a website to be fast, its pages need to be loaded in less than eight seconds. Ganns research (1999) found that 30 % of the users left the website when the website download time is 8 seconds and 70 % of the users switched to a different site when the download time is 10 seconds. Delays on the website may result from the network and software as well as uploaded images, graphics and links. Page editors or "Webmasters" who take charge of the website should regularly check the links within the page and delete or repair links that are not working because the competence of a website can be evaluated with fully working links (Sowards, 1997: 157). Software that will prevent the pages from loading should not be used. One of the methods used to provide accessibility to websites is search engine optimization (SEO). SEO is a system that allows the website to rank high on the search engine result page. When a word is written in the search engine, a comprehensive algorithm comes into play and starts working with all linked web pages. The websites that are most associated with the word group appear at the top of the page. There are some criteria for determining which sites of search engines will rank first. Search engine optimization optimizes these criteria, making it one of the top search engines (Giomelakis & Veglis, 2016: 380). At this stage, it is also very important to determine the domain that provides convenience for the user to browse the Internet and access a specific website. The domain should be simple and memorable, as it is used to make it easier for people to visit websites.

3.2 Content

The content of the website should cover all information about the event organization and be appropriate for the type of event and user needs. Also, the content of websites is an essential factor in determining the quality of the website (Bell & Tang, 1998; Scanlon et al., 1998).

Simplicity, clarity should always be taken into consideration when preparing content on websites (Yen, Hu & Wang, 2007: 164). If the content is complicated, the user can get bored and leave the website. The content should be interactive and attractive (Chu, 2001). Interesting posts that appeal to the ear and eye can be used in events. For instance, introductory short viral videos, photos of the previous years' events, introductory press releases, resumes of artists, descriptive information about the works to be presented, images of the artists during rehearsal, music sound recordings, various posts about the events can play a significant role in catching the attention of the potential visitors and users.

Information such as announcement texts, program contents, posters, services, transportation should be located on the website. If the information is not available, contact information such as an address, phone and e-mail should also be available for users. Due to the long duration of these events, it is necessary to include information about the event area, maps, eating-drinking-accommodation facilities and shopping information. There should also be information such as event's calendar, online event reservation, online event ticket sales, frequently asked questions, photo gallery, registration form, reports, in-site search engine, travel guides/brochures, trip planner, themed products, virtual tour (Wang & Fesenmaier, 2006: 863–875).

The availability of this information increases users' trust to the website and the event. However, since many people prefer to read from paper rather than computers, printable materials should be used on the website. Information should be available for fast printing or downloading to a computer (Hoyle, 2002: 61). The information should be clear and understandable (Patti, 2000). Confusing, misleading, using complex expressions, distorted websites negatively affect a people's decisions (Taylor & England, 2006: 83).

3.3 Update

Websites should be updated periodically (Madu & Madu, 2002: 251). Updates on the website should be announced to users, and the last update date should be stressed on the website. Daily, weekly, monthly and yearly routine work should be done on the websites. Daily routines include responding to e-mails, communicating with site visitors; these routines are important in terms of professionalism. As a weekly routine job, adding new pages ensures that the website is always up to date and gained dynamism. Monthly routine covers activities such as attempting to take place in new search engines and preparing articles about the destination in order to include electronic journals (Hoyle, 2002: 80–81). Also, the daily currency exchange information and daily weather information on the website will ensure the current perception of the site.

The content of the website can be checked by using CMS (Content Management System) without the need for a webmaster, without dealing with codes in terms of updates and changes. You can check the content, make corrections and even eliminate the content via CMS to make changes. It is a safer option than changing the system code, which saves time and money.

3.4 Interactivity

In addition to successful SEO studies, it is essential to organize websites with superior interactive designs and present them to users. Interactive websites cover many online criteria to attract users. In other words, there should be highly interactive web designs. Such websites enable users to visit the site in a short time and increase their loyalty. One of the methods used to make the site interactive is to put exciting videos of the past events or high-quality photos of the participating artists, to share exciting stories of the participants, audio animations, and messages. Adding dynamic and interactive elements will create long-term interaction. Interactive websites should pay attention to ensure bi-directional communication between the event organization and website users. Several methods can be used, such as live support services, chat rooms, guest books. Attention to these issues increases the trust of the participants in the organization (Peltekoğlu, 2009).

3.5 Security

The reliability of the website is also one of the reasons for event preference (Liu & Arnett, 2000: 24; Yoo & Donthu, 2001; Li, Tan & Xie, 2002). Many users are concerned about the privacy of their personal information (Yang et al., 2003).

In addition to personal information, the payment security provided for online ticket sales on the site is also essential. Websites should use a security program that protects personal information. Three main issues should be considered for privacy policies. These are information, security and consumer control. The information explains what data (email addresses, phone numbers, etc.) you collect and how you use it. When a person enters the system, it should be ensured that the information is collected automatically (IP addresses, browser type). Users should be guaranteed that the information will never be shared with others and sold to others. Security is how consumer information is stored. The security methods of the business, SSL (Secure Sockets Layer) encryptions and measures should be presented to the users in the outline. Users should feel comfortable that their information is safe. Consumer control expresses how the users of the site can see and change the information that they have previously provided to the site (Bayram & Yaylı, 2009).

Sharing information about the security company and contracts related to safeguarding personal information will change users' perceptions of security positively (Lee, 2002).

3.6 Usability

The increase in users' loyalty to a website depends on its usability and whether it is user-friendly (Nielsen, 2000: 388–389). The research shows that 35 % of users pay attention to the accessibility and ease of use of the site (Lee et al., 2002).

Websites have to be designed correctly in order to reach new users and protect existing users. Otherwise, potential users may choose another company's website (Taylor & England, 2006: 83). Therefore, the concept of usability has great importance for both service providers and users (Kent & Taylor, 1998).

According to Nielsen, usability is examined under five headings; learnability, efficiency, memorability, errors and satisfaction. Learnability means how easily the system can be learned, and efficiency indicates how efficiently the learned system can be used. Memorability looks at whether one can easily remember how to use the system after having used it for a certain period. Errors mean that users have a low error rate and can be easily corrected. Satisfaction refers to the thoughts of users regarding the usability. (Tsai et al., 2010).

The presence of the site map, navigation bar, directory and site search engine on the website will facilitate the usability. It is essential that the site map is on every page and the search option is accessible.

Also, there should be different language options, especially on the web pages of international events. For example, Rio Carnival attracts than

6 million people and has seven different language options on its website. To avoid misunderstandings in translations, verbatim translation should not be used.

3.7 Site Design

The design of the site attracts the attention of the user as well as the content of the site. Websites should have technical features (such as graphics, sound, animated animations) that will provide the visual richness for brand advertising.

In order to keep the users on the site and to ensure that they spend more time, the elements that will reflect the layout of the site in the most precise manner such as the colors, the fonts, the quality of the images, the placement of the sliders and the content have great importance (Geissler, 2001).

Colors that are memorable, impressive and suitable for the nature of the activity should be used to get the attention of the user. Excessive use of colors exhausts the eye and distracts people. Usually, the color of the event's logo is chosen as the primary color and complementary colors are preferred to match this color. Otherwise, the appearance of the site may become very complicated (Fırlar & Özdem, 2013).

The primary purpose of the websites is to provide information about the event. Therefore, aesthetic concerns should be a priority and fonts that are difficult to read or complex structures should not be used. Instead, plain, simple and most comfortable fonts should be preferred. The same applies to graphics, photographs and tables (Sowards, 1997: 157).

Pictures and videos that will be used in the event sites should be selected effectively, exciting and suitable for the purpose of the event. If necessary, animated figures should be used in small quantities (Gehrke & Turban, 1999). The animations can be distracting, making the site look amateurish and slow the page loading. Also, the music plays a significant role.

The use of intense background makes reading difficult. The background on all pages of the site should be consistent with each other. The website should comply with the nature of the event.

Conclusion

The Internet has become an indispensable part of our life with technological products such as a computer, mobile phone, and tablet. The Internet, which allows obtaining information in a short time and easily, increases the usage in the businesses. Especially with the transition to the Internet web 2.0 system, the second generation of Internet usage has started. One-way communication has been replaced by two-way communication. This system allows Internet users to be more active on the web, and it is used extensively by businesses that need bidirectional communication.

The websites, which are one of the communication tools of the Internet, are used effectively within the scope of event organization. The websites provide low-cost promotion and marketing, fast communication with customers, easy and continuous information and easy updating opportunities, detailed information without intermediaries, utilize personal products and facilitate direct booking opportunities.

Web designs, which have an essential function in the marketing of national and international events, must be meticulous, detailed and appropriate to meet the expectations. Many studies have been done on quality web evaluation. However, the diversity of websites, different expectations of businesses and users make it difficult to define the standard for successful websites.

Considering the concept of event marketing, success is based on three elements defined as 3E. These elements consist of Entertainment, Excitement and Enterprise. Entertainment includes providing a different, unique and unpredictable experience, excitement includes making the event memorable, and the enterprise involves being ready to try something new that has not been tried before (Hoyle, 2002). As Van der Wagen (2005) mentioned, the promotion efforts in event marketing should be designed in a way that ensures people to have fun and have a good time (Timur et al., 2014). A successful event website should be able to reflect 3E elements to users. Receiving a positive image from the website increases participation in the events. The quality of the website directly affects consumer satisfaction, and that consumer satisfaction positively affects purchasing behavior (Liu & Arnett, 2000; Bai, Law & Wen, 2008).

In the study, a satisfactory event's website design is addressed in seven main headings: accessibility, content, updating, interactivity, security, usability and site design. Paying attention to these criteria will increase the success of the website, and it will increase user loyalty, in other words, participation in the event.

The accessibility criterion refers to the speed of the website. The term can expand as having links to access information on the event web page, the fast opening of the website, fast access to video-images and links, no freezing, and correct planning of the main page.

The content criterion is quite extensive. In general terms, Availability of all necessary information on the site meets the needs of the participants. It is among the criteria that the website should be plain and original. The site should have before event information such as the program, participant list, event calendar, ticket sales information. The site content should also include information during the event such as accommodation, food and communication.

Update and sharing information is an indispensable criterion. Although it is challenging to update all websites frequently, especially the changes in the program are critical and need to be announced to the participants quickly. Changes, delays or additions can notify the participants with updates.

Being interactive covers several online criteria to attract users. Locating exciting videos from the past period of the event, high-quality photographs of the participating artists, sharing of exciting stories, audio animations and messages of the participants make up the interactive structure of the website.

Security concerns the protection of personal information shared by website users. The measures taken to protect privacy should be shared with the users.

Usability is the ability to access the information easily. The presence of the site map, navigation bar, directory and site search engine on the website will facilitate more comfortable use for the user within site.

Design of the website is also significant. It should be impressive, attractive, exciting, original and encouraging for event participation. Color usage, text format, background, videos and photos must be appropriate for the structure of the event organization.

Today, websites are a very preferred communication tool due to the advantages it provides to business owners and users. The important thing is to design websites that can be used effectively and that can meet the needs of the businesses and users. Activities that appeal to the masses should pay attention to the website designs. According to Penpece's (2014) research, only 24 % of international festivals held in Turkey have a corporate website. This situation reveals a vital deficiency in the announcement of the activities. Necessary attention should be paid to the designs of websites that play an active role in successful event organizations.

In the light of this subject, which is focused on website design, various suggestions can be made to enable event organizations to use websites efficiently and effectively. In this context;

- It should be avoided to receive professional support on graphics, design, web design, IT, which will be frequently needed in the production and dissemination of the content, and emphasis should be placed on the production of original content.
- Switching to applications that will enable the interaction of participants about the activities.

- Considering the feedback of the participants about the activities and establishing feedback channels for this.
- Both the event organizer and the design group should work carefully and meticulously to create a website. Communication between these two groups is critical.
- The social media accounts of the festival can be used for promotion all year. Festival promotion can be planned not only a few weeks before the festival but as soon as the program is announced.
- The feedback of the participants regarding the events should be taken into consideration in the next year's event and website design.
- Besides the event websites, the website of the destination where the event is located is also essential. Since 1995, almost all European Capitals of Culture have created their websites. After 1997, the majority of European Capitals of Culture used email messaging tools or electronic bulletins, and half of them had electronic ticketing services.

Bibliography

Akhter, S. and Bashır, M. K. (2008), *Importance of Web Site Design And Customer Support Services in Online Purchase* (Master Thesis), Lulea University of Technology. Department of Business Administration and Social Sciences. Division of Industrial Marketing and E-Commerce.

Aksu, A. A. and Tarcan, E. (2002), "The Internet and Five-Star Hotels: A Case Study from The Antalya Region in Turkey", *International Journal of Contemporary Hospitality Management*, 14 (2), pp. 94–97.

Bai, B., Law, R. and Wen, I. (2008), "The Impact of Website Quality on Customer Satisfaction and Purchase Intentions: Evidence from Chinese Online Visitors", *International Journal of Hospitality Management*, 27, pp. 391–402.

Bauer, H. H., Grether, M., and Leach, M. (2002), "Building Customerrelations over the Internet", *Industrial Marketing Management*, 31, pp. 155–163.

Bayram, M. and Yaylı, A. (2009), "Otel Web Sitelerinin İçerik Analizi Yöntemiyle Değerlendirilmesi", *Elektronik Sosyal Bilimler Dergisi*, 8 (27), pp. 347–379.

Bell, H. and Tang, N. (1998), "The Effectiveness of Commercial Internet Websites: A User's Perspective", *Internet Research: Electronic Networking Applications and Policy*, 8 (3), pp. 219–228.

Beri, B. and Singh, P. (2013), "Web Analytics: Increasing Website's Usability and Conversion Rate", *International Journal of Computer Applications*, 72 (6), pp. 35–38.

Chu, R. (2001), "What Online Hong Kong Travelers Look for on Airline/Travel Websites?" *International Journal of Hospitality Management*, 20 (1), pp. 95–100.

Cox, J. and Dale, B. G. (2001), "Service Quality and E-Commerce: An Exploratory Analysis", *Managing Service Quality*, 11 (2), pp. 121–131.

Cox, J. and Dale, B. G. (2002), "Key Quality Factors in Web Site Design and Use, an Examination", *International Journal of Quality & Reliability Management*, 19 (7), pp. 862–888.

Cudny, W. (2013), "Festival Tourism – The Concept, Key Functions and Dysfunctions in the Context of Tourism Geography Research", *Geographical Journal*, 65, pp. 105–118.

Çubukçu, M. İ. (2010), "Konaklama İşletmeleri Web Site İçeriklerinin Değerlendirilmesi", *İnternet Uygulamaları ve Yönetimi Dergisi*, 1 (1), pp. 39–59.

Fırlar, B., and Özdem, O. Ö. (2013), "Web Tasarımının Önemi: Destinasyon Web Sitelerinin Görsel Tasarımlarının Değerlendirilmesi", *ODÜ Sosyal Bilimler Araştırmaları Dergisi*, 4 (7), pp. 5–16.

Gann, R. (1999), "Every Second Counts", *Competing*, 28, pp. 38–40.

Gehrke D. and Turban, E. (1999). *Determinants of Successful Website Design: Relative Importance and Recommendations for Effectiveness.* In Proceeding Book of 32nd Hawaii Int. Conf. System Sciences.

Geissler, G. L. (2001), "Building Customer Relationships Online: The Website Designers' Perspective", *Journal of Consumer Marketing*, 18 (6), pp. 488–502.

Getz, D. (2007), *Event Studies, Theory, Research and Policy for Planned Events*, Oxford: Elsevier.

Getz, D. (2008), "Event Tourism: Definition, Evolution, and Research", *Tourism Management*, 29, pp. 403–428.

Giomelakis D. and Veglis A. (2016) "Investigating Search Engine Optimization Factors in Media Websites", *Digital Journalism*, 4 (3), pp. 379–400.

Haas, R. (2002), "The Austrian Country Market: A European Case Study on Marketing Regional Products and Services in A Cyber Mall", *Journal of Business Research*, 55 (8), pp. 637–646.

Hoyle, L. H. (2002), *Event Marketing: How to Successfully Promote Events, Festivals, Conventions and Expositions*, New York: John Wiley and Sons.

Karaosmanoğlu, E., Acar, R., and Uray, N. (2016), "Websiteleri Firmalara Ne Kazandırabilir? Websitesi Kalitesi, Tüketici-Odaklı Marka Değeri ve

Satınalma Eğilimi Arasındaki İlişki", *Anadolu Üniversitesi Sosyal Bilimler Dergisi*, 16 (1), pp. 159–174.

Kaynama, S. A. and Black, C. I. (2000), "A Proposal to Assse The Service Quality Of Online Travel Agencies: An Exploratory Study", *Journal of Protessional Services Marketing*, 21 (1), pp. 63–89.

Kent, M. L. and Taylor, M. (1998), "Building Dialogic Relationships through the World Wide Web", *Public Relations Review*, 24 (3), pp. 321–334.

Law, R. and Hsu, C. H. C. (2005), "Customers' Perceptions on the Importance of Hotel Web Site Dimensions and Attributes", *International Journal of Contemporary Hospitality Management*, 17 (6), pp. 493–503.

Li, Y. N., Tan, K. C., and Xie, M. (2002), "Measuring Web-Based Service Quality", *Total Quality Management*, 13 (5), pp. 685–700.

Liu, L. and Arnett, P. (2000), "Exploring the Factors Associated with Website Success in the Context of Electronic Commerce", *Information and Management*, 38 (1), pp. 23–33.

Madu, C. N. and Madu, A. (2002), "Dimensions of E-Quality", *International Journal of Quality & Reliability Management*, 19 (3), pp. 246–258.

Neırottı, L. D., Bosettı, H. A., and Teed, K. C. (2001), "Motivation to Attend the 1996 Summer Olympic Games", *Journal of Travel Research*, 39, pp. 327–331.

Nielsen, J. (2000), *Designing Web Usability*, Indianapolis: New Riders Publishing.

O'Cass, A. and Carlson, J. (2012), "An Empirical Assessment of Consumers' Evaluations of Web Site Service Quality: Conceptualizing and Testing a Formative Model", *Journal of Services Marketing*, 26 (6), pp. 419–434.

Onat, F. (2014), *Dijital Çağda Halkla İlişkiler Yazarlığı*, Ankara: Nobel.

Onat, F. and Gülay, G. (2015), "İnternetin Sanat Festivallerinde Halkla İlişkiler Amaçlı Kullanımı: 20. ve 21. İzmir Avrupa Caz Festivalleri İzleyici Araştırması", *Selçuk Üniversitesi İletişim Fakültesi Akademik Dergisi*, 8 (4), pp. 49–72.

Peltekoğlu, F. B. (2009), *Halkla İlişkiler Nedir?* İstanbul: Beta Basım Yayım Dağıtım.

Penpece, D. (2014), "Festivallerin Pazarlanması: Türkiye'deki Uluslararası Festivaller Üzerinde Bir Araştırma", *Ç.Ü. Sosyal Bilimler Enstitüsü Dergisi*, 23 (1), pp. 193–210.

Perdue, R. (2001), "Internet Site Evaluations: The Influence of Behavioral Experience, Existing Images and Selected Website Characteristics", *Journal of Travel & Tourism Marketing*, 11 (2/3), pp. 21–37.

Sarı, Y. and Kozak, M. (2005), "Turizm Pazarlamasına İnternetin Etkisi: Destinasyon Web Siteleri İçin Bir Model Önerisi", *Akdeniz İktisadi ve İdari Bilimler Fakültesi Dergisi*, 9, pp. 248-271.

Scanlon, J. M., Schroeder, W., Scanlon, T., and Snyder C. (1998), *Websites that Work: Designing with Your Eyes Open*, Tutorial, pp. 147-148.

Shock, P. J. (2000), "Effective Use of the World Wide Web", *Journal of Convention & Exhibition Management*, 2 (2), pp. 143-146.

Spool, J. M., Scanlon, T., Schroeder, W., Klee, M., and Landesman, L. (2000), *Web Sites that Work: Designing with Your Eyes Open*. CHI 2000 Tutorial.

Sowards, S. W. (1997), "Save the Time of the Surfer: Evaluating Web Sites for Users", *Library Hi Tech*, 15 (3-4), pp. 155-158.

Taylor, M. J. and England, D. (2006), "Internet Marketing: Web Site Navigation Design Issues", *Marketing Intelligence & Planning*, 24 (1), pp. 77-85.

Thampson, K. and Matheson, C. M. (2008), Culture, Authenticity and Sport: A Study of Event Motivations at the Ulaanbaatar Naadam Festival. Mongolia, In J. Cochrane, (Ed.), *Asian Tourism: Growth and Change*, Oxford: Elsevier.

The "Digital in 2019" Report Published by We Are Social and Hootsuite. https://dijilopedi.com/2019-internet-kullanimi-ve-sosyal-medya-istatistikleri/ (04.04.2020).

Timur, M. N., Çevik, S., and Kıyık K. G. (2014), "Etkinlik Turizmi: Kültür Başkenti Etkinliklerinin Başarı Unsurları Üzerine Bir Değerlendirme", *The Journal of Academic Social Science*, 2 (1), pp. 56-83.

Tsai, W. H., Chou, W. C., and Lai, C. W. (2010), "An Effective Evaluation Model and Improvement Analysis for National Park Websites: A Case Study of Taiwan", *Tourism Management*, 31, pp. 936-952.

Van Der Wagen, L. (2005), *Event Management for Tourism, Cultural, Business and Sporting Events* (2nd Ed.), New South Wales: Pearson Education Australia.

Wang, Y. and Fesenmair, D. (2006), "Collaborative Destination Marketing: A Case Study of Elkhart Country, Indiana", *Tourism Management*, 28 (3), pp. 863-875.

Yang, X., Ahmed, Z., Ghingold, M., Sock Boon, G., Su Mei, T., and Hwa, L. (2003), "Consumer Preferences for Commercial Web Site Design: An Asia Pacific Perspective", *Journal of Consumer Marketing*, 20 (1), pp. 10-27.

Yen, B., Hu, P. J.-H., and Wang, M. (2007), "Toward an Analytical Approach to Effective Web Site Design: A Framework for Modelling, Evaluation and Enhancement", *Electronic Commerce Research and Applications*, 6, pp. 159-170.

Yoo, B. and Donthu, N. (2001), "Developing a Scale to Measure the Percieved Quality of an Internet Shopping Site (SITEQUAL)", *Quarterly Journal of Electronic Commerce*, 2 (1), pp. 31–46.

Zhao, Z. and Gutierrez, J. (2001), The Fundamental Perspectives in E-commerce. In M. Singh and T. Teo (Eds.), *E-Commerce Diffusion: Strategies and Challenges* (pp. 20–3), Melborne: Heidelberg Pres.

Fatma Doğanay ERGEN
Information and Communication Technology for Event Tourism Management

1 Introduction

Information communication technologies (ICT) are altering tourism and many sectors globally. ICT offers a series of opportunities by reforming the industry structure through restructuring. Information and communication technologies define and customize products by supplying tools to develop and manage proposals around the world to strengthen them by supporting the globalization of industries (Bethapudi, 2013: 67). ICT tools are employed in almost every field of events in event tourism management. It is observed that various applications are extensively used in activities, especially in the scope of the Internet of things. For instance, self-service software, artificial intelligence, augmented and virtual reality technologies, and their usage areas are increasing nowadays. Thanks to the use of these technologies, the participants go through distinctive experiences through personal service. Moreover, speed and convenience are presented and lastly, solutions are found immediately. Event organizers, on the other hand, can reduce costs, capture marketing opportunities, and manage a successful event in a shorter time period. Therefore, it is anticipated that the use of ICT will continue to develop in the future due to its contributions to the event industry.

In this context, the basic objectives of the chapter include giving information about the scope of ICT, ICT tools, ICT products and services. Besides, the section provides information on the applications of information and communication technologies used in various activities such as festivals, exhibitions, ceremonies, sports events, meetings and conferences as well as ICT event organizers, exhibitors, sponsors, etc. Explanation of the contributions made to all event stakeholders is among the objectives of the chapter. It is also aimed to present information about trends for the future of ICT in the frame of event tourism management.

2 Information and Communication Technology for Event Tourism Management

Swift development has been observed in information and communication technologies, starting especially from the second half of the 20th century. ICT has found its place in almost all areas of human life, especially in the military, political and social areas (Sevim, 2019: 48). UNESCO (2019) defines the ICT concept as "a set of various technological tools and resources used to transmit, store, create or share information." These technological tools and resources include computers, the Internet (websites, blogs and emails), live broadcast technologies (radio, television and webcast), recorded broadcasts (podcasts, audio and video players and storage devices), and phones (fixed or mobile, satellite, vision/ video conferencing etc.).

When the tools of ICT are examined, it is identified that the essential tools are the Internet, computer and telephone. The most influential developments in ICT came out after the Second World War. The first computer was used by the US army in 1946, and later this technology was improved, and the first personal computer was created and introduced in 1965 (Sevim, 2019: 48). On the other hand, it needs to be stated that the most imperative reason for the developments in ICT is the birth and development of the Internet. The Internet and computer have created a revolution that has never been experienced in the world of communication. These unique talents were combined after the invention of tools such as telephones, radio and computers (Leiner et al., 1997: 2). ICT offers abounding products and services using the Internet, computer and mobile phone tools. Augmented reality (AR) express the combination of real scene and a virtual scene (Hjalager, 2015: 311), virtual reality created by the restructuring of the real environment (Coşkun, 2017: 64) Internet of objects that are a network of interconnected objects and people sharing data (Kuriş, 2020: 1), self-service software (Smart Hotel Technology Guide, 2018: 32; Dalgıç & Birdir, 2020: 334) that offers electronic support without the need to interact with the service representative artificial intelligence (British Government Industrial Strategy White Paper, 2017: 37). All these mentioned technologies are capable of performing tasks that require human intelligence such as visual perception, speech, recognition and language translation. At the same time, these applications are among the most used applications in event tourism management. Events take place as a result of people's acting together and are organized for some purposes such as meeting some social or individual needs or reaching common goals as a group and organization, sharing a common feeling or thought and being a social entity in social life (Ekin, 2011: 3). It is clear that it is very difficult to manage large crowded events successfully. Successful and effective management of large-scale events mostly depends on the use of ICT (Yamin et al., 2008: 1293).

ICT is increasingly critical in constructing competitiveness in tourism organizations. ICT is becoming the key determinant of a competitive advantage. Along with the speedy developments in the field of ICT and the role of ICT in reducing costs, it has promoted many areas. In this context, ICT is considered as a powerful tool in strategy (Bethapudi, 2013: 67–68).

a Virtual and Augmented Reality

One of the chief products of ICT is the augmented and virtual reality. An augmented reality (AR) system is a combination of the real scene displayed by the user, creating a unified view for the user, and a virtual scene containing computer-generated pictures, additional information, animation or entertainment (Hjalager, 2015: 311). Virtual reality is the restructuring of the real environment in the virtual environment. However, when new sounds and formats are added to the existing reality environment, AR takes place (Coşkun, 2017: 64). Virtual reality simulates images, sounds and feelings and puts users into an imaginary 3D environment via using devices such as VR headsets (Smart Hotel Technology Guide, 2018: 32). AR and VR technology are services designed to meet the needs of tourists and provide a better tourist experience and use multiple different tools such as cameras, smartphones and glasses. Although both technologies are used in destinations, museums and many other areas, it is noticed that they are utilized as a vital tool in the event sector (Hjalager, 2015: 311).

As the events provide us with information about different cultures, the event sector is regarded as an extensive area, and is a powerful aid in promoting our culture. For example, activities such as carnival, official receptions, shows, national days, music, theater and art festivals can be shown as examples (Jenny, 2017: 38). It is salient that the event sector is carried to new dimensions with technology every year. Thanks to AR and VR applications in events, the way the organizers, sponsors and vendors interact with the participants change. With these applications, improvements in event tools and data security have achieved a collective progress in the event industry. It is also worth noting that these systems have become crucial determinants in terms of innovation among the event trends that continue to shape the spaces (Levine et al., 2018). It is known that the basic rule of event management and marketing is to make the participant experience something unforgettable. The increasingly crowded special events and the increasing competitive environment bring with difficulties in management. Therefore, originality created by providing a unique experience that will be unforgettable to participants will be critical for success (Hoyle, 2002: 22). For example, if a venue hosting birthday parties can offer immersive digital technology by incorporating VR and AR applications for entertainment purposes, it can increase potential demands and provide a competitive advantage to the business. The use of these applications is recognized as an opportunity to increase competitiveness (Thurlby, 2018: 22). At the same time, it is almost impossible for the tourist to experience all the different events and celebrations organized at once so that the systems can provide tourists with an unforgettable experience in this respect (Jenny, 2017: 38).

In addition to the benefits of AR and VR technology providing a unique experience to the participants, it is observed that virtual reality (VR) technology provides benefits in terms of revenue generation and competitive advantage in terms of management (Tussyadiah et al., 2018: 152). Event tourism management companies generally appreciate that their competitive advantages lie in their ability to maximize the opportunities of the venue and in designing events. In this context, event organizers are looking for new and various ways to adjust their activities and create a competitive advantage (Chaturvedi, 2009: 299). It can be said that VR technology offers many opportunities to event organizers as digital spaces can be formed anywhere and, in any size, (Thurlby, 2018: 22) and enable interesting virtual meeting tools (Howland, 2016). VR technology offers many opportunities to event organizers (Thurlby, 2018: 22). Thanks to VR, it is possible to create virtual spaces and contribute in terms of cost. Event planners allocate a large part of their budgets to speakers, accommodation, food and beverages and various other logistics. On the other hand, most of these costs are eliminated with the virtual spaces created when a conference or trade fair is organized through virtual reality (Levine et al., 2018). Meanwhile, the personnel costs will be decreased, and some activity costs will be advantageous, including event equipment since there will be no installation making. Therefore, it will be more attractive to use AR and VR technologies for event organizers (Thurlby, 2018: 22). On the other hand, event organizers can make it easier for people to attend an event remotely thanks to the developments in virtual participation telepresence tools. For example, a remote control equipped with iPad developed by Double Robotics facilitates the remote interaction of the participants with other event participants. Event organizers can make these robots available to VIP guests who cannot attend an event because the business procedures conflict with their schedule. It enables participation in these events without the need for long flight times and inappropriate time zone changes.

Event attendees can perform enhanced experiences using an AR headset, glasses or contact lenses paired with special event planning software. At the same time, it is predicted that in events that require hundreds of thousands of square feet, such as automobile shows, it will leave the need for giant venues (Levine et al., 2018). However, in the event tourism management sector, there are other creative, interesting, and awareness-raising applications that can be effective and high-budget, especially used by large companies. Within these mentioned applications, the applications known as video projection mapping 3D or 4D video mapping come to the fore. These applications are works to attract the attention of the audience by reflecting three- or four-dimensional images on large wall surfaces such as buildings that are the symbol of historical buildings or the city. For example, in 2012, Samsung brand reflected the introduction of new models with 3D and 4D video on the Rumeli Fortress walls of Turkey by performing the mapping application to create awareness of the country's borders. This may be noticed as a means to attract attention. As a global brand, Samsung has implemented this method in many countries simultaneously or at different times. Mapping applications may also focus on promoting brands or models within the scope of social responsibility studies and efforts are drawn to attract attention (Akay, 2014: 60).

It is notable that AR and VR technologies are also used effectively in sports events. When the use of these applications is explored, it is detected that there are studies aimed to increase the efficiency of physical activities in sports, to manage the games more fairly and to follow the competitions more easily by the audience. However, it seems that most of the advanced AR applications in sports are related to presenting sports competitions to the audience. The Eyeply system draws attention among many applications made to present sports competitions to the audience better. The Eyeply (Hurwitz & Jeffs, 2009: 55) is a system designed under three layers where different services are offered: stadium layer, player layer and friend layer. In the stadium layer, viewers can see various parts of the stadium, such as restaurants, cafes, lodgings and exit gates, and are directed to the seat where they sit on via mobile devices. At the player layer, viewers, previous statistics, performance tables and real-time data about the players can be followed on the mobile device. Socialization options are offered in the friend layer (Bozyer, 2015: 319–320). Another example where these systems are used in sports events is the American Cup, where sailing races are held. Augmented reality makes the sailing race an exciting spectator sport. With automatic tracking systems placed on each sail, the location data of the sails are continuously transferred to high-performance computers on the shore. With the specially developed augmented reality software, location data is transformed into informative screen graphics in real-time. It is much easier and interesting to watch and interpret with an augmented reality system, as viewers can see all the information they need at once, from boats to seafarers and even the effects of wind on water (Honey & Milnes, 2013). On the other hand, since television can provide a satisfactory substitute for individuals who perceive significant restrictions to participate in a sports event, it can also be offered as a satisfactory substitute for individuals who perceive restrictions related to various experiences with VR technology (Guttentag, 2010: 646). It can also be an alternating option for individuals who perceive certain risks in real life. For example, since there are many risks in the real world due to cost and security constraints, Red Bull introduced the extreme sports platform that includes exciting outdoor experiences such as cliff jumping, skiing and helicopter aerobatics designed for seat enthusiasts with VR. Therefore, it is viewed that businesses create events as a gripping place with VR and AR applications and present them as entertainment platforms (Yeoman et al., 2015: 17 akt., Thurlbly, 2018: 14). Another application offered as an entertainment platform is location-based VR applications. It is regarded that these applications are used in meetings, special events, corporate trips, team building and many more. For example, Kalahari Resorts and Conventions, America's largest indoor water park business, offer a VR experience called Arena, which offers three virtual games that can accommodate eight players at a time (Heilman, 2017).

It is spotted that the way we communicate and interact with VR and AR technologies has started to shift. For example, it is known that preparations for celebrating the festival Diwali, known as India light festival, in a virtual world in 2030. Preparations are being made for families in two different locations, one in Bangalore (India) and the other in New Jersey (USA) to celebrate Diwali in a virtual world. Thanks to these technologies, it is aimed to ensure that families live the festival face to face and interact as if they were in the same room. With the new integrated technology platform, a creative seating arrangement that allows families to experience the same room interaction is designed and families are planned to meet, greet and perform rituals together. It is attempted to connect all distances and experience real life situations in events such as AR and VR applications and Diwali. At the same time, it is recognized that organizations are preparing to offer new virtual and augmented reality platforms by changing their business strategy models and customer experience design by utilizing these new technologies (Benjamin, 2020). Therefore, if event organizers want to compete, they have to be innovative (Yeoman et al., 2015: 17 akt., Thurlbly, 2018: 14).

Information and Communication Technology 293

b Internet of Things (IoT) and Self-Service Software

The Internet of Things (IoT) is defined as a "simple, inconspicuous and cost-effective item identification system (use) to connect everyday objects and devices to large databases and networks" (Giusto et al., 2010: 12). In another definition (Kuriş, 2020: 2), it is explained as a network of interconnected objects and people that data sharing to perform various tasks.

The main purpose of this network is to increase our daily life quality by performing various tasks. Basic concepts that make up the IoT system can be listed as objects, sensors, RFID (Radio Frequency Identification), controller, activator, cloud computing, fog computing and big data. Objects are physical devices that are able to connect to the Internet and with each other to make smart decisions. Sensors, another type of device called sensor in IoT, must be connected to the data network. Sensors are objects that can be used to measure physical properties and convert information into electrical or optical signals (Kuriş, 2020: 3). RFID is an automatic recognition system that operates with radio frequencies and allows tracking of the movements of an object carrying a tag equipped with a microprocessor and the information it carries on a tag. There are different software and hardware requirements for the formation of an RFID system. The hardware required for RFID is described as RFID tags, RFID readers, frequencies and standards. Moreover, interfaces are required as software. One of the most paramount components of RFID systems are labels. In its simplest sense, the label is a microchip with an antenna inside (Yüksel & Zaim, 2009: 1). The controller is responsible for collecting data from the sensors and providing an Internet connection. The activator is a basic motor that can be used to move or control a mechanism or system based on a specific set of instructions. Cloud computing is a technology that enables widespread networks with optional service and a pool of common resources by providing unlimited space. Sis computing creates a distributed computing infrastructure closer to the edge of the network, which performs easier tasks that require a quick response, reducing the data load on networks. Big data in IoT is potentially one of the biggest data sources in smart cities, industry, agriculture, transportation, health, retail, etc. (Kuriş, 2020: 9). Another technology element operated in this field is self-service software. Self-service software offers the user electronic support without the need to interact with a service representative. The usage areas of the software and the formats practiced are varied such as kiosks, mobile phones, etc. (Smart Hotel Technology Guide, 2018: 32; Dalgıç & Birdir, 2020: 334). It is sighted that the use of self-service is not a novice concept, and banks have used this system for many years. The ATM concept, which is employed to reduce costs and provide better services for customers, is an example of these systems. Self-service is utilized for two main reasons in other industries: increasing productivity and eliminating labor costs and decreasing costs. It is realized that generally self-service kiosks are used. Such kiosks include a computer with the software installed in an area accessible to customers and inside a protective case. Customers can make use of a self-service kiosk to interact with an employee from the company or meet their needs (Bodendorf, 2009: 40; Kaur, 2015: 1234). According to IoT, smart sensors need to be developed with the ability to connect locally to existing networks in order to interact with the real world. Access to real-time data and proper action is allowed through the use of smart distributed "objects" such as RFID tags and sensors (Giusto et al., 2010: 12). With the Internet of Things, devices are enabled to function online. Taking the case from the tourism industry, the tourism company offers real-time customer experience monitoring and the ability to extend its customer base (EY Turkey, 2019: 7). Recently, many types of identification and sensor devices, especially RFID tags, have been developed in the tourism and event industry. There are RFID tags, RFID-enabled tickets, smart maps and applications such as transportation, mobile geographical guidance, effective information retrieval and mobile ticketing (Yamin et al., 2008: 1294). RFID technology has great potential to facilitate and improve the management of social activities where people interact with each other over time and in different places (Ertek et al., 2017: 3). The main use of RFID technology in events is to facilitate the establishment of a quality standard. It is possible to facilitate the participation and participation of the participant in the activity with a chip-containing material (wristband, tag, card, etc.) given to the participant. The significant thing here is to ensure that the chip is defined following the permission of the participant, and accepts it while performing its operations. The participant can take part in different activities within the activity area with the material that has the feature of a chip, can announce place notifications, post twit shares in exchange for surprise gifts at certain points, and have the opportunity to enter areas that cannot be entered without chips (Akay, 2014: 66).

There are several ways event creators can use RFID applications to facilitate the event experience and collect data. When the main applications of RFID for events are examined, it is remarkable that RFID entry is faster and easier than other types of electronic scanners at the entrance point of the event. The staff does not need training to use, and the RFID equipment lights green to go and red to stop. It maintains flexible access control. Fast and controlled access can be made into certain areas where access is required, such as VIP areas or conference sessions (Stein, 2017). (VIP alanları veya konferans oturumları gibi giriş izni gereken belirli bölgelere hızlı ve kontrollü giriş yapılabilir). It is planned to create a completely smart exhibition space for EXPO 2020 to be held in Dubai. In this context, mobility and technological connections that will be established in order to regulate the densities at the entrance gates of the fair area can be shown as an example of the use of ICT within the scope of the events. (Yalçınkaya et al., 2018: 46). Another usage area of RFID technology is payment. RFID allows cashless payments by enabling participants to pay for food, beverage and commercial products with their wristbands instead of carrying cash or credit cards. Payment transactions are fast, easy and reliable. Research also shows that people spend up to 20 % more when using RFID instead of cash at events. It is also possible to collect better data with RFID. For instance, RFID technology gives real-time information about how participants move between events. If visitors register their wristband with an e-mail address, this data that can be used in future marketing activities may also have been acquired. Saving the information is not a concern, on the contrary, it seems to be safer than other systems. With RFID chips, each ticket is associated with a separate ID. When the participant's wristband is lost, the wristband can be easily disabled, and a new wristband can be arranged (Stein, 2017). It is deemed that this application is also used in the Exploratorium, an applied science museum in San Francisco. Although the card with RFID technology is a unique identity in the museum, visitors (at the beginning of their visit) have to enter their e-mail addresses in the registration kiosk to register their cards. In case of loss of the cards, records are taken by e-mail for the privacy of the images in the museum and to protect the information regarding the visitor (Hsi & Fait, 2005: 62).

Thanks to RFID, a personalized service can also be administered to the participants. At a conference, visitors were given an RFID-enabled name card upon arrival, and they were asked to touch the screen to check-in. A personalized welcome message with the names of the visitors is displayed on the screen, along with the sessions suggested to the visitors and personalized recommendations for the participants. However, printing and coding of each badge was carried out in a short time, taking 15 seconds. With this service, a great experience was created for visitors and real-time arrival data were adjusted for the event organizer (Hardie, 2017). RFID systems, when used in social events, obtain data for each participant of the event. This data is created with meaningful information about the structure and quality of the social network and the event, when combined with the data of the participants' features, locations and event schedule. Improved conference management information systems and management practices can enable participants to find sessions and other people they are interested in, minimize timing conflicts, increase participation in sessions, and improve the overall quality of the event (Ertek et al., 2017: 4). RFID technology also helps with cloakroom tickets, which seems simple in events, but is one of the issues that may be encountered. Thanks to RFID technology, it is perceived that there are no lost cloakroom ticket problems. With this system, which was used in the event with more than 3,000 visitors, the visitors were automatically assigned to the next available cloakroom number when they put their RFID name cards on the cloakroom points. The number of the cloakroom staff is displayed on the screen, which was also done at the end of the day when collecting their coats. Thanks to RFID, it is realized that waiting times can be shortened and can reduce confusion on missing tickets. Thus, the whole process was carried out quickly. It is understood that another area where the system is used is table plans and dinner arrivals. When the guests come to the area and touch their RFID-enabled invitations, they encounter a screen that shows the table they will be placed, who they are sitting with and how to go to their table. Thanks to this method, it was possible to host more than 1000 guests in less than an hour with less staff. It is also a system that allows organizers to make last-minute changes in the seating plan until the last moment each guest arrives. It is seen that RFID technology provides speed and convenience in beverage orders. It is shown that it is an acceptable service for VIP visitors as well. At the same time, RFID technology makes relevant contributions to control bar expenditure, to use the dashboard to add beverage tokens instantly when needed and to see costs (Hardie, 2017).

RFID technology is used in the exhibitions held at the museum with a handheld device as a museum guide system. Each exhibition has an RFID tag, and the RFID reader receives information such as the number of exhibition indexes on the RFID tag. The system distributes detailed information from the database to handheld devices over the wireless network. Visitors can listen to the audio description, watch videos, view the colorful picture of the items on display, and read the text from these devices. Thus, visitors can have enough time to learn the information and enjoy their tour (Wang et al., 2007: 1). RFID technology was also witnessed in the Exploratorium, an applied science museum in San Francisco. The museum allows visitors to continue their scientific discoveries beyond the walls of the museum with a prototyped system called eXspot. The EXspot system consists of a small RFID reader package, a card or necklace with an RF tag carried by visitors, a wireless network, a recording kiosk and dynamically generated Web pages to place it on the works exhibited in the museum. The RFID card, which is given to the visitors at the beginning of the museum visits, is designed to clearly show the graphics about the Exploratorium on one side and the RFID chip and the external antenna on the other side to satisfy the visitor curiosity. The EXspot system allows visitors to learn about the exhibitions they visit and take souvenir photos while in the museum. They can then access to the exhibition information on personalized Web pages. In addition to the Exploratorium, RFID technology can be found in many museums around the world such as Chicago Science and Industry Museum, Vienna Technology Museum, Aarhus, Danish Museum of Natural History for visitor (Hsi & Fait, 2005: 62–63).

It is stated that sensor devices can also be used in the event industry. In the activities of large communities where religious rituals are held, the events can be managed safely with sensor devices. For example, overcrowding during some mandatory rituals of religious organizations such as Hajj or Kumb Mela can give organizers hard time to manage. In order to prevent situations such as accidents, fire and terrorist attacks during the rituals, sensors can be activated in the activity areas and these areas can be monitored. As all area will be established with surveillance systems, managers will be able to monitor the situation continuously and take preventive measures. At the same time, in order to prevent illness and to make the pilgrimage duty a health-safe religious activity, the pilgrim must undergo a full medical examination and the data must be stored in the database. Thus, people with health risks can be tracked through the RFID and "smartwatch" system throughout the pilgrimage. When there is a negative situation in the health of the persons, the help signal can be sent to the central control station with the smartwatch. Therefore, since the systems will know the exact location of the patient, it will be ensured that medical aid will be organized and sent urgently and situations such as delays will be prevented (Yamin et al., 2008: 1295).

c Artificial Intelligence Applications

Another technology product effectively employed in the field of ICT is the artificial intelligence (AI). The first developments about artificial intelligence were put forward by Alan Turing in an article published in 1950 when he brought up whether the machines could think. With the Turing Test he proposed, it was possible to distinguish whether a machine is intelligent or not. However, John McCarthy is considered as the real name father of artificial intelligence. John McCarthy organized an academic conference on the subject in 1956 (Şener, 2019). In the summer of 1956, McCarthy, Minsky, Rochester and Shannon (1955: 5) proposed a 2-month artificial intelligence study at Dartmouth College in Hanover, New Hampshire in the summer of 1956. It was to attempt to find out how machines use language, how abstractions and concepts are created, how to solve the types of problems that arise for people and how to improve themselves. Today, technology promises to help cope with the challenges of humanity by entering our daily lives from research laboratories. It is predicted that there will be a rapid progress in the development of artificial intelligence and this speed will continue exponentially (International Telecommunication Union, 2019: 5). In this context, artificial intelligence is defined by the UK Government Industrial Strategy White Paper (2017: 14) as "Technologies capable of performing human intelligence tasks such as visual perception, speech, recognition and language translation." It is also added to the definition that artificial intelligence systems today often can learn or adapt to new experiences or stimuli. Artificial intelligence is a new form of intelligence that can synthesize several different ideas at once (Zsarnoczky, 2017: 89). When the concepts used in artificial intelligence are examined, algorithms need to be discussed. An algorithm is a set of instructions to calculate or solve problems and forms the basis of artificial intelligence systems, an expert system that imitates a person's decision-making ability. Additionally, an algorithm gives the ability to learn from experience, machine learning, data exchange and neural network of artificial neurons. It is deep learning, a new variation of neural networks by connecting many layers of artificial neuron and used to solve problems (Şener, 2019).

It is noticed that artificial intelligence applications are cognitive robotic process applications and managers are working to integrate these applications into their value chains more successfully. While robotic process automation applications provide digitalization in manual progressing processes such as finance, accounting, support functions, etc., artificial intelligence applications appear with smart assistants where chatbots and voice calls are performed. Using artificial intelligence and robotic process technologies in processes such as sales and customer relations by contacting the customer may make customers feel more special (EY Turkey, 2019: 7). The best artificial intelligence in the event industry should be allowing us to customize customer engagement and improve business processes so as to increase revenue. The artificial intelligence used in the event industry in 2019 is believed to have employed the technology to implement personalized participant experiences that satisfy event organizers, exhibitors and sponsors, with more meaningful interactions. Event organizers can also gain competitive advantage thanks to their personalized participant experiences (Groot, 2017).

Robots have made their solid places within the scope of event tourism management. Robots are electromechanical units programmed or controlled by someone and they perform certain actions and tasks. Robots are used in many areas from the bartender robots, those preparing the cocktail of the participants during the event to the saluting robots, those showing up as holograms and welcoming and informing the participants at the event entrance (Novak, 2017). An example of this situation is the robot named as "Mario," which can speak 19 languages in the Marriott hotel in Belgium, which gives information about the activities both inside and outside the hotel (tuyed.org.tr, 15.02.2020). At the same time, there are SoloShot robot photographers who take photos of the event with a camera programmed to move the participants' movements on the tripod depending on the follow-up. These robot cameramen, who are programmed to take the best photos, can also group photos according to their themes and distinguish non-aesthetic photos. On the other hand, it is exceptional that robots also take part in activities like in entertainments. For example, The Toyota brand was awarded the best marketing campaign award by recruiting the robot, produced in England, eight meters tall and weighing 770 pounds to introduce the new car model. Titan robot took the award for contributing to the winning of this prestigious award (Novak, 2017).

Today's technological revolution also demands efficiency, sustainability and productivity. The use of artificial intelligence, which can be reached in many different fields, is expanding into every sector (Zsarnoczky, 2017: 89). In this regard, it may be highlighted that many technology giants are making investments into artificial intelligence. The annual total cost savings that Chatbots will provide for companies in 2020 is estimated to be 8 billion dollars, and the new business opportunity created by artificial intelligence is expected to be 2.3 million dollars (EY Turkey, 2019: 32). It is purposeful that in the event planning industry, 80 % of businesses planned to use chatbot by 2020 (Arnold, 2019). Chatbots developed for automatic communication have improved greatly since the first model was released (Zsarnoczky, 2017: 89). Chatbot software is a speech tool that provides diversity in a chat-oriented process and its primary function is to answer questions programmatically and consistently. It often appears as pop-ups that appear on the bottom corner of the site that helps its visitors on a website. While the chat support is used as a new communication tool in the event industry, it is unique that the levels of artificial intelligence used differ from each other (Cooper, 2018). For example, chatbot software, which has many functions such as ordering food, planning a meeting, giving information about the weather, is expected to become one of the most crucial parts of the brand–customer relationship (Kayıkçı & Bozkurt, 2018: 57). Therefore, event planners can automate a series of tasks and serve their customers better by making use of chatbot software in more than one different area (Arnold, 2019). Today, chatbots that can answer the question asked as the most successful chatbot software are shown. For example, they are the consultancy and personal assistant applications (Zsarnoczky, 2017: 88), which are among the most popular versions of artificial chatbot software that can answer the question when asked for a place at the event venue (Cooper, 2018). There is no price or technical barrier for event planners to apply artificial intelligence-based chatbot software. Chatbot can be easily implemented for any event at a lower cost. At the same time, artificial intelligence matching engines can help participants experience better effectiveness. However, participants using a matching engine that analyzes their goals, interests, social media data, and participant profile can only receive specially selected suggestions for them. It is also observed that artificial intelligence contributes directly to the event organizers. Event organizers should find event locations, connect with vendors, and reach out to people they think are best suited to attend their events. This is a comprehensive task that includes manually searching for various sources, sending outreach emails, etc. Processes that will take months for event organizers can be carried out in seconds with artificial intelligence (Arnold, 2019). At the same time, chatbot software helps participants faster than people can deal with queries. For example, when the event attendees want to learn the wireless password, simply ask "What is the password for Wi-Fi?" They are provided to receive an instant response by writing. A simple question can be answered every time with a timely and simple answer (Cooper, 2018). On the other hand, chatbot software, Facebook Messenger, etc. can also be used through other popular applications. Applications are easier to use for participants who do not want to download applications only for the event (Arnold, 2019). There are separate applications such as Drift and Help Scout that serve like Chatbot. Drift application is used to answer the questions of people browsing your website, to collect e-mail addresses and to give support to everyone. Help Scout is another chat-based platform that focuses more on customer support and follow-up. The platform tries to answer questions directly by using bot-like instant answers and can suggest articles and resources to answer the questions to the customer (Myers, 2020).

It can be noticed that artificial intelligence can be used as chatbots and robots in exhibitions and events held in museums. It is increasingly used in wide range of areas, from developing artificial intelligence websites to analyzing visitor data and collections and determining visitor policies and exhibition content. In 2015 and 2016, the Chicago Institute of Arts managed to calculate how much time visitors spent in galleries by using artificial intelligence. In this way, they started to plan small exhibitions that focused on the permanent collection. With the help of artificial intelligence applications, museums not only show how visitors behave but also assist them develop museum collections and understand the visitors better (Gümüş, 2019: 26).

Deep learning, which is one of the most critical parts of artificial intelligence in event tourism management, is an extensive area. Through deep learning and adding a reasoning component to create new actions based on reasoning, it can become a system that is able to read the recording data in the event. During the event, it can estimate what logistic requirements are required in terms of beverage, food and crowd control at the venue. Although deep learning practice is available today, yet it also requires complex systems since it demands expertise, it is predicted that this system may take several years to become popular in the event industry (Groot, 2017). Using artificial intelligence for activities is regarded as a gain due to some simple benefits for the environment. Chatbots and event apps are only available online. Therefore, this technology can be used for participants, calendars, maps, survey cards, etc. because it is much greener and more sustainable than printed materials (Cooper, 2018).

Conclusion

The developments in information and communication technologies have affected many fields along with the event tourism industry. The fact that it is difficult to manage large-scale events and to carry out large-scale events with big budgets are among the fundamental reasons why the event organizers use ICT. At the same time, the intense competition environment and rapid changes in consumer expectations have made it imperative to use technological tools that the event tourism management will benefit. It is observed that today these technologies are taking place in different services and activities day by day and their usage areas are increasing and diversifying. There are many different applications in the fields of augmented and virtual reality, Internet of things, self-service software, artificial intelligence used in event tourism management and these are used in the event industry. These technologies produces vital contributions to event organizers, such as collecting data, protecting them, presenting primary reports on the event and performing transactions in a short time. Thus, the Internet of objects, such as artificial intelligence technologies retention of data generated, operation and management has become easier than ever before (EY Turkey, 2019: 27) and they can provide very useful information to event organizers and attendees and create better experiences for participants (Arnold, 2019).

It is considered that it is of great importance to pay attention to the following recommendations for the successful use of ICT in event tourism management.

In this context, suggestions that are thought to contribute to the relevant literature, researchers and industry are presented:

- The event audience should be analyzed well before event organizers use ICT applications. Davidson et al. (2002: 24) also stated that the targeted participants in the MICE sector should have the necessary skills and equipment to enable them to access the new technology. In the MICE sector, it is imperative to examine the profile of potential participants before selecting these tools, and to assess whether the participants are sufficiently ICT-oriented to take advantage of ICT methods.
- It is predicted that if artificial intelligence and robot applications are not managed well, it will bring with some problems (Kayıkçı & Bozkurt, 2018). In this context, there should be a legal regulation due to the importance of issues such as determining the usage areas of artificial intelligence and restricting certain areas.
- It is determined that AI system developed by Microsoft called as "Tay" learned racism and hate speech from Internet trolls (Groot, 2017). As a result of these behaviors learned from Internet trolls, the system was shut down within 24 hours (CNNTURK, 2016). Since artificial intelligence continues its development by learning from people's behavior, they can make sexist and racist discourses. It is thought that it is valuable to pay attention to such issues in artificial intelligence because these robots meet the participants and provide support service by answering the questions of the participants. Even in short-term activities, artificial intelligence systems need to be constantly controlled because it is deemed that artificial intelligence can learn quickly even in a short time like 24 hours.
- Although ICT applications contribute to event management, it also appears to have some threats. As for the upcoming studies to be carried out on this subject, the detailed investigation of the threats of ICT applications that may bring to the industry may also contribute to minimizing the negative situations that will be encountered in both literature and event tourism management.

Bibliography

Akay, A. Rafet, (2014), "Etkinlik Yönetimi Uygulamalarinda Yaratici Rekabet ve Sosyal Medyanın Entegrasyonu", *The Turkish Online Journal of Design, Art and Communication - TOJDAC* October 2014 Vol. 4, N. 4, pp. 55–70.

Arnold, Andrew, (2019), *How AI is Transforming the Event Planning Space One Conversation at a Time.* https://www.forbes.com/sites/ andrewarnold/2019/02/25/how-ai-is-transforming-the-event-planning-space-one-conversation-at-a-time/#303aa94510b3, (1.5.2020).

Benjamin, Sanjay, (2020), *The Future of Experience with Virtual and Augmented Reality.* https://www.happiestminds.com/whitepapers/the-future-of-experience-with-virtual-and-augmented-reality.pdf, (1.5.2020).

Bethapudi, Anand, (2013), "The Role of ICT in tourism industry", *Journal of Applied Economics and Business*, Vol. 1, N. 4, pp. 67–79.

Bodendorf, Freimut, (2009), Self-Service e-Transactions for Citizens Concept and Case Study. *2009 Third International Conference on Digital Society*, A.B.D., pp. 40–45.

Bozyer, Zafer, (2015), "Augmented Reality in Sports: Today and Tomorrow", *International Journal of Science Culture and Sport*, Vol. 4, pp. 314–325.

British Government Industrial Strategy White Paper, (2017), *Industrial Strategy Building a Britain Fit for the Future.* https://assets.publishing. service.gov.uk/government/uploads/system/uploads/attachment_data/ file/664563/industrial-strategy-white-paper-web-ready-version.pdf, (5.5.2020).

Chaturvedi, Ashutosh, (2009), *Event Management A Professional and Development Approach.* Retrieved from https://books.google.com.tr/books ?id=VXWJIkGxLpcC&pg=PT4&lpg=PT4&dq=Chaturvedi+event+manag ement&source=bl&ots=VmKS4yANaQ&sig=ACfU3U202f4jy7NnuJ_TJW I7LC0BJNLdcQ&hl=tr&sa=X&ved=2ahUKEwi80aeL96vqAhXSeZoKHR 3kAq8Q6AEwAXoECA8QAQ#v=onepage&q=Chaturvedi%20event%20 management&f=false, (6.5.2020).

CNNTURK, (2016), *Microsoft Tay Neden Kapatıldı?* https://www.cnnturk. com/teknoloji/microsoft-tay-neden-kapatildi, (28.04.2020).

Cooper, Kristen, (2018), *AI for Events: The Future Is Here.* https://helloendless. com/ai-for-events/, (3.4.2020).

Coşkun, Cumhur, (2017), "Bir Sergileme Yöntemi Olarak Artırılmış Gerçeklik", *Sanat ve Tasarım Dergisi*, N. 20, pp. 61–75.

Dalgıç, Ali and Birdir, Kemal, (2020), Smart Hotels and Technological Applications. In Evrim Çeltek (Eds.), *Handbook of Research on Smart Technology Applications in the Tourism Industry.* IGI Global, pp. 323–344.

Davidson, Rob, Alford, Philip and Seaton, Tony, (2002), "The Use of Information and Communications Technology by the European Meetings, Incentives, Conferences and Exhibitions (MICE) Sectors", *Journal of Convention & Exhibition Management*, Vol. 4, N. 2, pp. 17–36.

Ekin, Yakın, (2011), *Etkinlik Turizmi Kapsamında Festivaller ve Antalya Altın Portakal Film Festivali'nin Yerel Halk Üzerindeki Sosyal Etkileri Konulu Bir Araştırma* (Unpublished Doctoral Dissertation), Akdeniz Üniversitesi, Antalya.

Ertek, Gurdal, Chi, Xu and Zhang, N. Allan, (2017), A Framework for Mining RFID Data from Schedule-Based Systems. *IEEE Transactions on Systems, Man, and Cybernetics: Systems*, Vol. 47, N. 11, pp. 2967–2984.

EY Türkiye, (2019), Turizm Sektörü Dijitalleşme Yol Haritası Seyahat Acentaları Dijital Dönüşüm Raporu. TÜRSAB & TBV.

Giusto, Daniel, Iera, Antonio, Morabito, Giacomo and Atzori, Luigi, (2010), *The Internet of Things 20th Tyrrhenian Workshop on Digital Communications*, New York, Dordrecht, Heidelberg, London: Springer.

Groot, Tim, (2017), *AI in 2019: How Artificial Intelligence Is Impacting Events*. https://www.eventbrite.com/blog/ai-in-2018-artificial-intelligence-events-ds00/, (22.4.2020).

Guttentag, A. D. (2010), "Virtual Reality: Applications and Implications for Tourism, *Tourism Management*, 31 (5), pp. 637–651.

Gümüş, Fatih, (2019), *Müzelerde Yapay Zeka Uygulamaları, Etkileri ve Geleceği* (Unpublished Master Dissertation), İstanbul Üniversitesi Sosyal Bilimler Enstitüsü, İstanbul.

Hardie, Clemi, (2017), *10 Cool Things We've Done with RFID for Events*. https://www.noodlelive.com/blog/10-cool-things-weve-done-with-rfid-for-events/, (28.2.2020).

Heilman, Jeff, (2017), *How Virtual Reality and Augmented Reality Are Transforming Event Venues*. https://www.bizbash.com/local-venues-destinations/media-gallery/13482500/how-virtual-reality-and-augmented-reality-are-transforming-event-venues, (28.3.2020).

Hjalager, M. Anne, (2015), *Turizmi Değiştiren 100 Yenilik*, (O. Güler, G. Akdağ, A.C. Çakıcı and S. Benli Trans.), *Anatolia: Turizm Araştırmaları Dergisi*, Vol. 26, N. 2, pp. 290–317.

Honey, Stan and Milnes, Ken, (2013), *The Augmented Reality America's Cup*. http://spectrum.ieee.org/consumer-electronics/audiovideo/the-augmented-realityamericas-cup, (5.3.2020).

Hoyle, H, Leonard, (2002), *Event Marketing: How to Successfully Promote Events, Festivals, Conventions & Expositions*. New York: John Wiley & Sons.

Howland, Alex, (2016), *6 Ways Virtual Reality will Provide Competitive Advantage*. https://www.linkedin.com/pulse/6-ways-virtual-reality-provide-competitive-advantage-howland-ph-d-, (28.2.2020).

Hsi, Sherry and Fait, Holly, (2005), "RFID Enhances Visitors' Museum Experience at the Exploratorium", *Communications of the ACM*, Vol. 48, N. 9, pp. 60-65.

Hurwitz, Austin and Jeffs, Alistair, (2009), *Eyeply: Baseball proof of concept - Mobile Augmentation for Entertainment and Shopping Venues.* Mixed and Augmented Reality – Arts, Media and Humanities, 2009. ISMAR-AMH 2009. IEEE International Symposium on. Retrieved February 3, 2020, from http://doi.org/10.1109/ISMAR-AMH.2009.5336723.

International Telecommunication Union, (2019), United Nations Activities on Artificial Intelligence (AI) 2019. International Telecommunication Union, Geneva-Switzerland.

Jenny, Sandra, (2017), *Enhancing Tourism with Augmented and Virtual Reality* (Unpublished Bachelor Dissertation), Hame University of Applied Sciences, Finland.

Kaur, Gurvir, (2015), Role of Information Technology in Tourism Study of Airport Self Service Technology. *Advances in Economics and Business Management (AEBM)*, Vol. 2, N. 13, pp. 1234-1242.

Kayıkçı, M. Y. and Kutluk Bozkurt, A. (2018), "Dijital Çağda Z ve Alpha Kuşağı, Yapay Zeka Uygulamaları ve Turizme Yansımaları", *Sosyal Bilimler Metinleri*, 1, pp. 54-64.

Kuriş, Umur, (2020), *Nesnelerin İnterneti Ekosisteminde Yapay Zeka Tabanlı Saldırı Tespit Sistemi Geliştirilmesi* (Master Thesis). İstanbul Üniversitesi Cerrahpaşa Lisansüstü Eğitim Enstitüsü, İstanbul.

Leiner, M. Barry, Vinton, G. Cerf, Clark, D. David, Kahn, E. Robert, Kleinrock, Leonard, Lynch, C. Daniel, Postel, Jon, Roberts, G. Larry and Wolff, Stephen, (1997), *Brief History of the Internet.* https://www.internetsociety.org/wp-content/uploads/2017/09/ISOC-History-of-the-Internet_1997.pdf, (6.4.2020).

Levine, Josh, Schutz, Stephanie and Epstein, David, (2018), Virtual and Augmented Reality: What Event Organizers Must Know. https://blog.bizzabo.com/virtual-and-augmented-reality-for-event organizers, (4.2.2020).

McCarthy, John, Minsky, L. Marvin, Rochester, Nathaniel and Shannon, E. Claude, (1955), *A Proposal for the Dartmouth Summer Research Project on Artificial Intelligence. August 31.* http://jmc.stanford.edu/articles/dartmouth/dartmouth.pdf, (1.2.2020).

Myers, Ben, (2020), *10 Resources & Platforms to Create a Self-Service Conference or Event.* https://web.pheedloop.com/blog/10-resources-platforms-create-self-service-conference-event, (1.2.2020).

Novak, Milica, (2017), *Robots and Event Planning*. https://www.seebtm.com/en/robots-and-event-planning/, (3.3.2020).

Sevim, F. Muhammed, (2019), *Bilgi ve İletişim Teknolojilerinin Ekonomik Kalkınma Üzerindeki Etkileri: Türkiye Örneği* (Unpublished Master's Thesis), University of Sivas Cumhuriyet, Sivas.

Smart Hotel Technology Guide, (2018), *Using Technology to Navigate the Guest Experience Journey*. Singapore Hotel Association. Singapore Tourism Board. https://www.stb.gov.sg/content/dam/stb/documents/mediareleases/Smart%20Hotel%20Technology%20Guide%202018.pdf, (3.3.2020).

Stein, Madison, (2017), *What is RFID and Why Should Event Creators Care? Resources and Tools, Nov 7*. https://www.eventbrite.com/blog/what-is-rfid-ds00/, (1.2.2020).

Şener, Semih, (2019), *Yapay Zeka, Makine Öğrenimi ve Derin Öğrenme Arasındaki Farklar*. https://www. endustri40.com/yapay-zeka-makine-ogrenimi-ve-derin-ogrenme-arasindaki-farklar/, (22.2.2020).

Tussyadiah, P. Lis, Wang, Dan, Jung, H. Timothy and Dieck, T. M. Claudia, (2018), "Virtual Reality, Presence, and Attitude Change: Empirical Evidence from Tourism", *Tourism Management*, Vol. 66, pp. 140–154.

Thurlby, Shannon, (2018), *Virtual Reality & Augmented Reality: The Future of Events Management* (Unpublished Master Dissertation), Cardiff Metropolitan University, United Kingdom.

TUYED, (2020), *Turizm Yazarları ve Gazeteleri Derneği. Turizmde İnsansı Robotlar Devri. 13 Mart 2016*. Retrieved February 23, 2020, from http://www.tuyed.org.tr/turizmde-insansi-robotlar-devri/.

UNESCO, (2019), *Information and Communication Technologies (ICT) Definition*. http://uis.unesco.org/en/glossary-term/information-and-communication-technologies-ict, (5.4.2020).

Wang, Yafang, Yang, Chenglei, Liu, Shijun, Wang, Rui and Meng, Xiangxu, (2007), *A RFID & Handheld Device-Based Museum Guide System*. 2nd International Conference on Pervasive Computing and Applications, Birmingham, United Kingdom.

Yalçınkaya, P., Atay, L., and Karakaş, E. (2018), "Akıllı Turizm Uygulamaları", *Gastroia: Journal of Gastronomy and Travel Research*, 2 (2), pp. 34–52.

Yamin, Mohammad, Mohammadian, Masoud, Huang, Xu and Sharma, Dharmendra, (2008), *RFID Technology and Crowded Event Management*. CIMCA, IEE Computer Society. pp. 1293–1297.

Yeoman, Ian, Roberton, Martin, Beattie, M. Una, Backer, Elisa and Smith, A. Karen, (2015), *The Future of Events & Festivals*. London: Routledge.

Yüksel, M. Erkan and Zaim, A. Halim, (2009), Otomatik Nesne Tanımlama Teknolojisi Olarak RFID ve RFID'nin Faydaları. 5. Uluslararası *İleri Teknolojiler Sempozyumu (IATS'09)*, Karabük, Türkiye.

Zsarnoczky, Martin, (2017), How does Artificial Intelligence affect the Tourism Industry?. *Journal of Management*, Vol. 31, N. 2, pp. 85-90.

Özlem ŞEN

The Role of ICTs for Effective Marketing Activities in Event Tourism

1 Introduction

Tourism, which is the international industry and one of the largest employment industries, has more heterogeneous stakeholders than many other industries. The biggest reason for the energetic growth and development of the tourism industry is the development of ICTs. The recent synergistic interaction between technology and tourism has brought radical changes in the tourism industry. Communication that is universal and easy to access has brought the entire tourism industry to a new level of interaction. ICT is dramatically playing a significant role in the competitiveness of tourism activities, destinations and for other industries (UNWTO, 2001).

Tourism is a knowledge-intensive industry. Travelers take a lot of risks when they invest money and time on a travel experience. The higher the perceived risk level, the more travelers tend to try to find information. Marketing activities in the tourism industry are seen as an opportunity to offer a mix of value for money and decrease risk perceptions.

Increased interest in online technologies means that consumers' willingness to have reach to product selection and their views on these products has improved. Reliability is a substantial element of tourism marketing, and consumers avoid marketing messages developed by the brand and instead request information from reliable people.

Since consumers collect data from multi-channel platforms, tourism marketers must follow a logical communication strategy. Marketing communication synchronization between platforms provides efficiency and decreases external costs.

Tourism companies and destinations should invest in the right resources to improve the correct tourism marketing strategy. In order for the action plan created to be successful, the targets, responsibilities, and criteria must be clearly stated. Besides, this plan can make the organization realistic in terms of managing effective resources correctly. ICT is one of the most suitable environments to enable information or create information about different components. Thus, businesses can access a large number of significant and up-to-date information rapidly, easily and inexpensively (Requena et al., 2007).

The Internet is easy to access for travelers, representatives of the tourism industry, as well as researchers worldwide. Besides, the Internet has become increasingly complicated and dynamic. Therefore, the use of digital technologies that increase effective communication and articulation among different tourism actors in target networks and markets are considered worthy for business success (Law et al., 2014; Dexeus, 2019).

Events are convenient environments for the use of different senses and the knowledge and feelings of different people. Event marketing is a big part of a marketing communication organization that consists of a wide variety of dimensions. Events can also be a part of communication as their presence in their channels, media, and editions. Event marketing ensures that a certain message is delivered efficiently and memorably. As a versatile marketing and communication component, event marketing can be integrated efficiently with a variety of other actions and projects.

The event industry, which includes conferences, festivals, exhibitions, meetings and sports activities with significant contributions to tourism, is growing and developing rapidly. Therefore, event managers and marketers must identify consumer demands and needs. Through developing technology, effective marketing and management of events are possible with the effective use of ICTs.

Events are taken into consideration in the development and planning of destinations as well as being an important source of motivation for tourism. The roles and effects of events in tourism are also determined in destination planning and action plans are prepared accordingly. Events are increasingly major for destinations to survive in a competitive environment. In addition to the increase in the infrastructure quality of touristic events thanks to tourism activities, it has a positive impact on the local and the regional economy with the income from the sale of goods and services (Hernández-Mogollón et al., 2014).

To create a successful event and interact with consumers, event marketing needs to harmonize between the target group and the brand (Wohlfeil & Whelan, 2006). Developments in ICT make it easier for event marketers to advance and apply segmentation, targeting and positioning strategies. How and how much the event managers will use, rather than whether they use information technology (IT), has become a more significant issue.

Most events are not for the whole society, so it is significant to identify consumer segments that most closely match the event in question. Successful market segmentation and targeting at the events can increase the satisfaction of the participants and increase the motivation and success of the event managers. Businesses apply various strategies to achieve a competitive advantage. In a modern marketing approach, businesses want to gain a competitive advantage through consumer satisfaction. In addition to increasing consumer satisfaction, ICT supports businesses to measure this.

The characteristics of consumers such as demand, needs and expectations, the resources they have, and their purchasing behaviors differ from each other. Mass marketing approaches are not enough to meet these different demands, needs, expectations, resources, and behaviors. This variety of customer demands and expectations can be turned into an opportunity through market segmentation activities that serve to group customers in terms of similar needs, characteristics, and purchasing behavior. It may be more profitable for firms to perform their marketing events with a differentiated marketing approach rather than determining the whole market as a target market with a good proposal. Through differentiated marketing, it is possible to carry out a more effective event marketing activity. ICT strengthens its activities to define, customize and purchase goods and services.

Businesses want to improve their offers and to be preferable to increase their satisfaction by determining the needs of their customers and thereby gain a competitive advantage. In this context, it is necessary to exactly determine consumer needs. The process of determining consumer needs is a very costly, time-consuming and carefully managed process. When businesses determine their target group, they can also determine their marketing strategies to meet the expectations of the target group and increase consumer satisfaction. At this phase, the analysis and commentary of the data collected are used more through ICT.

At present, as a natural result of online marketing being more interactive than traditional marketing, ICT is increasingly interactive in market segmentation, targeting and positioning. The use of ICT provides great opportunities for marketers in creating detailed and complicated market models, sizing and estimating goods and services, and preparing offers according to needs.

2 Background

In recent years, we have observed large changes in economy in general and marketing in particular as at present, information exchange is spreading rapidly because of the increased use of smartphones and tablets. Also, tourists can readily reach information on any topic related to tourism. Therefore, tourists choose destinations by accessing better information infrastructures, Internet services, and mobile with high quality (Li et al., 2016).

ICT has enabled the development of the service industry with the enhanced security, convenience, accuracy, flexibility, diversity and reliability of the internal processes of the enterprise. Considering the important effects of ICTs on tourism businesses, some studies in the literature investigating the management experience/design of ICT, other studies have examined its effects on tourism marketing. Maskell et al. (2005) stated that the definition and selection of new technologies are complicated and heterogeneous, as well as difficult and costly processes. There is a lot of research that highlights the consumer perspective. Gretzel and Jamal (2009) predicted in their study the emergence of a new "Creative Class" of tourists using technology to mediate, create, and restructure their experiences. For instance, McCarthy and Wright (2004) have suggested that technology serves as an essential component for tourist experiences. Neuhofer et al. (2014) argue that tourism firms using technology can create a "technology-enhanced experience" for tourists. Yoo and Gretzel (2010) conducted research on the change of marketing functions due to the increasing use of Web 2.0. Tussyadiah (2015) made evaluations in the design of tourism experience of increasing technology mediation. Besides, it is offered by Benckendorff et al. (2014) that an overview of the effects of ICT applications in all sectors of the tourism industry. These studies give an idea about both the role of ICT in tourism and also especially its effects on tourist experience. There is a lack of literature investigating the importance and use of ICT in tourism marketing (Yoo & Gretzel, 2016).

MacKay and Vogt (2012) concluded that the rate of using the Internet in travel planning and decision-making processes is increasing because of the use of wireless instruments more in the study of ICT spreading from daily use to travel activities. Ho et al. (2012) analyzed the behavior patterns of users searching for tourism information. The study revealed findings that would provide managers with a source of information to make search engine marketing activities more functional. They found out how kind of behaviors tourists do while they seek, process and use travel knowledge offline and online. Ghaderi et al. (2019) concluded that the attitudes of tourists against smartphones positively affect their intention to join in regional tourism experiences. This finding supports the positive effect of ICT tools on tourism experience.

3 ICTs and Marketing

The emergence of Web 2.0 has created significant changes in consumers' behavior, such as creating, acquiring, and exchanging information (Cantallops & Salvi,

2014). The modern information society has made tourism an information-intensive industry as ICT has a potential impact on tourism. The role of ICT in the tourism industry cannot be ignored, and ICT is a major driving force in the current knowledge-driven society. Technological advances have created a new business environment by providing new components and distribution channels. ICT tools have facilitated commercial transactions in all sectors by communicating with commercial partners, distributing products and services and informing consumers around the world. On the other hand, tourism consumers also use online communication to get information and plan their travels and holidays. Information is an important component of the tourism industry.

ICT covers pretty much all areas of the tourism industry. For example, while determining and developing the tourism area and destinations itself, ICT tools are adhered to. The use of geospatial IT for various purposes worldwide has been widely known in recent years, and the tourism industry is no exception to take advantage of these advantages of ICT. ICT can be used by tourist professionals to identify the borders, surroundings, and communities living in the proposed tourist destinations. Also, the roads connected to the sites and water, power, market, etc. can get information about the availability of other services. These technologies are also beneficial monitoring and managing the website. The role of ICT tools in the tourism industry for marketing, operation, and management is known. Marketing activities are more effective thanks to ICT components (Shanker, 2008).

Airline companies and travel agencies/tour operators, who use all the components of ICT as important innovation, have been studied by researchers in the past decade (Buhalis, 2004; Bowden, 2007). Moreover, ICT is creating a growing innovation that affects wider areas in tourism and there is not much research on this subject yet. Online auctions, which allegedly transform business models, instead of simply transfer commerce to a dissimilar environment (Ho, 2008). It is also the use of iPhone, Android phone, GPS and other solutions for guidance services and comments (Liburd, 2005). ICT quickly enters these areas and museums and other attractions quickly increase comprehension, attractiveness, and accessibility for tourists and provide higher organizational efficiency (Go et al., 2003; McLoughlin et al., 2007; Nielsen & Liburd, 2008).

Researchers examining the effects, importance and development of ICTs in tourism marketing have mainly examined the effectiveness of online marketing activities in various scopes, including social media, search engines and websites (Law et al., 2014). For instance, in their study, Wang et al. (2012a) investigated the role of mobile technologies in tourism marketing while examining social technologies.

In particular, there is a lack of information about how marketers can convincingly present tourists' tourism experiences acquired using these technology tools. In their study, Pine and Gilmore (2011) examined the issue that the competitive battlefield for firms will have challenging and memorable experiences.

Creating a meaningful and memorable experience in tourism marketing has been a core value for marketers. As a result of this, various ICTs have been used to support and develop tourism experiences (Neuhofer et al., 2014). With the advancement of social technologies and, in parallel, mobile technologies and the increasing integration into the tourism industry Understanding how to Use ICT effectively is always more important to marketing tourism experiences. (Yoo & Gretzel, 2016).

Marketing focuses on establishing, developing and maintaining continuous relationships between the buyer and seller, where both parties benefit (AMA, 2004). Given the wide and heterogeneous ICT solutions in marketing channels, various technological solutions can have a dramatically dissimilar effect on customer value in the situation of retail-consumer relations (Saura & Molina, 2009).

Looking at technological innovations, it is clear that ICT caused changes in consumer behavior in accommodation and tourism in recent years. Consumer behavior has changed and facts have emerged in the market. Adoption of technology, especially the use of mobile platforms such as tablets, smartphones and other handheld devices is very major. Also, the use of ICT is serious to establish a dynamic dialogue between suppliers and consumers (Law et al., 2014).

ICT is recognized as a set of technology tools, such as industrial components such as business applications and strategies, and Internet and user-developed content (Web 2.0), especially smartphone applications. ICT has been identified as a crucial tool for further innovation in all industries and also in the tourism industry. For instance, the application of Web 2.0 tools, digital-based guides, geolocation technologies (by smartphone or GPS) or the mobile Internet is becoming more and more significant (Slokum & Lee, 2014).

Mobile and wireless technologies are among the tools that are rapidly adopted to better serve the tourism industry's customers (Buhalis & Law, 2008; Dimanche & Jolly, 2009). Most of the tourism firms have adopted various ICTs, mostly with web-based services (O'Neil, 2002; Presbury & Edwards, 2005).

The use of ICT has transformed all the functions of businesses and the marketing function has been affected by this transformation and change. In the literature, it is accepted that the use of ICT in the management of marketing processes provides a real competitive advantage for businesses (Roberts, 2000; Prasad et al., 2001; Bond & Houston, 2003; Tatikonda & Stock, 2003).

Literature has shown that all ICT components in this area can be divided into two basic types, roughly visitor-centered applications, and administrative applications, according to the main purpose (Lee et al., 2016). Visitor-centered applications often aim to serve a variety of pre-festival and festival experiences, which enable visitors to develop an experience of interacting, identifying, processing and creating ideas together. For instance, the tickets of Tomorrowland (Boom, Belgium) are in the form of a radio frequency identification (RFID) wristband that uses personal ticket information and facilitates a "cashless" environment. Besides, the users can connect on social media sites by using the wristbands, and crowded LEDs can be triggered remotely by the festival and produce light shows of the crowd (Rubinstein, 2015).

Management practices are generally related to the organization of visitors at the event organization, preparation stage, on-site applications and analysis stage. Also, the RFID wristband in the example of Tomorrowland provides real-time process reporting and mass information for personalized communication after the festival. Another example is the innovative use of ultra-fast wireless access with networked storage devices at the Vieilles Charrues Festival in Carhaix, France, to support organizers, press representatives, and volunteers. Equally a massive event, this three-day festival also faces a wealth of data to store and share using a stable and easy-to-use temporary infrastructure (Stankov et al., 2018).

In this chapter; in the events, the things that can be done to carry out more effective marketing events by using ICT have been gathered together.

4 Using ICTs in Event Tourism Marketing

Hoyle (2002) explains that there are three "E" in event marketing: entertainment, excitement, and enterprise. "Entertainment, excitement, and enterprise" are what should be included to create a successful marketing event. Whether a congress or an independent festival is marketed, all three of these items are critical to the ongoing success of any event. ICT facilitates their work by supporting event managers in this sense and also supports the experiences of the participants.

The online travel market is growing up dramatically and the websites of destinations are playing a significant role in information for consumers and suppliers as an effective marketing tool. Kasavana et al. (1997) claims that the Internet, a network of global networks, has changed the way the hotel industry plan, control, manage and integrate most of its commercial events, including

Tab. 16-1: The Main Effects of ICT on Tourism Events

Managing information and Flows	Communication and transaction costs reduce by using ICT tools
	ICT tools make changes traditional tourism distribution channels
Shaping consumer behavior	Changing consumer perceptions
	To influence consumer decision-making processes
New product development	Advanced price discrimination
	Networking among industry staff
	Product development flexibility Increases by using ICT tools
Small and medium tourism enterprises (SMTE)	Provides a balanced playground for SMEs
	Using ICT needs appropriate levels of skills, resources, and infrastructure
Disintermediation	ICT causes changing the role of intermediaries
	Using ICT tools can create job losses and restructuring
Job security, education, and skills	Advanced service quality and training opportunities
	Workers become active participants in decision-making
Planning and participation	Creates new forms of community participation

marketing events. In parallel with this change, tourism companies, like all other sectors, have rapidly started to use the tools of the Internet and ICTs.

Developments in ICT undoubtedly changed both business practices, strategies and industry structures (Porter, 2001). ICTs have many effects on events. Tab. 16-1 shows the effects of ICTs on events Buhalis and Zoge (2007).

Many ICT tools are used to organize and manage events effectively. The main ICT tools and initiatives used in the events can be summarized as follows (Gün & Sayar, 2015):

4.1 Online Branding

Online branding is the starting point of event tourism marketing. Given how intense time consumers spend online today, branding is essential in this environment.

Brand building and building relationships are the two most significant content for the virtual communities of tourism organizations. Brand communities;

it is characterized by a structured set of social relationships with members or fans of the brand, a collective consciousness, common rituals and traditions, and expert communities with a moral sensation of liability for the group and not geographically connected. A brand's online communities have become areas where organizations and consumers share information and sensation about brands. Thus, these online brand communities have the potential to increase consumer loyalty and increase brand awareness beyond the limited geographical location (Hede & Kellet, 2012). Therefore, it can be thought that branding online will increase the event consumers' loyalty to the event and the brand.

Because it is an extension and activation of a strong brand that can be activated online and offline, digital tourism marketing is an effective form of marketing (European Travel Commission and World Tourism Organization, 2014). In this case, online branding is inevitable for event managers.

When planning an event, the event brand; it should have a basic description that directs all other marketing messages, immaterial, unique attributes or points of sale that distinguishes the brand from its competitors, and an identity consisting of visual elements, including the brand's logo, colors, name and design.

4.2 Content Development and Management

Visual content is becoming dramatically importance for online users. According to the Cognitive Brain Imaging Center, the human brain absorbed the visual messaging faster and deeper through the cables. A press material's total number of online views is directly related to the number of multimedia items it contains. It should be noted, however, that visual content must be part of a major content strategy. Content is the major driving force in tourism marketing. Creating, collecting and distributing creative, new and interesting texts, images and videos is at the heart of accomplished ICT-used tourism marketing.

Traditionally, tools such as user tracking, website hits and queries are used to evaluate websites, but doubts have emerged regarding the effectiveness of these measures in evaluating information content, which is considered a significant component of website quality (Perdue, 2001). While the event consumers are turning to the Internet for information search, it is major to provide comprehensive information thanks to the website content. The website experience is more satisfactory for event consumers than others when they participate in guided information retrieval (Filo et al., 2009).

It is aimed to attract the participants and convince them about the participation through online sharing of the events, which can be achieved by presenting original content. There is no chance of success that repeats and makes no difference. Especially, with Web 2.0, Internet users play the roles of co-marketers, co-designers, and co-creators of tourism information by producing substantial content (Lee & Wicks, 2010).

4.3 Website Management

Some research in the literature suggests that visiting a website is a self-selection behavior. Besides, only consumers with interest or appeal to a particular good or service will spend time to visit this website (McQuitty & Peterson, 2000). Therefore, it is essential to emphasize the necessity of creating an event website for event managers and marketers.

Their website is defined by researchers as an important marketing tool for tourism businesses (Gretzel et al., 2000). Therefore, website design has become an important part of tourism marketing. Furthermore, Website design is very major for tourism marketing. In their study, Luna-Nevarez and Hyman (2012) found that user-friendly and especially the website design components are the most important factors affecting the credibility of users' target websites. Lee et al. (2010) commit that sensory comments on target Websites are a path to make marketers' Websites more convincing. Lee and Gretzel (2012), who investigated the effect of the Website in the destination selections of travelers, concluded that the images on the destination Website create mental images that significantly affect the attitudes and expectations of the website visitors.

Event marketers try to reach the target audience by creating a website and social media accounts before the events. In this context, it should be noted that personalized communications are also effective on the website where landing pages are widely used to make the page more relevant to what the customer is looking for. Popular websites are used to create in-depth text content, rich media, etc. to build strong links with visitors. It contains video and audio as well as participating in customer communities (Chaffey & Chadwick, 2016).

4.4 Social Media

Today, consumers can easily obtain information about goods or services are both from those they know and from large groups of people who have relevant experience without geographical and temporal limitations (Ladhari & Michaud,

2015). Therefore, the number of platforms offering online consumer reviews or goods and services ratings has increased considerably (Duan et al., 2008).

Businesses can send free and zero-latency messages on social media platforms instead of paying for advertising in traditional media. Second, social media can provide a two-way transmission platform that allows participation and communication and between message creators and recipients. This interactive feature of social networks differs from previous Internet approaches (e.g., websites), which narrowly focus on the provision of information in many models that hardly allow feedback, sharing and shopping (Li & Duan, 2018).

Social media has the power to completely change marketing practices. The larger effects of this significant technology on tourism, destinations, and innovations proceed to be explored (Werthner & Klein, 2006; Schegg et al., 2008).

Social media marketing is defined as the use of social media for institutions and organizations to encourage shopping with consumers (Tuten & Solomon, 2015). Social media is an online communication tool between interactive human and organizational networks and is a major technique for research, promotion, customer service, and relationship building organizations (Tuten & Solomon, 2015). It is also an inexpensive strategy that provides a variety of ways for organizations to reach customers and interact and engage their consumers at different stages in the purchasing process (Tuten & Solomon, 2015). Cant (2016) describes social media marketing as companies using existing network tools to expand their customer base and increase brand exposure. Tiwari et al. (2018), users on social media often prefer content that is considered more attractive that can be obtained from both textual and visual content or a combination of both. Zhu and Chen (2015) state that marketers should give target consumers an idea about products in social media marketing. Also, Cant (2016) states that social media is a vital phenomenon in the lives of people access to many people. So, it recommends that organizations continue to focus on improving communication with consumers on social media platforms such as Facebook or Twitter. Furthermore, social media marketing should be the focus of a brand's promotional strategies (Cant, 2016).

Event marketing in social media has been used recently to promote its brands and products among businesses and is considered an increasing trend (Li & Duan, 2018). It is becoming more efficient and easier for businesses and individuals to share information while publishing events on social media (Li & Duan, 2018). Tiwari et al. (2018) can be used to provide updates from content events on social media. Events on social media should post content variety and avoid similar messages that may disturb viewers (Li & Duan, 2018). It is considered that promoting the events by providing frequent updates to the viewers will increase the use of social media, and may also raise the brand value (Walsh, Clavio, Lovell & Blaszka, 2013). However, Zhao et al. (2015) state that the studies on event marketing in social media are not sufficient.

E-WOM is more manageable than WOM, because messages about destinations or events or facilities are posted online and readily accessible for everyone. However, the proliferation of the social media platform and the transparentation of opinions make eWOM management very complicated. Consequently, various studies show that hospitality and tourism practitioners can use social media to monitor and correct their target and company images in a timely and cost-effective way (Law et al., 2014).

There are case studies that suggest that the realization of a sports event does not provide significant benefits before and after the actual event without effective marketing strategies. For example, Banyai and Potwarka (2011) evaluated the image aim of an Olympic host city using social media, and it was concluded that the Olympic image did not have a significant effect and the rate of return to the target was low. Lee et al. (2012) analyzed the use of social media in the promotion of events. They stated that emotional comments on Facebook had a great impact on the effectiveness of this type of marketing in their study.

4.5 E-Mail Marketing

E-mail is the oldest form of communication. The change in the use of this medium from the 1980s to the present is that e-mail is no longer a medium of communication (Gajek, 2014).

In e-mail marketing, it is one of the most effective communication channels as messages are sent to customers who allow an existing customer base. Many of the digital communication techniques are similar to their traditional equivalents; for example, display ads are broadly similar to print or display ads, while e-mail marketing is equivalent to direct mail. However, the approaches used to target the online audience are potentially very different from each other, providing customization based on customer profile and previous interactions, many options for conveying more relevant messages (Chaffey & Chadwick, 2016).

4.6 Wearable Devices

The Internet offers a platform that enables new products to be developed faster, as it is possible to test new ideas and concepts and explore different product options through online market research. Businesses can use their consumer panels to test their ideas faster and often at lower costs than traditional market research. Although Google has many successes and failures in this regard, it is a very innovative company. For example, Google ad-words (success) and Google Buzz and Wave (failure), Google Glass (advanced wearable technology and augmented reality glasses (Gibbs, 2015). Internet network enables businesses to partner more easily to launch new products (Chaffey and Chadwick, 2016).

4.7 Augmented Reality

Augmented Reality (AR) is a technology that mediatizes digital information with live video screens in the real-time environment of the user. Technology as Augmented Reality is implemented in tourism centers, such as theme parks, museums, around the world, to improve visitor experiences.

AR uniquely blends computer-simulated images in a real environment is an advanced Virtual Reality (VR) stage that (Yovcheva et al., 2012). VR, which is often linked to 3D and visuals (Dadwal & Hassan, 2015), can be more broadly defined as a direct or indirect, vivid image of a physical, real-world environment. Components of these views can be enhanced by computer-generated sensory inputs such as video, audio, GPS data or graphics. VR is based on a widely known concept of "Mediator Reality," where a real appearance is altered, enhanced, or possibly absent by a computer. Thus, the technology improves the user's perception of reality while consuming the product. The adjacent real world of the user becomes digitally interactive by Supported by computer vision and sophisticated VR technology with object identification. VR can simulate information about the environment and objects integrated into the current world (Hassan & Ramkissoon, 2016).

VR is constantly gaining considerable interest in various areas of the tourism industry, especially in museums. It can provide a memorable experience as they are associated with the sharing of history, knowledge or learning experience (Morabito, 2014).

Tourists increasingly prefer to travel independently and demand updated information. The tourism industry has quickly adopted Augmented Reality technology. There are many downloadable tourism applications such as printed tourism guides, places of interest, attractions, transportation links, and directions. They use a handheld device to scan the Quick Response (QR) code of tourists. Installing QR codes in places of interest that do not allow users to view site-specific content can be given as an example of tourism-based marker Augmented Reality.

Participants can increase their participant experience with VR as well as create entirely new attractive virtual tourist events both indoor and outdoor, regardless of the charm of the event. Besides, the use of VR in events can provide businesses with a competitive advantage.

4.8 Internet of Things (IoT)

The Internet of Things (IoT) has become an increasingly significant matter with the increasing number of devices connected to the Internet. IoT includes various devices such as sensors, smartphones, and resources such as mass sensors. The availability of data produced by these various devices has created new opportunities for innovative applications in different areas such as supply chain management systems, smart transportation systems and smart cities (Akbar et al., 2015). Tourism is one of the sectors expected to increase the development of IoT applications to a great extent (Car et al., 2019). IoT applications help travelers connect with various devices and increase their tourism experience.

In events, IoT can be of great benefit in enhancing the customer experience. For example; In large events, if there are activities in more than one area, reminders can be sent to the participants about the program using these sensors. Again, through the sensors, the staff can be used to alert and automatically direct them to the right area when the guest arrives.

Moreover, while enabling customization, IoT can offer businesses financial benefits either through automated or intelligent energy savings. For instance, in one event, Internet-enabled devices and sensors can allow the temperature to be adjusted continuously, i.e., heating or cooling is only used when it is really necessary, which can save energy. A similar situation applies to light. The sensors automatically detect the natural light levels in the event area, thus reducing the power of the lighting devices and preventing the waste of energy.

5 Solutions and Recommendations

All tourism firms and destinations must invest in time and the right resources to improve a detailed and exhaustive tourism marketing strategy. To create an action plan that expresses targets, competency, responsibilities, and metrics for achievement can provide that the organization is realist in the state of managing the available resources. A detailed and exhaustive action plan also enables these resources to be spent cost-effectively and creates a platform to adapt marketing activities to what is successful. This is especially significant for digital marketing while considering the trends changing and the speed at which new platforms become available.

Tourism is a knowledge-intensive industry. Tourism consumers take a lot of risks when they invest time and money on a travel experience. When the perceived risk level increases, travelers tend to seek more information. Tourism marketing is seen as an opportunity for tourism firms to minimize consumer's risk perceptions and offer a mix of value for money.

Increased interest in online technologies means that consumers' willingness to have access to product selection and their views on these products has improved. Reliability is a substantial element of tourism marketing, and consumers avoid marketing messages developed by the brand and instead request information from reliable people.

Tourism consumers collect data from different platforms. For that reason, tourism marketers must follow a consistent communication strategy. Messaging synchronization across platforms provides efficiency and minimizes external costs.

Managers should implement a more effective management method by including all the advantages of technology and ICT tools in their routine workflows in order not to lose their competitive advantage and increase customer satisfaction.

All managers in the tourism and hospitality industry must ensure the sustainability of information and communication technology tools. Managers of the department should follow the developments regarding the use of ICT tools and employees should organize continuous training and measurement events to compensate for the lack of information on the subject. All managers and employees are required to make professional suggestions about the capabilities of ICT. As they continue to develop their knowledge, they should inform ICT suppliers and other stakeholders about the nature of the events.

Conclusion

At present, as a result of online marketing is more interactive than traditional marketing, ICT is becoming more interactive in event tourism marketing. The use of ICT ensures great opportunities for marketers to create detailed and complex market models, to size and estimate goods and services, and to prepare proposals according to needs.

It supports IT businesses to be flexible and market-oriented. Commercial success depends on responding to rapidly changing customer needs using

ICT to deliver products that meet the targeted segments. It can be provided by tourism companies to develop their flexibility, interaction, efficiency, and competitiveness by using Internet and ICT tools.

ICT provides rapid growth of event marketing with more security, convenience, precision, flexibility, diversity, and reliability in processes, thereby increasing total efficiency.

It is unavoidable to use ICT tools in tourism businesses as in all other businesses. To take full advantage of the Internet; a company should consider the informative content of the good / service, the entire value proposition, its communication and the interactive relationships that can be developed with its customers. It is also significant to note that "conventional" (offline) and electronic (online) events and applications should be planned and implemented completely (Qirici et al., 2011).

Developments in ICT make it easier for event tourism marketers to develop and implement all the rebuke strategies. At the moment, it is more significant how and how much event managers will use, rather than whether they use information technology.

Considering that the one-way communication form is becoming increasingly ineffective in all sectors, companies should see the opportunity to communicate with their customers as an opportunity thanks to ICT tools. Event managers and marketers should strive to improve the quality of interaction and create appropriate environments that can benefit mutually.

Key Terms and Definition

Internet of Things (IoT): Although it does not have a common definition yet, the Internet of Things, it was defined as the inclusion of physical objects in business processes by connecting to the Internet by Haller et al. (2008).

Information and Communication Technology (ICT): Information and communications technology is a term that includes all technologies for the communication of information telecommunications, recording information, broadcast media, intelligent building management systems, audiovisual processing and transmission systems, and network-based monitoring and control functions.

Wearable Devices: "Wearable technology" and "wearable devices" are used to describe electronic connections and computers integrated into clothes and other accessories. Examples of these devices are glasses (Google Glass etc.), watches (IWatch etc.) and similar accessories. These technologies are especially used in the fields of medicine and health, but they are used in the fields of fashion, entertainment and fitness (Wright & Keith, 2014).

Bibliography

Akbar, A., Carrez, F., Moessner, K., & Zoha, A. (2015), *Predicting Complex Events for Pro-active IoT Applications*. In 2015 IEEE 2nd World Forum on Internet of Things (WF-IoT) (pp. 327-332). IEEE.

AMA (2004), "AMA adopts new definition of marketing", *AMA Newsletter*, 1 (5).

Banyai, M. and Potwarka, L. R. (2011), "Assessing Destination Images of an Olympic Host City Using Social Media", *European Journal of Tourism Research*, 5 (1), pp. 6-18.

Benckendorff, P. J., Sheldon, P. J., and Fesenmaier, D. R. (2014), *Social Media and Tourism*. Tourism information technology, (Ed. 2), 120-147.

Bond III, E. U. and Houston, M. B. (2003), "Barriers to Matching New Technologies and Market Opportunities in Established Firms", *Journal of Product Innovation Management*, 20 (2), pp. 120-135.

Bowden, J. (2007), "The Rise of the ICT-dependent Home-based Travel Agents: Mass Tourism to Mass Travel Entrepreneurship", *Information Technology & Tourism*, 9 (2), pp. 79-97.

Buhalis, D. (2004), "eAirlines: Strategic and Tactical Use of ICTs in the Airline Industry", *Information & Management*, 41 (7), pp. 805-825.

Buhalis, D., and Zoge, M. (2007), The Strategic Impact of the Internet on the Tourism Industry. In Proceedings book of *Information and Communication Technologies in Tourism* (pp. 481-492), Springer, Vienna.

Buhalis, D. and Law, R. (2008), "Progress in Information Technology and Tourism Management: 20 Years on and 10 Years after the Internet—The State of eTourism Research", *Tourism Management*, 29 (4), pp. 609-623.

Cant, M. C. (2016), "Using Social Media to Market a Promotional Event to SMEs: Opportunity or Wasted Effort? *Problems and Perspectives in Management*, 14 (4), pp. 76-82.

Cantallops, A. S. and Salvi, F. (2014), "New Consumer Behavior: A Review of Research on eWOM and Hotels, *International Journal of Hospitality Management*, 36, pp. 41-51.

Car, T., Stifanich, L. P., and Šimunić, M. (2019), Internet of Things (Iot) in Tourism and Hospitality: Opportunities and Challenges. In Prooceddings book of *Tourism in South East Europe: Creating Innovative Tourism Experiences: The Way to Extend the Tourist Season* (pp. 163-175), Opatija, Croatia.

Chaffey, D. and Ellis-Chadwick, F. (2016), *Digital Marketing: Strategy, Implementation and Practice* (6th Ed.), Pearson, United Kingdom.

Dadwal, S. and Hassan, A. (2015), The Augmented Reality Marketing: A Merger of Marketing and Technology in Tourism. In N. Ray (Ed.), *Emerging Innovative Marketing Strategies in the Tourism Industry* (pp. 78-96),IGI Global, Hershey, PA.

Dexeus, C. R. (2019), The Deepening Effects of the Digital Revolution. In *The Future of Tourism* (pp. 43-69). Springer, Cham.

Dimanche, F. and Jolly, D. (2009), "Tourism, Mobility, and Technology: Perspective and Challenges", Selected Papers from the Annual Conference of the Travel and Tourism Research Association (TTRA) Europe, Nice, France, April 2007. *Tourism Analysis*, 14 (4), pp. 421-513.

Duan, W., Gu, B., and Whinston, A. B. (2008), "Do Online Reviews Matter?—An Empirical Investigation of Panel Data", *Decision Support Systems*, 45 (4), pp. 1007-1016.

Edwards, D. and Presbury, R. (2005), "BEST Sustainable Tourism Think Tank III "The Role of Tourism in Community Development, Culture and Environmental Stewardship" Costa Rica, July 8-11, 2003", *Journal of Teaching in Travel & Tourism*, 4 (3), pp. 95-100.

Filo, K., Funk, D. C., and Hornby, G. (2009), "The Role of Web Site Content on Motive and Attitude Change for Sport Events", *Journal of Sport Management*, 23 (1), pp. 21-40.

Gajek, K. (2014), "New Internet Communication Technologies in Polish Sports Organizations", *Informatyka Ekonomiczna*, (34), pp. 37-46.

Ghaderi, Z., Hatamifar, P., and Ghahramani, L. (2019), "How Smartphones Enhance Local Tourism Experiences?", *Asia Pacific Journal of Tourism Research*, 24 (8), pp. 778-788.

Gibbs, W. J. (2015), Interface Technology Trends Implications for News and Information Services. In *Encyclopedia of Information Science and Technology*, (3rd Ed.; pp. 2069-2078), IGI Global, USA.

Gil-Saura, I. and Ruiz-Molina, M. E. (2009), "Customer Segmentation Based on Commitment and ICT Use", *Industrial Management & Data Systems*, 109 (2), pp. 206-223.

Go, F. M., Lee, R. M., and Rosso, A. P. (2003), "E-heritage in the Globalizing Society: Enabling Cross-cultural Engagement through ICT", *Information Technology & Tourism*, 6 (1), pp. 55-68.

Gretzel, U. and Jamal, T. (2009), "Conceptualizing the Creative Tourist Class: Technology, Mobility, and Tourism Experiences", *Tourism Analysis*, 14 (4), pp. 471-481.

Gretzel, U., Yuan, Y. L. and Fesenmaier, D. R. (2000), "Preparing for the New Economy: Advertising Strategies and Change in Destination Marketing Organizations", *Journal of travel Research*, 39 (2), pp. 146-156.

Gün, T. and Sayarı, B. (2015), Placemaking Betwixt and between Festivals and Daily Life. In *Information and Communication Technologies in Tourism* (pp. 374–384), Taylor and Francis, New York.

Haller, S., Karnouskos, S., and Schroth, C. (2008), The Internet of Things in an Enterprise Context. In *Future Internet Symposium* (pp. 14–28), Springer, Berlin, Heidelberg.

Hassan, A. and Ramkissoon, H. (2016), Augmented reality application to museum visitor experiences. *Visitor Management in Tourist Destinations*, 3, p. 117.

Hede, A. M. and Kellett, P. (2012), "Building Online Brand Communities: Exploring the Benefits, Challenges and Risks in the Australian Event Sector", *Journal of Vacation Marketing*, 18 (3), pp. 239–250.

Hernández-Mogollón, J. M., Folgado-Fernández, J. A., and Duarte, P. A. O. (2014), "Event Tourism Analysis and State of the Art", *European Journal of Tourism, Hospitality and Recreation*, 5 (2), pp. 83–102.

Ho, L. (2008), "What Affects Organizational Performance?", *Industrial Management & Data Systems*, 108 (9), pp. 1234–1254, doi:10.1108/02635570810914919.

Ho, C. I., Lin, M. H. and Chen, H. M. (2012), "Web Users' Behavioural Patterns of Tourism Information Search: From Online to Offline", *Tourism Management*, 33 (6), pp. 1468–1482.

Hoyle, L. H. (2002), *Event Marketing*, Wiley, Canada.

Irvin, R. (2007), "Information and Communication Technology (ICT) Literacy: Integration and Assessment in Higher Education", *Journal of Systemics, Cybernetics and Informatics*, 5 (4), pp. 50–55.

Kasavana, M. L., Knutson, B., and Polonowski, S. (1997), "Netlurking: The Future of Hospitality Internet Marketing", *Journal of Hospitality & Leisure Marketing*, 5 (1), pp. 31–44.

Ladhari, R. and Michaud, M. (2015), "eWOM Effects on Hotel Booking Intentions, Attitudes, Trust, and Website Perceptions", *International Journal of Hospitality Management*, 46, pp. 36–45.

Law, R., Buhalis, D., and Cobanoglu, C. (2014), "Progress on Information and Communication Technologies in Hospitality and Tourism", *International Journal of Contemporary Hospitality Management*, 26 (5), pp. 727–750.

Lee, B. C. and Wicks, B. (2010), "Tourism Technology Training for Destination Marketing Organisations (Dmos): Need-Based Content Development", *Journal of Hospitality, Leisure, Sports and Tourism Education* (Pre-2012), 9 (1), pp. 39–52.

Lee, W. and Gretzel, U. (2012), "Designing Persuasive Destination Websites: A Mental Imagery Processing Perspective", *Tourism Management*, 33 (5), pp. 1270–1280.

Lee, S. S., Boshnakova, D., and Goldblatt, J. (2016), *The 21st Century Meeting and Event Technologies Powerful Tools for Better Planning, Marketing, and Evaluation*, Apple Academic Press, Waretown.

Lee, W., Gretzel, U., and Law, R. (2010), "Quasi-trial Experiences through Sensory Information on Destination Web Sites", *Journal of Travel Research*, 49 (3), pp. 310–322.

Lee, W., Xiong, L., and Hu, C. (2012), "The Effect of Facebook Users' Arousal and Valence on Intention to Go to the Festival: Applying an Extension of the Technology Acceptance Model", *International Journal of Hospitality Management*, 31 (3), pp. 819–827.

Li, X. and Duan, B. (2018), "Organizational Microblogging for Event Marketing: A New Approach to Creative Placemaking", *International Journal of Urban Sciences*, 22 (1), pp. 59–79.

Li, Y., Hu, C., Huang, C., and Duan, L. (2016), The concept of smart tourism in the context of tourism information services. *Tourism Management*, 58, pp. 293–300.

Liburd, L. J. (2005), "Sustainable Tourism and Innovation on Mobile Tourism Services", *Tourism Review International*, 9 (2), pp. 107–118.

Luna-Nevarez, C. and Hyman, M. R. (2012), Common Practices in Destination Website Design", *Journal of Destination Marketing & Management*, 1 (1–2), pp. 94–106.

MacKay, K. and Vogt, C. (2012), "Information Technology in Everyday and Vacation Contexts", *Annals of Tourism Research*, 39 (3), pp. 1380–1401.

Maskell, P., Bathelt, H., and Malmberg, A. (2005), Building Knowledge Pipelines: The Role of Temporary Clusters. DRUID working paper 05-20. Copenhagen Business School, Department of Industrial Economics and Strategy.

McCarthy, J. and Wright, P. (2004), "Technology as Experience", *Interactions*, 11 (5), pp. 42–43.

McLoughlin, J., Kaminski, J., and Sodagar, B. (2007), Assessing the Socio-economic Impact of Heritage: From Theory to Practice. In J. McLoughlin, J. Kaminski and B. Sodagar (Eds.), *Technology Strategy, Management and Socio-Economic Impact* (pp. 17–42), Archaeolingua, Hungary.

McQuitty, S. and Peterson, R. T. (2000), "Selling Home Entertainment on the Internet: An Overview of a Dynamic Marketplace", *Journal of Consumer Marketing*, 17, pp. 233–248.

Morabito, V. (2014), *Trends and Challenges in Digital Business Innovation*, Springer, Heidelberg.

Neuhofer, B., Buhalis, D., and Ladkin, A. (2014), "A Typology of Technology-enhanced Tourism Experiences, *International Journal of Tourism Research*, 16 (4), pp. 340–350.

Nielsen, N. C. and Liburd, J. J. (2008), "Geographical Information and Landscape History in Tourism Communication in the Age of Web 2.0", *Journal of Travel and Tourism Marketing*, 28 (3&4), pp. 282–298.

O'neil, D. (2002), "Assessing Community Informatics: A Review of Methodological Approaches for Evaluating Community Networks and Community Technology Centers", *Internet Research*, 12 (1), pp. 76–102.

Perdue, J., Woods, R. H., and Ninemeier, J. (2001), "Club Management Competencies 2005: Updated Information for the Classroom", *Journal of Hospitality & Tourism Education*, 13 (2), pp. 20–33.

Pine, B. J. and Gilmore, J. H. (2011), *The Experience Economy*, Harvard Business Press.

Porter, M. (2001), "Strategy and the Internet", *Harvard Business Review*, 79 (3), pp. 62–78.

Prasad, V. K., Ramamurthy, K. and Naidu, G. (2001), "The Influence of Internet-marketing Integration on Marketing Competencies and Export Performance", *Journal of International Marketing*, 9 (4), pp. 82–110.

Qirici, E., Theodhori, O., and Elmazi, L. (2011), "E-Marketing and ICT-Supported Tourist Destination Management. Implications for Tourism Industry in Global Recession", *International Journal of Management Cases*, 13 (3), pp. 152–158.

Roberts, J. (2000), "From Know-How to Show-How? Questioning the Role of Information and Communication Technologies in Knowledge transfer", *Technology Analysis & Strategic Management*, 12 (4), pp. 429–43.

Rubinstein, P. (2015), The Use of Technology to Enhance Music Festivals. Retrieved from https://www.youredm.com/2015/03/07/the-use-of-technology-to-enhance-music-festivals/ (12.10.2020).

Schegg, R. and Stangl, B. (2018), "Special Section on Recommendations and Analytics in Tourism", *Information Technology Tourism*, 18, pp. 1–4. doi:10.1007/s40558-018-0109-8.

Schegg, R., Liebrich, A., Scaglione, M., and Ahmad, S. F. S. (2008), An Exploratory Field Study of Web 2.0 in Tourism, In *Information and Communication Technologies in Tourism* (pp. 152–163), Springer, Vienna.

Shanker, D. (2008), Ict and Tourism: Challenges and Opportunities. In *Conference on Tourism in India – Challenges Ahead*, 15–17 May.

Slocum, S. L. and Lee, S. (2014), "Green ICT Practices in Event Management: Case Study Approach to Examine Motivation, Management and Fiscal Return on Investment", *Information Technology Tourism*, 14, pp. 347–362. doi:10.1007/s40558-014-0019-3.

Stankov, U., Pavluković, V., Alcántara-Pilar, J. M., Cimbaljević, M., and Armenski, T. (2018), Should Festival Be Smarter?: ICT on Mass Events – The Case of the Exit Festival (Novi Sad, Serbia). In *Handbook of Research on Technological Developments for Cultural Heritage and Etourism Applications* (pp. 245–263), IGI Global.

Tatikonda, M. V. and Stock, G. N. (2003), "Product Technology Transfer in the Upstream Supply Chain", *Journal of Product Innovation Management*, 20, pp. 444–67.

Tiwari, A., Weth, C., and Kankanhalli, M. (2018), *Multimodal Multiplatform Social Media Event Summarization. ACM Transactions on Multimedia Computing, Communications, and Applications (TOMM)*, 14 (2s), pp. 1–23.

Tussyadiah, I. P. (2015), *Personal Technology and Tourism Experience.* Iscontour.

Vilaseca-Requena, J., Torrent-Sellens, J., and Jiménez-Zarco, A. I. (2007), "ICT Use in Marketing as Innovation Success Factor", *European Journal of Innovation Management*, 10 (2), pp. 268–288.

Walsh, P., Clavio, G., Lovell, M. D., and Blaszka, M. (2013), "Differences in Event Brand Personality between Social Media Users and Non-Users", *Sport Marketing Quarterly*, 22 (4), pp. 214–223.

Wang, D., Park, S., and Fesenmaier, D. R. (2012a), "The Role of Smartphones in Mediating the Touristic Experience, *Journal of Travel Research*, 51 (4), pp. 371–387.

Wang, W., Chen, J. S., Fan, L., and Lu, J. (2012b), "Tourist Experience and Wetland Parks: A Case of Zhejiang, China", *Annals of Tourism Research*, 39 (4), pp. 1763–1778.

Werthner, H. and Klein, S. (2006), ICT-enabled Innovation. In *Innovation and Product Development in Tourism* (pp. 71–83),Erich Schmidt Verlag, Berlin.

Wohlfeil, M. and Whelan, S. (2006), "Consumer Motivations to Participate in Event-marketing Strategies", *Journal of Marketing Management*, 22 (5–6), pp. 643–669.

Wright, R. and Keith, L. (2014), "Wearable Technology: If the Tech Fits, Wear It", *Journal of Electronic Resources in Medical Libraries*, 11 (4), pp. 204–216.

Yoo, K. H. and Gretzel, U. (2016), Use and Creation of Social Media by Travellers. In E. Christou and M. Sigala (Eds.), *Social Media in Travel, Tourism and Hospitality: Theory, Practice and Cases*, (Chapter 15; pp. 189–205), United Kingdom: Ashgate.

Yovcheva, Z., Buhalis, D., and Gatzidis, C. (2012), "Smartphone Augmented Reality Applications for Tourism", *e-Review of Tourism Research (eRTR)*, 10 (2), pp. 63–66.

Additional Reading

Aldebert, B., Dang, R. J., and Longhi, C. (2011), "Innovation in the Tourism Industry: The Case of Tourism@", *Tourism Management*, 32 (5), pp. 1204–1213.

Karimi, K. and Atkinson, G. (2013), *What the Internet of Things (IoT) Needs to Become a Reality*. White Paper, FreeScale and ARM, 1–16.

Munar, A. M. and Jacobsen, J. K. S. (2014), "Motivations for Sharing Tourism Experiences through Social Media", *Tourism Management*, 43, pp. 46–54.

Palen, L. (2008), "Online Social Media in Crisis Events", *Educause Quarterly*, 31 (3), pp. 76–78.

Raun, J., Ahas, R., and Tiru, M. (2016), "Measuring Tourism Destinations Using Mobile Tracking Data", *Tourism Management*, 57, pp. 202–212.

Tuten, T. L. and Solomon, M. R. (2017), *Social Media Marketing*. UK: Sage.

UNWTO (2001). *eBusiness for Tourism: Practical Guidelines for Destinations and Businesses*. Madrid: World Tourism Organisation.

Yoo, K. H. and Gretzel, U. (2010), "Antecedents and Impacts of Trust in Travel-Related Consumer-Generated Media", *Information Technology & Tourism*, 12 (2), pp. 139–152.

Zhao, Z., Resnick, P., and Mei, Q. (2015), *Enquiring Minds: Early Detection of Rumors in Social Media from Enquiry Posts*. In Proceedings of the 24th International Conference on World Wide Web (pp. 1395–1405).

Zhang, P. (2008), "Technical Opinion Motivational Affordances: Reasons for ICT Design and Use", *Communications of the ACM*, 51 (11), pp. 145–147.

Zhang, Z. K., Cho, M. C. Y., Wang, C. W., Hsu, C. W., Chen, C. K., and Shieh, S. (2014), IoT Security: Ongoing Challenges and Research Opportunities. In *7th International Conference on Service-Oriented Computing and Applications*, pp. 230–234.

Zhu, Y. Q. and Chen, H. G. (2015), "Social Media and Human Need Satisfaction: Implications for Social Media Marketing", *Business Horizons*, 58 (3), pp. 335–345.

İbrahim ÇETİNTÜRK

A Study on Brand Loyalty in Accommodation Establishments[1]

1 Introduction

Meeting people's need for transportation, accommodation, catering, entertainment, and recreation etc. on temporary travels, the tourism industry is billed as a global sector that stretches all over the world. Accommodation, one of the sub-sectors of tourism, is a major industry that offers great inputs to the global economy and creates jobs for millions of people around the globe (Tanford et al., 2012). The ever-growing competition is one of the top problems that hotels face in today's world. Concentration on brand loyalty is a prerequisite for hotels to survive in the future (Kandampully & Suhartanto, 2000). As a theme that is particularly emphasized and studied, brand loyalty is considered to be one of the keys to an ever-lasting and long-term success for companies (Eren & Erge, 2012).

The provision of higher-quality and lower-cost products/services to customers through reduction of costs based on brand loyalty provides establishments with a major competitive edge over their counterparts. Repeated exchange makes loyal customers less likely to choose rival establishments. This positively and strongly affects the profitability of an establishment (Selvi & Ercan, 2006). Studies report that a 5% rise in brand loyalty leads to 25 % to 85 % of profit growth to an establishment (Kandampully & Suhartanto, 2000). Brand loyalty to accommodation establishments stands for a customer's tendency to frequent a certain hotel or a hotel chain. A customer's choice of the same accommodation establishment in a non-coincidental way is an indicator of brand loyalty. Brand loyalty has a positive effect on a hotel's financial and non-financial performance. In addition, customers with high brand loyalty are less likely to be price-sensitive.

It has become more and more important to focus on challenges of the tourism industry and study them accordingly. This study is intended to help

---

[1] This chapter has been generated from the Master's Thesis of İbrahim Çetintürk titled "Brand Loyalty in Hospitality Industry, the Case of Antalya" written in Suleyman Demirel University, Social Sciences Institute.

establishment gain further insight into brand loyalty and contribute to academic studies over brand loyalty. This study aims at determining the perception of customers of five-star accommodation establishments over brand loyalty. It is also intended to help accommodation establishments adopt a strategy to boost their potential for loyal customers.

2 Literature Review

Only by gaining further insight into demands of their customers, businesses can keep their current customers and have a sustainable competitive edge in such a fiercely competitive market. It is an undeniable fact that a business that makes a good analysis of demands of its customers will have a competitive edge. Studies over brand loyalty are usually focused on service quality, customer satisfaction, corporate image, brand value, and factors that affect brand loyalty. As a part of their study, Demirağ and Durmaz (2019) analyzed the effect of perceived service quality on perceived value and attitudinal brand loyalty. Rather and Camilleri (2019) studied the effect of service quality and harmony between consumers and brands on brand loyalty while Yang et al. (2017) studies the effects of service quality on customer satisfaction, brand loyalty and brand image. Tayfun and Yayla's (2014) study was carried out over five-star hotels based in Ankara. As a part of their study, they analyzed the brand loyalty in dimensions of behavioral and attitudinal loyalty. The study suggested that behavioral and attitudinal loyalty is affected the most by brand image. In another study, Tayfun and Yayla (2013) analyzed the importance of factors that affect brand loyalty in hotel choices of tourists from the perspective of demographic variables. Nam et al. (2011) conducted a study to examine the intermediary effects of customer satisfaction between brand value and brand loyalty in hotel and restaurant business. They reported that customer satisfaction plays a partial intermediary role between brand value and brand loyalty. Wu et al. (2011) studies the effects of service quality, brand image, and fair price on customer satisfaction and loyalty. Wilkinsvd's (2009) study contributed to insight into correlations among service quality, perceived value, customer satisfaction, brand trust, brand attitude, and behavioral loyalty in service business. Kandampully and Hu's (2007) study was intended to explain the correlation between service quality and customer satisfaction, and their effects on corporate image and customer loyalty. As a result, they reported that corporate image and customer loyalty are affected by both service quality and customer satisfaction. The studies of Kandampully and Suhartanto (2000, 2003) are among the studies on brand loyalty and corporate image. It is reported that hotel image and customer satisfaction are significant factors that play into a customer's repurchase behavior and loyalty to the hotel. In addition, studies are significant in terms of offering methods that would boost customer loyalty for hotels.

While there are many definitions of the term loyalty in literature, researchers, and practitioners do not have a consensus (Kandampully & Suhartanto, 2003). Oxford dictionary describes the meaning of the word loyalty as commitment or sacrifice for one person, country, group or cause (Cuong et al., 2015). Oliver (1997) described loyalty as "a profound commitment to purchase or demand a product or a service to be one's choice on a regular basis in the future, and a strategic action to purchase the same brand repeatedly."

Brand loyalty is a subjective term on account of some challenges that arise from the way it is measured (Eren & Erge, 2012). Jacoby and Kyner (1973) suggested that brand loyalty is not a random action. Brand loyalty is the way consumers tend to purchase the same product/service in a repeated fashion and behavior. It is a consumer's attitude to support a company. It is the tendency to purchase an already-known or recommended product/service again in the future. It is also a profound commitment to re-purchase a product/service (Skogland & Siguaw, 2004). Brand loyalty is a testament to the commitment of a customer to a brand. It is the likelihood of a customer to switch to another brand once there is something different about the price or qualities of a product. The more the brand royalty is, the less interested customers are in products of rival brands (Aaker, 1992).

In literature, loyalty is analyzed in three different dimensions: Behavioral approach, attitudinal approach, and causal behavior theory (hybrid approach). First of all, the behavioral approach stands for actual purchasing patterns of customers (Back & Parks, 2003). Among methods to be adopted to measure a behavioral approach are repeated purchase rates, purchase ratio, and number of brands purchased (Aaker, 1992). Chen and Gürsoy (2001) strongly criticize the behavioral approach, and argue that it would be more pertinent to adopt an attitudinal approach for studies over tourism industry. The major problem with the behavioral approach is that repeated exchanges do not result from psychological commitment to a brand. For instance, if a new hotel launched across the street offers more value to a customer, he/she may choose the new hotel. In addition, repeated exchanges cannot be always construed as a customer's commitment (Bowen & Chen, 2001). Attitudinal approach, which is the second dimension of loyalty, is described as a positive reference of customers about a product/service and its recommendation to others even if they do not purchase it themselves repeatedly (Bowen & Chen, 2001; Huang & Zhang, 2008). The third one is the causal behavior theory (Pritchard & Howard, 1997).This approach analyzes brand loyalty based on both behavioral and attitudinal criteria. A customer's repeated purchasing patterns are reinforced by attitudes. This enables to move from artificial loyalty to goods and services into true loyalty. The causal behavior theory combines the first two dimensions (behavioral and attitudinal), and measures a customer's product choices, brand-changing trends, purchase frequency, purchase patterns, and total purchase amount (cited by Bowen & Chen, 2001).

Literature reviews were utilized to designate brand loyalty factors. Wilkins et al. studied the dimensions of loyalty in hotel and service industry: Service quality, perceived value, customer satisfaction, and brand trust. Çakıroğlu and Galai (2019) reported that among determinants of brand loyalty are brand image, brand trust, brand adequacy, perceived quality, and customer satisfaction. Among widely acknowledged determinants of brand loyalty are service quality, perceived value, brand image, and customer satisfaction (Suhartanto, 2011; Suhartanto et al., 2013). The determinants of brand loyalty to be addressed in the study based on the literature review are as follows:

Physical adequacy: Colors, lighting, size, width, sound, music, temperature, and humidity etc. are among the evidences concerning the internal positions of tourism establishments (Kozak, 2008: 244). An accommodation establishment's internal and exterior architectural style, circulation areas, decor, landscape, and recreational areas etc. constitute physical evidences. A large and spacious lobby with a soothing effect is significant for the first impression of a customer to be positive.

Trust: Customers would like any product they purchase to be unproblematic. Unproblematic products generate royalty of customers to a product and/or a firm. Trust in a product is not built only at the moment a customer purchases a product. It predates any purchase and remains in effect during and after (Gilbert et al., 2004). For instance, customers have their personal belongings, notes, and secrets in a hotel room. Customers leave their room with confidence built by the hotel, and know that no harm would be done to their personal belongings. This builds a customer's trust in an establishment.

Service Quality: The perception of customers about service quality has to do with what they get (physical output) and how they get it (service process) (Kandampully & Hu, 2007: 438). Most of the studies over service quality were analyzed by SERVQUAL method. This method enables to measure the difference between a customer's expectation and perception about service quality, and service performance. Canoğlu (2008) described main benefits of service quality for a hotel as customer loyalty, positive image, and maximization of its potential revenue. The studies are in agreement with the fact that service quality is a major factor that affects customer satisfaction, perceived value and brand loyalty.

Price: Zeithaml (1988) describes price from a consumer's perspective as "anything that is given or sacrificed to get a product." Price is a marketing tool that enables a business to generate revenue based on its costs. In other words, price is an effort to get some value back from consumers in exchange for a value offered for them (Altunışık et al., 2016). Should customers consider price an element of value for their choice of an accommodation establishment, they may choose a low-cost establishment.

Brand Image: An establishment's image is a major factor for potential customers to assess one (Kandampully & Hu, 2007). A brand image with a positive or negative effect on a market is a significant value for an establishment. For instance, hotel chains such as Hilton and Holiday Inn appeal to various categories of customers based on their positive image (Kozak, 2008: 130). Club Med, one of the leading international companies in accommodation business, has a positive image in the eyes of customers thanks to its successful animation shows (Hacıoğlu et al., 2003: 124, 125).

Perceived Value: Perceived value stands for a general assessment of consumers about the use of a service based on what they purchase and what they give in exchange for it. In addition, perceived value is described from four standpoints: (1) Value is low price. (2) It is what a consumer wants to see in a product. (3) It is the quality that consumers get for the price they pay. (4) It is what consumers get in exchange for what they give (Zeithaml, 1988).

3 Method

The population of the study comprises five-star accommodation establishments based in Antalya. A questionnaire was adopted as the data collection tool of the study. The questionnaire form was built on Selvi and Ercan's (2006) questionnaire drawn up to analyze brand loyalty to hotels.

The aim of the study is to determine the perception of customers of five-star accommodation establishments over brand loyalty. The study was based on sampling rather than reaching out to the entire population for time, coast, and accessibility purposes. Cluster sampling was adopted as a method to determine the sample out of the population. Nine five-star accommodation establishments based in various attraction centers of Antalya were selected as a cluster, and a total of 600 questionnaires were disseminated among the hotels that constituted the population. 390 of the questionnaire were taken into consideration. Prior to the study, a pilot study was performed to avoid any potential error in practice. The pilot study was performed for 30 guests who stayed in the town of Alanya, Antalya. The structure, validity, and reliability of the questionnaire were tested as a part of the pilot study. Cronbach Alpha coefficient was computed as an analytical method of internal consistency. The computation revealed that Cronbach Alpha coefficient was 0.84, and it was considered to be a highly reliable questionnaire.

Developed as a data collection tool, the questionnaire form consisted of two sections: The first section includes closed-end questions about gender, age, marital status, educational background, and occupation to determine demographic characteristics of the respondents. The second section includes 21 questions of 5-Likert scale to determine brand loyalty of customers to accommodation establishments. The test of normalcy revealed that the data did not have a normal distribution. Descriptive statistical methods (number, percentage, mean, median, and standard deviation) were adopted for data analysis. Based on the reliability table of statistics, Cronbach Alfa coefficient equaled to 0.93.

4 Practice

The results on the demographic characteristics of the respondents are presented in Tab. 17-1.

The demographic characteristics of the respondents show that 46.9 % of them are women while 53.1 % of them are men. The gender breakdowns of the respondents are similar to one another. The age of the respondents ranges from 18 to 30 for 41.50 % of them, and 31 to 40 for 29.5 % of them. The study reported that young subjects were more involved in the study. As for the variable of marital status of the respondents, 48.50 % of them are married and 51.50 % of them are single. Of the respondents, 47.9 % of them are international tourists and 42.1 % of them are domestic guests. Of the international tourists, 6.9 % of them are from Britain, with 13.8 % from Germany, 10.5 % from Russia, 3.6 % from the Netherlands, 1 % from the United Nations, 5.4 % from Belgium, 1 % from France, 3.3 % from Switzerland, 5.6 % from Norway, 1.8 % from Denmark and 4.9 % from other countries (China, Ukraine, Romania, etc.). As for the variable of occupation, 23.1 % of the respondents are students, with 29,1 % of them public officers, 10.8 % businesspersons, 7.9 % housewives, 9.7 % unemployed, 9 % pensioners, and 10.5 % other professionals. As for the educational background of the respondents, 4.1 % of them are elementary school dropout, with 26.2 % high school dropout, 52 % holding a bachelor's degree,

Tab. 17-1: Results on Demographic Characteristics of the Respondents

Gender	N	%	Nationality	N	%
Woman	183	46,9	Domestic	164	42,1
Man	207	53,1	International	146	47,9
Age			**Profession**		
18–30	162	41,5	Student	90	23,1
31–40	115	29,5	Public Officer	113	29,1
41–50	62	15,9	Businessperson	42	10,8
51–60	29	7,4	Housewife	31	7,9
61 and over	22	5,6	Unemployed	38	9,7
			Pensioner	35	9,0
			Other Professionals	41	10,5
Marital Status			**Education**		
Married	189	48,5	Elementary School	16	4,1
Single	201	51,5	High School	102	26,2
			Bachelor's	204	52,3
			Degree Master or Secondary Bachelor's Degree	68	17,4

Tab. 17-2: Respondents by Centers of Attraction

Destination	N	%
Alanya	128	32,8
Side	159	40,8
Kemer	21	5,4
Kaş	13	3,3
Belek	69	17,7

and 17.4 % holding a master's or a secondary bachelor's degree. The breakdown of respondents by centers of attraction is presented in Tab. 17-2.

Of the respondents, 32.8 % of them stayed in Alanya, with 40.8 % in Side, 5.4 % in Kemer, 3.3 % in Kaş, and 17.7 % in Belek. The breakdown of the replies of the respondents to the questions about travel and accommodation is presented in Tab. 17-3.

Based on how often the respondents travel, 9.7 % of them travel once every three months, with 31.50 % of them once every six months, 46.4 % of them once

Tab. 17-3: Results on Travel and Accommodation Experiences of the Respondents

		N	%
How often do you go on holiday?	Once every three months	38	9,7
	Once every six months	123	31,5
	Once a year	181	46,4
	Once every two years or less often	48	12,3
The reason for your (going on) holiday or vacation?	Recreation and Entertainment	238	61,0
	Business Trip	25	6,4
	Congress and Meeting	24	6,2
	Education and Seminar	29	7,4
	Historical and Cultural Trip	47	12,1
	Health	21	5,4
	Others	6	1,5
How many times did you come to this hotel?	First time	260	66,7
	Second time	82	21,0
	Third time	21	5,4
	Fourth time and more	27	6,9
Will you choose this hotel in the future?	Yes	313	80,3
	No	77	19,7

a year, and 12.3 % of them once every two years or less often. The respondents reported that they travel once a year at most. Recreation and entertainment rank at the top of the list with 61 % among the reason why the respondents travel. The fact that the respondents consider Antalya a destination of entertainment and recreation is an indicator of how the city's identity is built on entertainment and recreation. 12.1 % of the respondents visit Antalya for its diverse history and cultural assets. Based on how often the respondents visit the hotel, 66.7 % of them visited for the first time while it was more than once for 33.3 % of them. The majority of the respondents reported that they would visit the same hotel again. This shows that customers are pleased with hotels to a large extent, and they tend to choose the same hotel repeatedly. Behavioral loyalty is the most distinct aspect of brand loyalty to a certain product and service. However, behavioral dimension must be consolidated by attitudinal loyalty (emotional attachment). The interpretation and grading of the mean scores generated in line with five-point Likert scale in the final section of the questionnaire form were built on the criterion of scoring limits set out in Tab. 17-4.

Study on Brand Loyalty 341

Tab. 17-4: Scoring Limits of the Questionnaire Questions

Statements	Point	Mean Scores
Very high	5	4.20–5.00
High	4	3.40–4.19
Medium	3	2.60–3.39
Low	2	1.80–2.59
Very low	1	1.00–1.79

The scoring limits presented in Tab. 17-4 are five equal ranges of the arithmetic means of the replies for the extent that the respondents agree with the statements in the questionnaire form. 1.00 to 1.79 means very low, with 1.80 to 2.59 low, 2.60 o 3.39 medium, 3.40 to 4.19 high, and 4.20 to 5.00 very high for agreement. The attitudes of the respondents about brand loyalty are presented in Tab. 17-5.

The extent of agreement of the respondents with brand loyalty is presented in Tab. 17-5. Some descriptive analyses were performed on indicators of sub-dimensions of brand loyalty (physical competence and trust, service quality, price, brand image and perceived value). The descriptive analyses include mean value and standard deviation of the aforementioned dimensions. The respondents reported positive views in the questionnaire over brand loyalty. In addition, it was reported that their arithmetic mean for "physical competence and trust" ($\bar{x}$=4.00), service quality ($\bar{x}$=3.99), price ($\bar{x}$=3.65), brand image ($\bar{x}$=3.92), and perceived value ($\bar{x}$=3.72) are similar to one another.

Physical competence and trust was computed to have the highest arithmetic mean score among the sub-dimensions of brand loyalty ($\bar{x}$=4.00). This figure shows that the respondents have positive views on physical competence of accommodation establishments and trust in them. The respondents reported that they feel safe and peaceful at their hotel ($\bar{x}$=4.06), and customer safety is adequate ($\bar{x}$=4.14) and their trust in the hotel ($\bar{x}$=3.97) its diversity of services ($\bar{x}$=3.93) would be effective in staying at the hotel again.

Most of the respondents reported that staff are ready to help and willing to provide services ($\bar{x}$=4.10), and the fact that staff are well educated and well qualified is a significant factor ($\bar{x}$=3.96) for them to stay at the hotel again, and they would choose the same hotel again as they are pleased with the staff ($\bar{x}$=4.20). The respondents also reported that the hotel met their expectations more than enough ($\bar{x}$= 4.13), and the attention paid to the quality of food and beverages played a role in their choice of the hotel ($\bar{x}$=3.90), and the promotions the hotel offers have partially improved their loyalty to it ($\bar{x}$=3.79).

Price was computed to have the lowest arithmetic mean as a factor among sub-dimensions of brand loyalty. The respondents reported that the price-performance ratio of the hotel (quality/price balance) is coherent ($\bar{x}$=3.91). They also reported that they are indecisive ($\bar{x}$=3.42) about whether they would stay at the same hotel despite an increase in rates of the hotel. Customers loyal to a brand are insensitive to price. The respondents reported that they would choose to stay at the same hotel even if other hotels offer the same service at a lower price ($\bar{x}$=3.65).

The respondents reported that the destination where the hotel is located in has an influential brand image ($\bar{x}$=4.01), and they already had a positive view on the hotel even before they arrived ($\bar{x}$=3.89), and they were impressed by the hotel's prestige and image as they made the decision to purchase a vacation there ($\bar{x}$=3.85).

The respondents partially agreed with the statement: "I'd choose the same hotel again even if I experience any failure and defect where I stay" ($\bar{x}$=3.65). In addition, the respondents reported that they agree with the statement: "The fact that the hotel communicates celebratory messages on special occasions forges an emotional bond between us" ($\bar{x}$=3.70). The respondents reported that they are indecisive about whether they would delay their vacation rather than replacing the hotel if the hotel they'd like to stay at is fully booked ($\bar{x}$=3.40). In addition, they reported that special services provided by a hotel would be influential in their choice of the same hotel in the future ($\bar{x}$=3.89). Last but not least, the respondents reported that they would recommend the hotel for their relatives and friends ($\bar{x}$=3.99).

Conclusion and Discussion

In a rapidly growing and changing world of standards, brand loyalty is a significant element for a company to sustain its presence, boost profitability, and have a competitive edge over other companies. The aim of the study is to determine the perception of customers of five-star Antalya-based accommodation establishments on brand loyalty. The fact that the study was carried out in Antalya made it even more valuable. Antalya is one of the world's rare attraction centers known for modern accommodation establishments and superior service quality.

The businesses on hospitality focus on fulfilling customer demands and expectations at a better level and increasing service quality and customer

Tab. 17-5: Extent of Agreement of Respondents with Brand Loyalty

Statements	X̄	S.S	
Physical Competence and Trust	4,00	,99	High
1. I feel safe and peaceful in this hotel	4,06	,99	High
2. I believe that the security of the customers in this hotel is sufficient (food, fire, pool, etc.)	4,14	,98	High
3. My confidence to the hotel is effective in my coming to the same hotel	3,97	1,01	High
4. Product variety (entertainment, sports, food and drinks etc.) in the hotel is effective in my coming to the same hotel	3,93	1,06	High
Service Quality	**3,99**	**1,02**	**High**
5. The services I received from the hotel exceed my expectations	4,13	,99	High
6. I prefer this hotel because the personnel provide a professional service	4,20	1,01	Very high
7. The personnel in this hotel are well qualified, I would like to come again	3,96	1,08	High
8. I prefer this hotel because of the attention paid to the quality of food and drinks	3,90	1,13	High
9. The promotions provided by the hotel management (reward, discount, gift) increase my devotion to the hotel	3,79	1,08	High
10. In the hotel the personnel is willing to serve and help	4,10	1,02	High
Price	**3,65**	**1,12**	**High**
11. The quality-price balance in this hotel is coherent	3,91	,99	High
12. I prefer the same hotel in spite of the increase in price	3,42	1,30	High
13. I prefer this hotel although the other hotels in this area have low prices	3,65	1,15	High
Brand Image	**3,92**	**1,05**	**High**
14. If the destination center where the hotel is located has a good brand image, I will choose the same hotel in the future	4,01	,99	High
15. Before coming to this hotel, I had a positive idea about it	3,89	1,10	High
16. When choosing the hotel, I was affected by the fame and image of the hotel	3,85	1,11	High
Perceived Value	**3,72**	**1,15**	**High**
17. Although I observe some faults and deficiencies in the hotel, I prefer the same hotel in the future	3,65	1,17	High
18. Hotel management's sending celebration messages on special days constitutes a strong emotional bond	3,70	1,19	High

(continued on next page)

Tab. 17-5: Continued

	X̄	S.S	Statements
19. If I am not able to find a room on a date that I want, I postpone my holiday instead of going to another hotel	3,40	1,30	High
20. I am going to choose this hotel in the future because of the private services the hotel provided for me	3,89	1,15	High
21. I am going to recommend this hotel to my relatives and friends	3,99	1,11	High

satisfaction in order to create brand loyalty. A business with loyal customers is able to reduce the costs of sales and marketing. Thus, the business increases its profitability accordingly. The aim of the study is to determine the perceptions of customers towards brand loyalty. The population of the study consists of the customers of five-star accommodation enterprises in Antalya.

The study was carried out in the centers of attraction in the towns of Belek, Alanya, Kaş, Kemer and Side, Antalya. The majority of the respondents reported that they would re-visit the hotel they stayed at. This is an indicator of loyalty to the accommodation establishments for most of the respondents. Recreation and entertainment rank at the top of the list among the reason why the respondents travel. In addition, the fact that the respondents consider Antalya a destination of entertainment and recreation is an indicator of how the city's identity is built on entertainment and recreation. On the other hand, the majority of the respondents reported that they would re-visit the hotel they stayed at. This shows that customers are pleased with hotels to a large extent, and they tend to choose the same hotel repeatedly.

The study suggests that the views of the respondents concerning the brand loyalty to the accommodation establishment are positive. Descriptive analyses were performed on all indicators about physical competence and trust, service quality, price, brand image, and perceived value. The most positive view out of sub-dimensions of brand loyalty was physical adequacy/trust as a factor. Price was computed to have the lowest arithmetic mean as a factor among sub-dimensions of brand loyalty. The respondents reported that they would partially choose the same hotel even if others offer services at a lower price. One of the most significant indicators of brand loyalty is the continuous purchase of the brand by consumers without considering it cheaper or more expensive than its rivals. Customers loyal to a brand are insensitive to price. Past studies were particularly focused on determinants of brand loyalty, pioneers of brand loyalty, and factors that affect brand loyalty (Wilkins et al., 2009; Chitty et al., 2007; Kandampully & Suhartanto, 2003; Bowen & Chen, 2001; Ekinci et al., 2011; Suhartanto, 2011; Çetinsöz & Artuğer, 2013; Suhartanto et al., 2013). The fact that the questionnaire was conducted in accommodation establishments based in Antalya is one of the limitations of the study. It is considered that researchers with interest in carrying out studies in this regard would offer significant inputs for the literature of studies that cover various centers of attraction.

Bibliography

Aaker, D. A. (1992), "The Value of Brand Equity", *Journal of Business Strategy*, 13 (4), pp. 27–32.

Al-Msallam, S. (2015), "Customer Satisfaction and Brand Loyalty in the Hotel Industry", *International Journal of Management Sciences and Business Research*, 4 (9), pp. 1–13.

Altunışık, R., Özdemir, Ş., and Torlak, Ö. (2016), *Marketing Principles and Management* (2nd Ed.), Istanbul: Beta Publishing.

Back, K.-J. and S. C. Parks (2003), "A Brand Loyalty Model Involving Cognitive, Affective, and Conative Brand Loyalty and Customer Satisfaction", *Journal of Hospitality & Tourism Research*, 27 (4), pp. 419–435.

Bowen, J. T. and Chen S. L. (2001), "The Relationship between Customer Loyalty and Customer Satisfaction", *International Journal of Contemporary Hospitality Management*, 13 (5), pp. 213–217.

Canoğlu, M. (2008), *Determination of the Relationship between the Image and Service Quality Perceptions and Repeat Buying Behaviours of Hotel Customers* (Master Thesis), Çukurova University Institute of Social Sciences, Adana.

Chen, J. S. and Gürsoy, D. (2001), "An Investigation of Tourists Destination Loyalty and Preferences", *International Journal of Contemporary Hospitality Management*, 13 (2), pp. 79–85.

Chitty, B., Ward, S., and Chua, C. (2007), "An Application of the ECSI Model as a Predictor of Satisfaction and Loyalty for Backpacker Hostels", *Marketing Intelligence & Planning*, 25, pp. 563–580.

Cuong, L. D., Quy, V. T. M., Anh, N. T. N., Chung, N. D., and Thang, N. M. (2015), *Assessing Service Quality to Enhance Customer Satisfaction and Loyalty in case of Thien Duc Hospital* (Doctoral dissertation), FSB.

Çakıroğlu, A. D. and Galaş, C. (2019), "The Factors Affecting on Brand Loyalt in Personal Care Products: A Study in Giresun", *Kafkas University* the *Journal of Faculty of Economics and Administrative Sciences*, 10 (19), pp. 214–235.

Çetinsöz, B. C. and Artuğer, S. (2013), "A Research on Determination of Brand Equity of Antalya", *Anatolia: Journal of Tourism Research*, 24 (2), pp. 200–210.

Çetintürk, İ. (2017), "Customer Value, Customer Satisfaction and Brand Loyalty A Research on University Leisure Centres", *JTHM-Journal of Travel and Hospitality Management*, 14 (2), pp. 93–109.

Demirağ, B. and Durmaz, Y. (2019), "Investigation of the Effects of Perceived Service Quality on Perceived Value and Attitude Brand Loyalty: Five-Star Resort and City Hotels Example", *OPUS International Journal of Society Researches*, 11 (18), pp. 693–726.

Ekinci, Y., Zeglat, D., and Whyatt, G. (2011), "Service Quality, Brand Loyalty, and Profit Growth in UK Budget Hotels", *Tourism Analysis*, 16 (3), pp. 259–270.

Eren, S. S. and Erge, A. (2012), "The Effects of Brand Trust, Brand Satisfaction and Customer Value on Brand Loyalty of Customers", *Journal of Yasar University*, 7 (26), pp. 4455–4482.

Faullant, R., Matzler, K., and Füller, J. (2008), "The Impact of Satisfaction and Image on Loyalty: The Case of Alpine Ski Resorts", *Managing Service Quality*, 18, pp. 163–178.

Gilbert, R., Veloutsou, C., Goode, M., and Moutinho, L. (2004), "Measuring Customer Satisfaction in the Fast Food Industry: A Cross National Approach", *Journal of Services Marketing*, 8 (5), pp .371–383.

Hacıoğlu, N., Gökdeniz, A., and Dinç, Y. (2003), *Boş Zaman ve Rekreasyon Yönetimi*, Ankara: Detay Yayıncılık.

Huang, J. and Zhang, D. (2008), Customer Value and Brand Loyalty: Multi-Dimensional Empirical Test. In *International Seminar on Future Information Technology and Management Engineering*, University of Leicester, Leicestershire, United Kingdom, IEEE. (pp. 102–106).

Jacoby, J. and Kyner, D. B. (1973), "Brand Loyalty vs. Repeat Purchasing Behavior", *Journal of Marketing Research*, 10 (1), pp. 1–9.

Jensen, J. M. and Hansen, T. (2006), "An Empirical Examination of Brand Loyalty", *Journal of Product & Brand Management*, 15 (7), pp. 442–449.

Kandampully, J. and Hu, H. H. (2007), "Do Hoteliers Need to Manage Image to Retain Loyal Customers?", *International Journal of Contemporary Hospitality Management*, 19 (6), pp. 435–443.

Kandampully, J. and Suhartanto, D. (2000), "Customer Loyalty in the Hotel Industry: The Role of Customer Satisfaction and Image", *International Journal of Contemporary Hospitality Management*, 12 (6), pp. 346–351.

Kandampully, J. and Suhartanto, D. (2003), "The Role of Customer Satisfaction and Image in Gaining Customer Loyalty in the Hotel Industry", *Journal of Hospitality and Leisure Marketing*, 10 (1/2), pp. 3–25.

Kozak, N. (2008), *Tourism Marketing*, Ankara: Detay Publishing.

Nam, J., Ekinci, Y., and Whyatt, G. (2011), "Brand Equity, Brand Loyalty and Consumer Satisfaction", *Annals of Tourism Research*, 38 (3), pp. 1009–1030.

Oliver, R. L. (1997), *Satisfaction: A Behavioral Perspective on the Consumer*, New York: McGraw-Hill.

Oliver, R. L. (2014), *Satisfaction: A Behavioral Perspective on the Consumer*, New York: Routledge.

Pritchard, M. P. and Howard, D. R. (1997), "The Loyal Traveler: Examining a Typology of Service Patronage", *Journal of Travel Research*, 35 (4), pp. 2–10.

Rather, R. A. and Camilleri, M. A. (2019), "The Effects of Service Quality and Consumer-Brand Value Congruity on Hospitality Brand Loyalty", *Anatolia: An International Journal of Tourism and Hospitality Research*, 30 (4), pp. 1–13.

Selvi, M. S. and Ercan, F. (2006), "Customer Loyalty Assessment in Hotel Enterprises: An Application in Five Star Hotel Enterprises in İstanbul", *Balıkesir University-The Journal of Social Sciences Institute*, 9 (15), pp. 159–188.

Shang, R. A., Chen, Y. C., and Liao, H. J. (2006), "The Value of Participation in Virtual Consumer Communities on Brand Loyalty", *Internet Research*, 16 (4), pp. 398–418.

Skogland, I. and Siguaw, J. A. (2004), "Are Your Satisfied Customers Loyal?" *Cornell Hotel and Restaurant Administration Quarterly*, 45, pp. 221–234.

Suhartanto, D. (2011), "An Examination of the Structure and Determinants of Brands Loyalty across Hotel Brand Origin", *Asean Journal on Hospitality and Tourism*, 10 (2), pp. 146–161.

Suhartanto, D., Clemes, M., and Dean, D. (2013), "Analyzing the Complex and Dynamic Nature of Brand Loyalty in the Hotel Industry", *Tourism Review International*, 17 (1), pp. 47–61.

Tanford, S., Raab, C., and Kim, Y. S. (2012), "Determinants of Customer Loyalty and Purchasing Behavior for Full-Service and Limited-Service Hotels", *International Journal of Hospitality Management*, 31 (2), pp. 319–328.

Tayfun, A. and Yayla, Ö. (2013), "Brand Loyalty Factors Affecting the Hotel Elections of Tourists Investigation with Respect to the Demographic Variables", *Journal of Business Research*, 5 (4), pp. 159–169.

Tayfun, A. and Yayla, Ö. (2014), "The Examination of the Factors that Affect Brand Loyalty of Tourists in Hotel Election in terms of Behavioral and Attitudinal Loyalty", *Journal of Tourism and Gastronomy Studies*, 2 (1), pp. 30–35.

Wilkins, H., Merrilees, B., and Herington, C. (2009), "The Determinants of Loyalty in Hotels", *Journal of Hospitality Marketing & Management*, 19 (1), pp. 1–21.

Wu, C. C., Liao, S. H., Chen, Y. J., and Hsu, W. L. (2011), Service Quality, Brand Image and Price Fairness Impact on the Customer Satisfaction and Loyalty. In *International Conference on Industrial Engineering and Engineering Management*, IEEE International Conference, (pp. 1160–1164).

Yang, K. F., Yang, H. W., Chang, W. Y., and Chien, H. K. (2017), The Effect of Service Quality among Customer Satisfaction, Brand Loyalty and Brand Image. In *International Conference on Industrial Engineering and Engineering Management*, IEEE International Conference, (pp. 2286–2290).

Zeithaml, V. A. (1988), "Consumer Perceptions of Price, Quality, and Value: A Means-End Model and Synthesis of Evidence", *Journal of Marketing*, 52 (3), pp. 2–22.

Ozan ÇATIR

Investigation of Brand Positioning in Hotel Managements by Text Mining: A Case of Izmir Province

1 Introduction

Positioning a hotel brand in a tourism sector strongly isn't as easy relating with a logo or a brand as creating a marketing drive (Cai & Hobson, 2004). Positioning by considering only the needs of customers isn't enough not only for differing from other rival hotels, but it is also insufficient for making perceived the brand to the customers (Park et al., 1986). Therefore, it should be made a unique and diversified hotel brand positioning (Choi & Chu, 2001). A good hotel brand positioning provides the high occupancy rate, revisits and high profitability. For these reasons, it is important to know how to position a hotel brand uniquely (Brown & Ragsdale, 2002). The most important problem that leads to complicate the strong positioning is difficulty in establishing the products and services that are difficult to be imitated by rival hotels and that provides unique advantage and difficulty for delivering these to the customer (Anderson et al., 1999).

Assessment of customers' expectations and analysis of market conditions are required for a strong market positioning (Plumeyer et al., 2017). Survey data have generally been used for studies related to market positioning in tourism sector (Brown & Ragsdale, 2002). Survey studies containing minor sample studies may be insufficient to represent the customers' experiences completely (Krawczyk & Xiand, 2016). The studies that have been conducted recently and that have been analyzing the on-line comments with large scale reflect the customers' opinion better (Chiu et al., 2015). Therefore, this study has served for brand positioning of hotels by providing more trustworthy, more significant and practical information in accordance with both concerning customers' demands and technics used. Brand positioning includes basic customer preferences and expectations of brand, key rivals and competitive groups, better and worse services of brand and brand performance subjects (Hu & Trivedi, 2020).

This survey includes three basic research questions. These questions are; a) What are the preferences and expectations of customers from hotels in terms of customer's point of view? b) What are the fundamental scopes and qualities identifying rival brands and rival in the market? c) What are the differences of brand positioning between 4- and 5-star hotels. In this study, content analysis has been conducted with 4- and 5-star hotels' online comments with large scale in Izmir province by using text-mining method to hotel brand positioning with text mining. The assessment of certain brand characteristics have been conducted in accordance with customers' preferences and their perceptions.

Prior studies related to hotel brand positioning have been reviewed in the first part of the article. Research design has been presented and text analysis process has been detailed in method part. Yielded data have been presented in the following part. Eventually, both academic and practical arguments and contributions have been stated.

2 Literature Review

2.1 Hotel Brand Positioning

Brand positioning is how the customer positions the brand in his/her mind according to main rivals (Brown & Ragsdale, 2002). The hotel managements who want to make a distinguishable positioning should change their actual services, should design their price possibilities and also should step forth with the products to be served to certain target market (Torres & Kline, 2006). Brand positioning studies made that are related to the tourism sector have generally been conducted by using data that are yielded from questionnaire technique and interview technique (Prasad & Dev, 2000; Chen & Uysal, 2002; Kim & Agrusa, 2005; Kim et al., 2007; Wen & Yeh, 2010).

As an example, Chen and Uysal (2002) have conducted a brand positioning study by using two-dimensional correspondence analysis according to customer perception. Whereas Dev et al. (1995) have used multi-dimensional scaling analysis to determine the perceptions of related customers with predetermined seven hotel brands. Factor analysis has also been used to determine related dimensions of the brand (Kim et al., 2007). Since large data analysis has come to agenda, users' on-line content analysis by text mining analysis and revealing of customers' experiences have been resorted (Chiu et al., 2015). It's thought that analysis of users' contents with questionnaire data is more objective, more neutral and that margin is lower (Yamanishi & Li, 2002; Chiu et al., 2015; Krawczyk & Xiang, 2016). Although recent studies for brand positioning in tourism sector provided significant contribution procedurally practically, these studies assess only certain dimensions of customers' experiences. Hananto (2015) has examined the customers' remarks about determined characteristics of the brand whereas Chiu et al. (2015) has researched perceived brand performance in accordance with chosen characteristics. It's thought that arbitrary examinations lead to limited perception of brand positioning. For this reason it can said that there is a literature vacancy for multidimensional studies of brand positioning. In this context customers' preferences and expectations for hotels and basic dimensions and qualifications identifying rival brands and rivalry and differences f brand positioning between 4- and 5-star-hotels have been tried to determine.

3 Method

3.1 Research Design

In the first step, a web incision program has been used to get comments about 4- and 5-hotels from a voyage website. The whole comments related to hotels have been transferred from web media to an excel environment. Perceived brand relations have been filtrated manually and they have been coded as qualifications in the second step. Next, Perceptions of 4-star hotels have been compared with the perceptions of 5-star-hotels according to inferred codes.

3.2 Data Acquisition

Voyage Websites is one of the significant marketing instrument, which contain many data including rich customer content and comments, general assessments, rating, hotel types and hotel identifications. A data incision program has been used to scan comments for 4- and 5-star hotels in Izmir that presented from January 10, 2020, to February 25, 2020. Turkish comments have been analyzed. 5369 comments of 4-star hotels and 2966 comments of 5-star hotels have been used to assess.

The 4- and 5-star hotels in Izmir have been used in this study since town hotels are active all year round and these hotels are more institutional. 3-star hotels and hotels with less stars have been excluded from this study. Since more comments can be found for 4- and 5-star hotels, this study used these hotels.

In the first step of the study, key brand characteristics have been determined according to the key rating in on-line assessments. RapidMiner program have been utilized to check out the gained data. This program separates all the comments word-to-word with text mining analyze module and it presents outcomes with used word combinations. Separated words have been rated according to frequency rate. These words have been related to brand characteristics.

Tab. 18-1: Data Set Sample of Hotel Comments

ID	Hotels	Comments
1	4-Star Hotels	"I stayed during my business journey, location is super, and it is within walking distance to Konak. Hotel is clean and well-cared. But we couldn't start the air conditioning."
2	5-Star Hotels	"Hotel is in the city center and it is in such a safe environment that you can walk at nights. My room was very good even I checked-in at the late hours of the night. Everything was brand-new. Tea, coffee and fruit plate services were very kind and good. It was hard to find a place in the breakfast. There is no satisfactory food like pastry or croissant but you can find better thing than the standard breakfast at home."

3.3 Data Preprocessing

Comments of hotels have been downloaded by a special program that is used for web mining. Then data have made ready for text mining application by transferring to Excel program. Tab. 18-1 presents the data set sample of hotel comments.

Following to downloading of hotel comments, text-mining step has been started. RapidMiner program has preferred for text mining analysis. RapidMiner, "is an open source software which was written in Java language and it is very easy to use and it is used for academic studies". There for this program has been used. RapidMiner is the software which works with drag and drop method by allied operator functions and required parameters are easily determined by clicking operator functions. Thus coding data aren't needed. Many data can be processed by this program and many significant data can be yielded by these data (Celik, Akcetin, Gok, 2017). Yielded excel file has been transferred to RapidMiner program. Model has been prepared for separating the words in comments one by one, for extracting word frequencies and for association analysis. Design view of the model that was formed during data preprocessing is presented in Scheme 18-1.

In Scheme 18-1, the operators of model that was formed in Rapidminer program are seen. These operators make certain processes including "transform cases," "tokenize," "filter stop words" and "stem."

Brand Positioning in Hotel Management 353

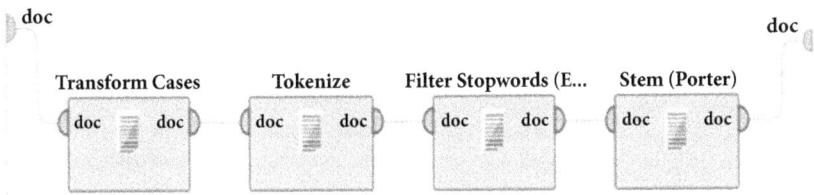

Scheme 18-1: Design View of Data Preprocessing

4 Research Findings

In this section, the findings obtained as a result of the analysis made with the data analysis program are presented. Tab. 18-2 presents word frequencies rating obtained from room comments related to the 4- and 5-star hotels in Izmir in data binning form.

In Tab. 18-2, it is seen that the word of room is most commonly used by customers (4243 comments and 1901 comments) for 4 and 5 star hotels when the key words about rooms are studied. It has established that the second most commonly used word for 4-star hotels is cleanness (816 comments) while view (816 comments) is the second commonly used word for 5-star hotels. It can be stated that customers of 5-star hotels give more importance to view than customers of 4-star hotels. The less used word for 4-star hotel is window (53 comments) and the decoration word is less used (51 comments) for 5-star hotels. It may be concluded that there is no problem about decoration in 5-star hotels. Although the comments are generally positive for hotels, noise is remarkable (99 comments) for 4-star hotels.

When customers' room preferences and expectations are searched, it may be stated that the customers of 4-star hotels give importance to the cleanness. Secondly, the view of the rooms is an important element for customers. The bed's properties should have enough quality to satisfy customers' expectations. Another expectation and preference reason is the rooms with enough width and size. The noise that 4-star hotels' customers experience has been a compliant subject. When an assessment is made for a 5-star-hotel managements, it may be stated that customers' have view expectation for their rooms as 4-stat hotels' customers. It has also been realized that customers take the width, cleanness, comfort and facility of the rooms into the consideration. Bed is another important subject for hotels' customers. It may be also said that 4- and 5-star hotels enter into rivalry with itself and each other. In this circumstance it may

Tab. 18-2: Word Frequencies Rating Obtained from Room Comments Related to the 4- and 5-Star Hotels

4-Star Hotel Words	Room f	5-Star Hotel Words	Room f
Room	4243	Room	1901
Cleanness	816	View	866
View	797	Big	490
Bed	781	Bed	353
Big	563	Bathroom	344
Bathroom	537	Cleanness	260
Small	476	Spacious	218
Spacious	436	Comfort	108
Voice	109	Relax	88
Noise	99	Furniture	67
Balcony	94	Decor	62
Hygiene	93	Terrace	53
Toilet	88	Decoration	51
Decoration	80		
Sleep	78		
Sheet	75		
Furniture	72		
Bathtub	61		
Terrace	61		
Aeration	58		
Design	57		
Well-kept	55		
Window	53		

be said that cleanness, bed quality, views, width and comfort factors should be assessed for related rooms' rivalry. There is no significant difference between 4- and 5-star hotels' customers' point of views for room factors. Similar comments have been made for the rooms in both types of hotels. Tab. 18-3 and 4 have presented comment frequencies for food and drink made for 4- and 5-star hotels.

When Tab. 18-3 is examined, it has been seen that the commonly used words in 4- and 5-star hotels' food and drink comments are breakfast (2302 and 1310 comments), food (1109 comments and 845 comments) and restaurant (454 and 575 comments). It can be inferred from the customers' comments that

Brand Positioning in Hotel Management   355

Tab. 18-3: Word Frequencies Related to Food and Drink Obtained from 4- and 5-Star Hotels' Comments

4-Star Hotel	Food and Drink	5-Star Hotel	Food and Drink
Words	f	Words	f
breakfast	2302	breakfast	1310
Food	1109	Food	845
Restaurant	454	Restaurant	575
Delicious	391	Diversity	239
Diversity	348	Delicious	156
Buffet	203	Buffet	145
Coffee	171	Coffee	106
Treat	105	bar	91
Fruit	104	beverage	60

food services served by 4- and 5-star hotel managements are diverse and delicious. Customers have also paid attention to the hotels with buffet (Buffet 203 comments; 145 comments).

When food-drink preferences and expectations of the customers are assessed, it may be stated that customers of 4-star hotel managements have an expectation for breakfast. Quality of the restaurant, taste of the foods and diversity of the served foods are also other important subjects. Customers prefer buffet service. It may also be stated that 5-star hotels' customer have similar first preferences for breakfast and they also give importance to the quality of the restaurants, taste of food and food diversity. These customers also prefer buffet service. Breakfast, quality of the restaurant, food and drink taste and diversity and buffet service should be assessed for rivalry of food-drink services. There is no significant difference in customers' point of views for food-drink factors between 4- and 5-star hotels. Similar comments have been made for the rooms in both types of hotels. Tab. 18-4 offers comment frequencies of adjectives for 4- and 5-star hotels.

When Tab. 18-4 is assessed, it can be seen that most commonly used adjectives for 4- and 5-star hotels are good (2250 comments, 1380 comments), relax (1069 comments, 533 comments), wonderful (898 comments, 920 comments), perfect (722 comments, 1042 comments) and high class (567 comments, 684 comments). It is seen that adjective of "old" is used for both 4- and 5-star hotel; hence, it can be understood that some parts of the hotels are worn out. It can be stated that customers of 4- and 5-star hotels have generally found these

Tab. 18-4: Comment Frequencies for Adjectives for 4- and 5-Star Hotels

4-Star Hotels	Adjectives	5-Star Hotels	Adjectives
Words	f	Words	f
Good	2250	Good	1380
Relax	1069	Perfect	1042
Wonderful	898	Wonderful	920
Perfect	722	relax	533
Quality	567	Old	452
Comfortable	553	Quality	684
Enough	518	Mangificient	249
Old	326	Free	181
Spacious	299	Small	169
Warm	285	Modern	169
Successful	263	Comfortable	167
Magnificient	243	Enough	140
Standard	223	Spacious	137
Far	219	Warm	134
Easy	218	Suit	133
Comfort	214	Kind	131
Modern	184	Successful	130
İdeal	172	Super	119
Kind	166	Easy	111
Pleasant	261	Rapid	110
Kind	156	Pleasant	108
Silence	152	Kind	100
Rich	152	beautiful	97
Bad	146	Silence	92
Missing	145	Rich	75
Super	144	Incredible	70
Rapid	129	Happy	69
Problem	128	Luxury	67
İnsufficient	121	exceptional	66
Unbelievable	108	important	59
Fresh	106	uncomfortable	56
Size	99		
Weak	94		
Cold	87		
Wretched	76		
Handy	70		

Tab. 18-4: Continued

4-Star Hotels	Adjectives	5-Star Hotels	Adjectives
Much	67		
Crowded	65		
Luxury	65		
Complaint	65		
Negative	63		
Positive	60		
Regular	59		
Decent	59		
Noisy	58		
Peaceful	54		
Pleasant	53		

managements good, comfortable, wonderful, perfect and high class. When it comes to negative adjectives it is seen that customers of 4-star hotels have used old (326 comments), terrible (146 comments), deficient (145 comments), insufficient (121 comments), bad (94 comments), awful (76 comments), crowded (65 comments), negative (63 comments), noisy (58 comments) adjectives. Whereas customers of 5-star hotels have used old (452 comments) and uncomfortable (56 comments) adjectives. It may be said that 5-star hotel customers have made more positive comments than 4-star hotel customers.

When customers' preferences and expectations including related adjectives are assessed it may be said that customers of 4-star hotel managements have found these hotels as good, comfortable, wonderful, perfect, high class and conventional. They have also used certain negative adjective for dissatisfactions as old, deficient, insufficient, terrible, awful, negative and noisy. However customers have generally made positive comments for 4-star hotels. On the other hand customers of 5-star hotel managements have found these managements good, perfect, wonderful, comfortable, high class, magnificent and modern. They have also used certain negative adjectives for dissatisfactions as old, small and uncomfortable. However, customers have generally made positive comments for 5-star hotels. It can be inferred that managements organize certain activities for the expectations of hotel customers. Negative aspects that have been highlighted by the customers of 4-star and 5-star hotels in Izmir are old, noisy and uncomfortable hotels. In this context, making physical arrangements and renewing work-outs are important for the rivalry of hotels.

Tab. 18-5: Word Frequencies Related Stuff That Is Obtained from 4- and 5-Star Hotel Comments in Izmir.

4-Star Hotel Words	Staff f	5-Star Hotel Words	Staff f
Staff	2781	Staff	1657
rReception	713	Helpful	455
Helpful	677	Reception	265
Smiling	555	Smiling	249
Friend	335	Concerned	187
Reservation	257	Professional	152
Concerned	225	Team	93
Team	187	Reservation	91
Jack	178	Friend	89
Manager	73	Jack	60
Understanding	56		

It can be inferred that brand perception of 4-star and 5-star hotels' customer are generally positive. There is no significant difference between brand positioning works of hotels according to point of views of hotel customers. Tab. 18-5 present word frequencies related stuff that is obtained from 4-star and 5-star hotel comments.

Tab. 18-5 shows that customers have used helpful (677 comments, 455 comments), jocund (555 comments, 249 comments), respectful (225 comments, 187 comments) and friendly (335 comments, 89 comments) adjectives in their comment about stuff in 4-star and 5-star hotel managements. It may be stated that customers have made positive comments about workers. No negative adjective has been found in the comments about stuff. Most commonly mentioned stuff is receptionists (713 comments and 265 comments).

As the customers' expectations and preferences about stuff are assessed it may be stated that customers have found the stuff in 4-star hotels as helpful, jocund, friendly, sympathizer and respectful. When it comes to stuff in 5-star hotels customers' comments about stuff are similar as helpful, jocund and respectful. One of the most commonly used words is reception. Therefore it can be inferred that receptionist have an important role in customer's satisfaction. It may be stated that there are certain teamwork applications for both 4- and 5-star hotel managements. Having helpful, jocund, friendly, sympathizer, respectful and teamwork skill stuff provides advantages for 4-star and

Tab. 18-6: Word Frequencies Related to Transportation Obtained from 4- and 5-Star Hotels Comments in Izmir

4-Star Hotel	Transportation	5-Star Hotel	Transportation
Words	f	Words	f
Distance	424	Centrum	561
Way	202	City	286
Airport	143	Distance	270
Metro	129	Walking	119
Car	120		
On foot	110		
Taxi	108		
Car	120		
Traffic	76		
Bus	62		

5-star hotel managements in rivalry. It can be said that customers have positive perceptions about stuff and that positive characteristics of stuff should be brought into the forefront. Tab. 18-6 shows word frequencies related to transportation obtained from 4-star and 5-star hotels comments in Izmir.

Tab. 18-6 shows that it can be said that the most commonly used words for 4-star hotels is distance (424 comments) while it is centrum (561 comments) for 5-star hotels. The bus word (62 comments) for 4-star-hotels and the walking word (119 comments) have been the least used words in hotel comments. It can be stated that most of the hotels are very close to centrum in Izmir and they are within walking distance. Besides, it may be added that customers are able to reach their hotel by metro, taxi or with their own vehicles. The traffic word (76 comments) has been found in comments about 4-star-hotels. It may be an evidence for the presence of the traffic jam in centrum.

Customers' preferences and expectations for transportation shows that customers' of 4-star hotel managements prefer airport, railway, highway, bus and taxi transportation to arrive their hotels. Indeed, it has also been stated that 4- and 5-star hotels are in the city-center and they are within walking distance. Since hotels are in the city-center, it may be said that transportation is easy to arrive. The transportation to the hotels may be provided by every type of vehicle. Accessibility may be an advantage in the rivalry. Centricity and accessibility are effective factor to use in brand positioning of both types of hotel for transportation. Due to the centricity of hotels, traffic jam in centrum

Tab. 18-7: Word Frequencies in Service Comments Obtained from 4- and 5-Star Hotel Comments in Izmir

4-Star Hotel Words	Service f	5-Star Hotel Words	Service f
Service	597	Service	427
Car park	782	Car park	171
Internet	197	Internet	105
Air conditioning	192	Lobby	101
Elevator	95	Air conditioning	77
Turkish Bath	89	Turkish bath	74
Sauna	77	Sauna	60
Wifi	77		
Music	73		

is reported as a negative factor. Nevertheless, hotel transportation is generally easy. Tab. 18-7 shows word frequencies of 4- and 5-star hotels about transportation that were obtained from hotel comments.

Tab. 18-7 shows that words of car park (782 comments, 171 comments), Internet (197 comments, 105 comments), air conditioner (192 comments, 77 comments), Turkish bath (89 comments, 74 comments) and sauna (77 comments, 60 comments) are mentioned for 4- and 5-star hotels. Since most commonly used word is car park, the service of this factor is significant. It may be inferred that air-conditioner is an obligatory service for air conditioning service of hotels because these hotels are in Izmir of which destination warm climate has. Nowadays Internet service is an inevitable service for hotels. Absence of Internet service may lead to customers' dissatisfaction.

When customers' preferences and expectations for services are searched, it may be inferred that customers prefer car park, Internet, air conditioner, Turkish-bath, sauna and Wi-Fi services. The customers of 5-star hotels also prefer car park, Internet, lobby, air conditioner, Turkish bath and sauna accordingly. In addition to accommodation and food-drink services of the hotels, other services that are served by the hotels are also important for customers. One of the most important services that are provided by the hotels in the city center of Izmir is car park. Internet and air conditioning services are also important factors for hotel rivalry. It can be stated that there is no significant difference for the provided services between 4- and 5-star hotels in Izmir. Hotels are

Tab. 18-8: Word Frequencies of Actions in Hotel Comments for 4- and 5-Star Hotels in Izmir

4-Star Hotel	Actions	5-Star Hotel	Actions
Words	f	Words	f
Recommending	846	Recommending	466
Acknowledging	679	Acknowledging	322
Be interested in	600	Prefer	319
Satisfaction	588	Satisfaction	272
Branding	53	Be interested in	187

Tab. 18-9: Word Frequencies about Price Comments of 4- and 5-Star Hotels in Izmir

4-Star Hotel	Price	5-Star Hotel	Price
Words	f	Words	f
Price	1287	Fee	126
Free	280	Expensive	76
Fee	162		
Money	122		
Expensive	74		

perceived by customers similarly. Tab. 18-8 shows word frequencies of actions in hotel comments for 4- and 5-star hotels in Izmir.

Tab. 18-8 shows that recommending and acknowledging are the mostly used actions in both 4-star hotels and 5-star hotels. In this context, it may be inferred that customers are satisfied with the services of the hotels. The word of satisfaction in the comments is also used commonly. The action of respect also proves that stuff concern customers closely. As the customers' preferences and expectation for actions show, 4-star hotel managements' customers are satisfied with the services and they state that they would recommend the hotels, they say thank you for the service and they are content with the services. It may be said customers' expectations were satisfied. Accordingly, it may be stated that 5-star hotel managements' customers are satisfied with the services and they state that they would recommend the hotels, they say thank you for the service and they are content with the services. It is also seen that hotel stuff is respectful to the customers. Tab. 18-9 shows word frequencies about prices that were obtained from 4- and 5-star hotels' comments.

Tab. 18-9 shows that the most commonly used word is price (1287 comments) for 4-star hotels while it is cost (126 comments) for 5-star hotels. Since expensive word (74 comments, 76 comments) is widely used for both 4- and 5-star hotels it is seen that customers perceive these hotels as expensive. The word of free (280 comments) is used for 4-star hotels; it may be a signal that shows certain services are provided freely.

Customers' preferences and expectations of 4- and 5-star hotels show that hotels' prices are expensive for customers. It is also seen that some of the services are free in 4-star hotels. Price rivalry always exists among hotels. Therefore it is seen that 4- and 5-star hotels suit high-income group. It is concluded that hotels should either screw down by lowering the cost to participate in the rivalry or they should diversify their services to sell customers on buying more precious services from their hotels. It is seen that brand perception of the customers about both types of hotel prices is stated as expensive. It may be stated that similar price policies are applied.

Conclusion and Recommendations

Hotel managements struggle with gaining interest by positioning themselves into the market. They sometimes diversify their services or they may sometimes screw down by decreasing their cost during these positioning activities. Hotel managements should provide their services by taking the target group's characteristics into consideration and by providing a product with demanded qualifications that customers expect. Therefore, hotel managements should follow customers' expectations, preferences, status of rivals and rivalry qualifications and brand positioning work-outs of rival managements closely.

The answers of three basic questions have been searched in this study for this purpose. a) What are the customers' preferences and expectations from hotels in accordance with customer's point of view? b) What are the rival brands in the market and what are the basic dimensions and qualifications that identify rivalry? c) What are the differences of brand positioning between 4- and 5-star hotels? In this study, content analysis has been conducted with 4- and 5-star hotels' online comments with large scale in Izmir province by using text-mining method to hotel brand positioning with text mining to realize this aim.

Customers' preferences, expectations, rivalry status and difference of brand positioning between 4- and 5-star hotels have been examined in accordance with customers' comments in eight different themes (**room, food-drink, adjectives, transportation, actions, stuff, service and price**). The results obtained according to defined study conclusions are presented below.

Brand Positioning in Hotel Management 363

a) **What are the customers' preferences and expectations from hotels in accordance with customer's point of view?**

In accordance with the customers' preferences and expectations about the room, it may be stated that customers' of 4-star hotel managements regard the room cleanness. Secondly, views of the rooms are important factor for customers. The quality of the beds should be in the quality that satisfies customers' expectations. The room with enough width and size is also another reason for customers' expectations and preferences. The customers of 4-star hotels complain about noise in the hotels. When it comes to 5-star hotel managements; it may be said customers also have expectations for rooms with view same as 4-star hotel customers. It has been seen that customers take room's size, cleanness, comfort and facility into consideration. Quality of the beds is another important factor for customers.

When customers' preferences and expectations for food-drink service are analyzed it may be indicated that customers of 4-star hotels have expectations for the breakfast service. The quality of the restaurant, foods' tastes and food diversity are also other important factors for customers. The buffet service is included in the customers' preferences. Similarly to former results it may be expressed that customers of 5-star hotel managements have also expectation for breakfast firstly and that quality of the restaurant, foods' tastes and food diversity are also other important factors for customers. The buffet service is included in the customers' preferences

When customers' preferences and expectations for adjectives are analyzed it may be indicated that customers of 4-star hotel managements have found these hotels as good, convenient, wonderful, perfect, high-class and comfortable. These customers have also used old, deficient, insufficient, terrible, awful, negative and noisy adjectives to express their dissatisfactions. However, customers have made positive comments for 4-star hotel managements in general. Whereas customers of 5-star hotel managements have found their hotels as good, perfect, wonderful, convenient, high-class, great and modern. These customers have also used old, small and inconvenient adjectives to express their dissatisfactions. However, customers have made positive comments for 5-star hotel managements in general.

When customers' preferences and expectations for adjectives are analyzed, it may be indicated that customers of 4-star hotel managements have found these hotels' stuff as helpful, jocund, friendly, sympathetic and respectful. Similarly to former results it may be expressed that customers of 5-star hotel managements have assessed the stuff as helpful, jocund and respectful. The most commonly used word is reception in the comments. Therefore it may be inferred that receptionist have an important role in customers' satisfaction. It may also be added that both of 4- and 5-star hotels have team works applications in their managements.

When customers' preferences and expectations for transportation are analyzed, it may be seen that customers of 4-star hotel managements prefer airport, railway, highway and bus, taxi to arrive their hotels. Besides it has been stated that 4- and 5-star hotels are in the city-center and they are within walking distance.

When customers' preferences and expectations for services are analyzed, it may be indicated that customers of 4-star hotel managements prefer car park, Internet, air- conditioning, Turkish bath, sauna and Wi-Fi. Similarly to former results it may be expressed that customers of 5-star hotel managements prefer car park, Internet, lobby, air-conditioning, Turkish bath and sauna.

When customers' preferences and expectations for services are analyzed, it may be indicated that customers of 4-star hotel management would recommend the hotels, they are content with the services and they say thank you to the managements. It may be inferred that the expectations of the customers have been fulfilled. Similarly to former results it may be expressed that customers of 5-star hotel managements would recommend the hotels, they are content with the services and they say thank you to the managements. It is seen that hotel stuff is respectful to the customers.

When customers' preferences and expectations for prices are analyzed, it may be indicated that customers of 4 and 5-star hotel management have found the prices as expensive. It is also seen that some of the services of the 4-star hotels are free.

b) **What are the rival brands in the market and what are the basic dimensions and qualifications that identify rivalry?**

It may be regarded that 4- and 5-star hotel managements are in rivalry within themselves and within each other. Thus cleanness, bed quality, view, size and comfort factor should be assessed for rooms. The breakfast, quality of restaurants, foods and drinks' taste and diversity and buffet services factors should be assessed for food-drink rivalry. The negative aspects of the 4- and 5-star hotel managements in Izmir are being old, being noisy and being inconvenient. In this aspect, physical arrangements and renewing work-outs are important for hotel rivalry. If the stuff is helpful, jocund, friendly, sympathetic, respectful and if stuff has team work skills, these factors may provide an advantage for 4- and 5-star hotel managements in the rivalry. It may be inferred that arriving to the hotels is easy since the hotels are in the city-center. The transportation may be provided to the hotels by all types of transportation vehicle. The accessibility may be used as an advantage in the rivalry. The additional services to accommodation and food-drink services are also important for hotel customers. One of the most important additional services for the hotels in the city center of Izmir is car park. Besides Internet and air-conditioning services are also important factor for hotel rivalry. The price rivalry always exists among the hotels. Therefore it is seen that 4- and 5-star hotel managements suit high-income group's fancy. It is concluded that hotels should either screw down by lowering the cost to participate in the rivalry or they should diversify their services to sell customers on buying more precious services from their hotels.

c) **What are the differences of brand positioning between 4- and 5-stars hotels?**

There is no significant difference about rooms between 4- and 5-stars hotels in accordance with the customers' point of view. Similar comments have been done for rooms in both types of hotels. There is no significant difference about food-drink service between 4- and 5-stars hotels in accordance with the customers' point of view. Similar comments have been done for rooms in both types of hotels. It may be expressed that brand perception of customers for 4- and 5-star hotels is positive. There is no significant difference about brand positioning workouts between 4 and 5-stars hotels in accordance with the customers' point of view. It can be stated that customers' perception about stuff is positive and that positive aspects of the stuff should be brought into the forefront in brand positioning workouts. Since both types of hotels are in the city-center and they have accessibility, this factor can be used in brand positioning. The traffic jam in the city-center has been stated as a negative aspect. However, hotels have accessibility in general. It may be stated that there is no significant difference about services between 4- and 5-star hotels in Izmir. The hotels are perceived as similar by their customers. It is seen that customers of both types of hotels found hotels' brand perception as expensive for their prices. It may also be added that hotels apply similar price policies.

Two theoretical contributions have been provided by this study. Firstly this study examines the content formed by customers in contrast to the previous studies that were conducted with questionnaire and interview techniques for brand positioning in the literature. Since this survey was made with text mining method, it also provides a contribution to the literature. Secondly, determination of customers' expectations, determination of basic qualifications for rivalry and comparison of 4- and 5-star hotels also provide another theoretical contribution to the brand positioning study.

In terms of pragmatists, this study may help to analyze the contents with large scale of the users and customers' experiences more systematically and scientifically, to be understood the brand positions in the market by hotel managements better and to improve the performances in basic brand qualifications against important rivals. Generally, the results of the study are efficient for hotel managements to confirm brand status according to the customers' perceptions, to determine the rivalry status and to specify rivalry strategies to improve.

This study has certain restrictions. This study only includes one-voyage website comments about 4- and 5-star hotels in Izmir. Different hotel comments about different hotel managements in different tourism destinations may be examined. The obtained comments are selected from Turkish language. Therefore the comments in different language can be used in forthcoming studies. All 4- and 5 star hotel managements have been selected for the study. How the brand of a unique hotel chain managements according to the customers' perception may be examined in forthcoming studies.

Bibliography

Anderson, R. I., Fish, M., Xia, Y., and Michello, F. (1999), "Measuring Efficiency in the Hotel İndustry: A Stochastic Frontier Approach", *International Journal of Hospitality Management*, 18, pp. 45–57, https://doi.org/10.1016/S0278-4319(98)00046-2.

Brown, J. R. and Ragsdale, C. T. (2002), "The Competitive Market Efficiency of Hotel Brands: An Application of Data Envelopment Analysis", *Journal of Hospitality and Tourism Research*, 26, pp. 332–360, https://doi.org/10.1177/109634802237483.

Cai, L. A. and Hobson, J. S. P. (2004), "Making Hotel Brands Work in a Competitive Environment", *Journal of Vocational Marketing*, 10, pp. 197–208, https://doi.org/10.1177/135676670401000301.

Chen, J. S. and Uysal, M. (2002), "Market Positioning Analysis: A Hybrid Approach", *Annals of Tourism Research*, 29, pp. 987–1003, https://doi.org/10.1016/S0160-7383(02)00003-8.

Chiu, C., Chiu, N. H., Sung, R. J., and Hsieh, P. Y. (2015), "Opinion Mining of Hotel Customer Generated Contents in Chinese Weblogs", *Current Issues in Tourism*, 18, pp. 477–495, https://doi.org/10.1080/13683500.2013.841656.

Choi, T. Y. and Chu, R. (2001), "Determinants of Hotel Guests' Satisfaction and Repeat Patronage in the Hong Kong Hotel İndustry", *International*

Journal of Hospitality Management, 20, pp. 277–297, https://doi.org/10.1016/ S0278-4319(01)00006-8.

Çelik, U., Akçetin, E., Gök, M. (2017), *Rapidminer ile Uygulamalı Veri Madenciliği*, Pusala Yayınevi, İstanbul.

Dev, C. S., Morgan, M. S. and Shoemaker, S. (1995), "A Positioning Analysis of Hotel Brands: Based on Travel-manager Perceptions", Cornell Hotel & Restaurant Administration Quarterly, 36, pp. 48–55, https://doi.org/10.1177/001088049503600617.

Hananto, A. (2015), "Application of Text Mining to Extract Hotel Attributes and Constructperceptual Map of Five Star Hotels from Online Review: Study of Jakarta and Singapore Five-star Hotels", *Asean Marketing Journal*, VII, pp. 58–80, https://doi.org/10.21002/amj.v7i2.5262.

Hu, F. and Trivedi, R. H. (2020), "Mapping Hotel Brand Positioning and Competitive Landscapes by Text-mining User-generated Content", *International Journal of Hospitality Management*, 84, https://doi.org/10.1016/j.ijhm.2019.102317.

Kim, D. J., Kim, W. G., and Han, J. S. (2007), "A Perceptual Mapping of Online Travel Agencies and Preference Attributes", *Tourism Management*, 28, pp. 591–603, https://doi.org/10.1016/j.tourman.2006.04.022.

Kim, S. S. and Agrusa, J. (2005), "The Positioning of Overseas Honeymoon Destinations", *Annals of Tourism Research*, 32, pp. 887–904, https://doi.org/10.1016/j.annals.2004.12.004.

Krawczyk, M. and Xiang, Z. (2016), "Perceptual Mapping of Hotel Brands Using Online Reviews: A Text Analytics Approach", *Information Technology & Tourism*, 16, pp. 23–43, https://doi.org/10.1007/s40558-015-0033-0.

Park, C. W., Jaworski, B. J., and MacInnis, D. J. (1986), "Strategic Brand Concept-image Management", *Journal of Marketing*, 50, pp. 135–145, https://doi.org/10.2307/1251291.

Plumeyer, A., Kottemann, P., Böger, D., and Decker, R. (2017), "Measuring Brand Image: A Systematic Review, Practical Guidance, and Future Research Directions", *Review of Managerial Science*, 13, pp. 227–265, https://doi.org/10.1007/s11846-017-0251-2.

Prasad, K. and Dev, C. S. (2000), "Managing Hotel Brand Equity", *Cornell Hotel & Restaurant Administration Quarterly*, 41, pp. 22–31, https://doi.org/10.1177/001088040004100314.

Torres, E. N. and Kline, S. (2006), "From Satisfaction to Delight: A Model for the Hotel Industry", *International Journal of Contemporary Hospitality Management*, 18, pp. 290–301, https://doi.org/10.1108/09596110610665302.

Wen, C.-H. and Yeh, W.-Y. (2010), "Positioning of International Air Passenger Carriers Using Multidimensional Scaling and Correspondence Analysis", *Transportation Journal,* 49 (1), pp. 7–23.

Yamanishi, K. and Li, H. (2002), "Mining Open Answers in Questionnaire Data", *IEEE Intelligent System,* 17, pp. 58–63, https://doi.org/10.1109/MIS.2002.1039833.

Berna KIRAN BULĞURCU

Determining the Relative Importance of the Factors for the Selection of Winter Tourism Centers

1 Introduction

Considering the global economy, we observe that the tourism industry comes after oil and automotive industries, and the tourism industry is on the rise. Turkey is one of such countries in the world that attracts the utmost attention with its cultural and regional characteristics. Given its rich capacity of serving domestic and international tourists from sea tourism, health tourism, congress tourism to mountain tourism, Turkey's potential to meet the expectations for winter tourism attracts attention. As the major centers of winter tourism, Bursa and Nevşehir are among the top five cities where the highest number of domestic and international tourists stayed overnight during January–March 2020 (Ministry of Culture and Tourism, 2020: 20). Ranking 13th in the world in respect of tourism revenues, Turkey aims at managing the growing demand for winter tourism and upgrading its current position, as substantiated by the special expertise commission report drafted by the Ministry of Development for 2023. The report explains the strategies for the development of winter tourism in detail (2018: 30). One such strategy is to remove the physical shortcomings of the provinces that are situated on the corridor of winter tourism and to strengthen their infrastructure. Intensifying the promotional and marketing activities for the winter tourism theme and actions to be taken are the diverse elements of the strategic plan. If the non-governmental organizations, local administrations, and local people may contribute to raising the mobility of winter tourism, Turkey may reduce the problem of seasonality, as defined in the same report as a part of this strategy. As Turkey is surrounded by sea on three corners, it inevitably stands out for vacation due to its sunny weather, seas, and sandy beaches. Besides, Turkey aims at using the advantage of being a country where four seasons may be lived concurrently, and strives to distinguish winter tourism from skiing tourism to keep the tourism dynamic throughout the year. For instance, the characteristics of winter tourism in Cappadocia, differentiated from skiing, attract more domestic and international tourists. The importance attributed by Turkey to winter tourism and the efforts to differentiate winter tourism from the tourism based on skiing and the resulting growth of international tourists have recently given Turkey a robust position in the annual barometer of the World Tourism Organization. This allowed Turkey to hold the 4th position amongst the European countries and the 6th position in the world, thereby strengthening Turkey's competitive framework in the tourism industry. Per the International Report on Snow and Mountain Tourism of April 2020, Turkey established 50 new mechanic ski lifts during the last decade. The report underlined that Turkey still did not have an adequate share in the winter tourism market to draw the attention of international tourists. The media outlets reported that the government was planning to implement a 50-million EUR project regarding winter tourism and the goal was to raise the number of winter tourism centers to 100 for 14 million skiers (Vanat, 2020). Therefore, the entire works that will add to the winter tourism centers in Turkey bear a great level of importance. Several national studies in the literature cover winter tourism in all aspects. The most striking point about such studies is on how winter tourism is perceived regionally in Turkey and how it should be developed. Per the data retrieved from the Ministry of Culture and Tourism in 2019, the focal point of the studies is the winter tourism centers, nine of which are active (Erzurum Palandöken, Bursa Ulu Mount, Bolu Köroğlu Mount, Kayseri Erciyes, Kastamonu Çankırı Ilgaz, Kars Sarıkamış, Kocaeli Kartepe, Isparta Davraz Mount, Sivas Yıldız Mount) and seven of which are partially inactive with a total bed capacity of 11,362. In examining the studies conducted to determine the locations with a potential for winter tourism (Altaş et al., 2015; Arslan-Muhacir & Yaman, 2016; Karaman & Gül, 2016; Yıldız & Kaya, 2016; Kadıoğlu, 2017; Evren et al., 2019) or development of the available winter tourism centers with the potential (İbret, 2006; Atay & Akyurt, 2009; Koca et al., 2011; Göktuğ & Arpa, 2015; Korgavuş, 2017; Eşitti, 2018; Özçoban, 2019), it can be observed as a commonality that the emphasis is mainly laid on how the winter tourism centers should be, under which conditions they should serve, and the requirement regarding the satisfaction of tourists who are accommodated in the tourism centers to raise the share of the winter tourism. Considered for competitiveness, Turkey has to encourage the production of academic studies on the selection of winter tourism centers and act in line with these studies to be able to compete with the global winter tourism centers and stand out in the field of winter tourism. A comprehensive survey of the national and international literature can show that one striking issue involves the criteria considered by tourists when making a selection out of the winter tourism centers. When these criteria are accurately assessed from the standpoints of tourists, it becomes clear that particular issues have to be taken into account by the winter tourism centers. When it comes to the assessment of the relevant criteria on this matter, the relative importance of the criteria brings into mind the integrity of the techniques which are successfully implemented. Here, techniques called Multi-Criteria Decision Making (MCDM) methods become important and preferred particularly during group decision-making processes with a scientific and analytical framework. The multi-criteria decision-making techniques that allow us to assess with the help of a small group of experts without the need for traditional statistical assumptions mostly aim at making assessments that bring together the objective research data with subjective judgments in a rational manner (Lo et al., 2020: 2). MCDM problems follow a procedure that starts with the determination of the factors that may be assessed, allowing the specialist decision-makers to determine the relative importance of these factors and to present the order of precedence of the relevant factors. Analytical Hierarchy Process (AHP), Analytical Network Process (ANP), ARAS, COPRAS, DEMATEL, ELECTRE, MOORA, MOOSRA, OCRA, TOPSIS, PROMETHEE, VIKOR are all MCDM techniques and may individually be implemented in various fields ranging from social sciences and positive sciences to engineering and medical sciences as regards the assessment of factors and the selection of alternatives.

There are limited studies conducted with MCDM techniques for winter tourism. One such research is on the selection of the most suitable one among the strategies stipulated about the overhauling Brezovica skiing center by Stojčetović et al. (2015) by employing the AHP method. By conducting a SWOT analysis, the skiing center management came up with a total of 16 factors on the strengths and weaknesses of the situation at hand related to the renovation of the center as well as the opportunities and threats that may be internal or external. Furthermore, both the SWOT factors and strategies determined by diversification, quality improvement, and customization strategies and the strategies were compared and analyzed through the AHP method, and a diversification strategy was picked up amongst the existing strategies. Erbas (2016) attempted to measure the IPA technique, a technique frequently used in measuring the tourism destination competition, in terms of its deficiencies in measuring the competitiveness with the help of the ADA (Attribute–Determinance Analysis) and AHP techniques, and also formed a new technique called CDPA (Competitive Determinance–Performance Analysis). This new and hybrid technique helped to assess increasing the competition activities at the Erciyes winter destination in Kayseri as a major winter tourism center. The manager of the skiing center contacted 20 individuals with expertise in different fields such as civil servants working at the Ministry of Tourism or freelance skiing trainers. It was noted that Erciyes should attribute precedence to the entertainment factor to outrun its rivals in winter tourism in the future. The major factors were cited as snow quality, cost value rate, and climate change.

Oğuz (2018) researched on the selection of the optimum center for accommodation as the demand for winter tourism in Turkey is on the rise. Oğuz thus used the fuzzy AHP method and conducted a verbal assessment. The study assessed price, accessibility, accommodation, characteristics of the facility, and alternative tourism as the main criteria along with 19 sub-criteria. The persons selected in the study were those who were accommodated at different winter centers in Turkey. The data obtained as a result of the assessment show that price was noted as the most important criterion. Price was followed by the characteristics of the facility. The least important criterion was the alternative tourism. The concept of fuzziness was used in this study particularly to prevent any uncertainties regarding verbal judgments.

Erzincanlı and Aksakal (2019) presented a study in which they analyzed the integrated MCDM problem with the help of Entropy and IRP models dedicated to the selection of the skiing centers in Turkey which draw further attention in the international arena every single day. The conditions of the ski-tracks of six different skiing centers such as Palandöken, Uludağ, Erciyes, Sarıkamış, Kartalkaya, Kartepe, and the quality of the services, general service quality, training status, snow quality, cost and health services and other criteria relating to these six skiing centers were assessed. As a result, it was noted that the Palandöken skiing center was the most ideal one.

Although several topics falling under the title of tourism were researched, the study particularly considers Turkey as an exemplary country where winter tourism should be attached importance. The main purpose of this study is to assess the main and sub-factors that are effective in selecting the winter skiing centers. The study also aims at developing suggestions as to the selection of winter tourism centers in Turkey and to increase the level of competition under the notion of competitiveness. MCDM methods are frequently used in places where taking a decision is equal to making a suggestion. This is the reason why the DEMATEL technique is used as it is capable of simultaneously reviewing and prioritizing the cause-and-effect relationship in assessing the main and sub-factors which are effective in the selection of the winter tourism centers. Furthermore, the AHP technique is used, as it can calculate the mutual and relative importance of the factors, to reinforce the results obtained by DEMATEL. A joint weighted calculation is conducted by merging and comparing the results obtained by these two techniques and the issue of how to turn the winter tourism centers in Turkey into an exemplary field of research will be elaborated.

Following the detailed literature review for reaching the main goal, surveys developed based on the techniques were conducted with the attendance of the persons who are knowledgeable about the winter tourism centers and who previously stayed in minimum three different winter tourism centers of Turkey and the world and performed winter sports and preferred the winter tourism regularly, that is, those who can be qualified as experts, to speak in terms of the MCDM methods. The results have been transformed into data and transferred to the Excel, and, accordingly, analyses have been made in line with the techniques used here. The secondary purposes of the study are to discuss the economic and social benefits of winter tourism by providing information on the concept of winter tourism and to explain how the final factors that should be deemed as significant in the selection of the winter tourism centers are identified through a scientific method.

After the introduction, the second part of the study dwells on the concept of winter tourism as well as its economic and social benefits. The literature review puts forth the studies regarding the criteria which are considered important in selecting the tourism center. The third part explains in detail the steps of the implementation techniques of DEMATEL and AHP which are the subjects of the methodology of this present study. On the other hand, the fourth part of the study, which focuses on the implementation process, analyzes the data of the survey for the selection of the winter tourism centers, and the findings will be presented and discussed with a reference to tables. The conclusion part, which covers a general assessment of the study, and presents suggestions for raising the competitiveness of the winter tourism centers, and discusses how any future research on this topic, may be conducted.

2 Literature Review

The birth of the winter tourism emanates from the idea of ensuring that the scope of tourism may be disseminated throughout the entire year. In this respect, a new tourism field was created as investments were made with the prediction that people going on vacation in the summer time will also prefer to have a vacation during winter. Several definitions of winter tourism were made. The broadest one of these definitions is the type of tourism that allows for the most appropriate conditions for skiing and refers to travels, accommodation, entertainment, and sight-seeing at various periods of the year (İlban & Kaşlı, 2011: 321). The definition of winter tourism also includes other concepts as speed skating, sled racing and ice hockey (İbragimov, 2001: 34).

The mountainous areas should be rationally arranged to do winter tourism at present. An important item here is to remove the shortcomings related to the infrastructure. Governments should, therefore, consider the social benefits of winter tourism and act accordingly. The success of the European countries that house the Alpes is evident. With the help of rational solutions, regional economic growth was assured and migration was eased in this region which previously experienced continuous migrations and a lack of economic growth (Price et al., 2011: 7). These regions are currently ranked in the first position in winter tourism, and as stated by Williams and Shaw (1988: 88) in their studies, winter tourism will bring along several economic and social benefits to the society with the creation of employment, contribution to the government revenues, and improvement of the infrastructure. On raising the tourism revenues, which decline during the winter and taking advantage of the inert labor force that worked during the summertime, Çakmak and Yılmaz (2018) compared winter tourism with other types of tourism and underscored that the government should provide further incentives to winter tourism by making positive discrimination since the investments into this particular tourism remain restricted. The study also focuses on the sustainability of winter tourism across the country without directly considering the high economic level of the people who prefer winter tourism, and concentrates on the efforts to improve the service quality.

The first pillar of the improvement of the service quality should be to eliminate the infrastructure shortcomings, if any. Therefore, to secure skiing and attract more tourists to a tourism center, the first task will be to support the development of the regions that have mountains conducive to winter tourism. Considering that Turkey still has several deficiencies in this particular issue, activities should be carried out to raise the service quality and to ensure the flux of tourists by raising specialized labor force in winter tourism, equipping accommodation centers with adequate bed capacities and services, resolving the transportation problems, and improving the mechanical facilities (Ministry of Culture and Tourism, 2007).

Wang (2011: 3) stressed that the interests of the society and the service providers and service recipients should match one another both economically and socially for a particular tourism center to be marketed in a manner that attracts tourists to the region. Wang also proposed that a tourist-centered approach should be adopted. In this context, the factors that may be effective in selecting the winter center should be searched for satisfying tourists and improving service quality. There are several studies in the literature regarding the issues that have to be considered in the selection of tourism centers. Price, accessibility, comfort, destination image, and alternative tourism resources emerge as five main factors that have been identified as a result of the literature review conducted to determine the selection factors of winter tourism centers. These five main factors and the explanatory sub-factors of these main factors are shown in Tab. 19-1.

If we are to list a part of the studies determining all these factors, the first will be Manap's study (2006) on which, from among alternative tourism centers such as Alanya, Belek, Beldibi, Bodrum, Çeşme, Fethiye, Kemer, Kuşadası and Marmaris, Side will be selected as the one where one can make the most of the trio of the sea, sand, and sun in Turkey. The tourists were recommended to decide on a system through the AHP method by considering both qualitative and quantitative criteria such as natural environment, entertainment, distance to the airport, price, number of blue flag bays, number of archeological sites, and intensity. Another study was conducted by Adıgüzel et al. (2009) regarding the customer preferences for accommodation spots in the Belek region of the Antalya province. This study was carried out using the AHP method. Pairwise comparison matrix was used to determine the relative importance of such specific criteria as space and price, customer safety, service quality, a carnet of the hotel, rate of liquidation, price/performance ratio, as responded by the customers who attended the survey on the selection of the best of the six five-star hotels. The study becomes different, as the data of the study attaching precedence to the customer choices were collected from the customer views from the website of the ETS tour company.

Prioritizing the selection of the choice of tourism centers by the tourists, Hsu et al (2009) determined a total number of 22 sub-factors as internal and external factors. The survey conducted via the AHP method into the tourists visiting Taiwan helped to select the best center out of 8 tourism centers. As social interaction is the main factor in the decision-making problem where factor weighting is done with the use of AHP, the suitability for family and friend visits and personal security problems were shown as the most important sub-factors. In the course of the assessment involving all these factors, the Fuzzy TOPSIS method was used for the ordering and selection of a tourism center, and tourism center was chosen as the best alternative.

In another study conducted based on the interviews with 351 foreign tourists who were accommodated in totally 53 vacation centers, Ateşoğlu and Bayraktar (2011) developed hypotheses to show how word of mouth will be effective on the selection of an appropriate vacation spot. They also made an inference regarding the analyses with a one-sample t-test. The inferences indicated that the criteria, which were effective in the selection of suitable vacation spots in Turkey, were

Tab. 19-1: The Main and Sub-Factors of Selection of Winter Tourism Centers

Main Factors	Sub-Factors	References
Price (C_1)	Accommodation fee (C_{11}) Transportation fee (C_{12}) Mechanical facility fees (C_{13}) The fees of training and hired equipment (C_{14})	Manap (2006), Chou vd. (2008), Hsu vd. (2009), Adıgüzel vd. (2009), Ateşoğlu ve Bayraktar (2011), Albayrak ve Özkul (2013), Doğan ve Gencan (2013), Memiş (2016), Oğuz (2018), Uğur vd. (2018)
Accessibility (C_2)	Distance to airport (C_{21}) Distance to city center (C_{22}) Distance to accommondation center (C_{23}) Distance to ski track (C_{24})	Manap (2006), Chou vd. (2008), Hsu vd. (2009), Vetitnev vd. (2013), Albayrak ve Özkul (2013), Oğuz (2018), Uğur vd.(2018)
Comfort (C_3)	Adequacy of the accommodation facilities (C_{31}) Adequacy of the mechanical facilities (C_{32}) Taking runway safety measures and presence of health teams (C_{33}). Adequacy of daily accommodation facilities (C_{34}) Presence of alternative ski track (C_{35}) Presence of ski training (C_{36})	Adıgüzel vd. (2009), Hsu vd. (2009), Vetitnev vd. (2013), Albayrak ve Özkul (2013), Memiş (2016), Ayaz ve Apak (2017), Oğuz (2018), Erzincanlı ve Aksakal (2019)
Destination image (C_4)	Recognition of the center (C_{41}) Environment-friendly approach of the center (C_{42}) Thoughts of people who previously visited the center (C_{43})	Ateşoğlu ve Bayraktar (2011), Baldemir ve Akyurt Kurnaz (2013), Memiş (2016), Uğur vd. (2018)
Alternative Tourism Opportunities (C_5)	The presence of the cultural tourism locations (C_{51}) The presence of the entertainment tourism locations (C_{52}) The presence of the nature tourism (C_{53}) The presence of the thermal tourism (C_{54})	Hsu vd. (2009), Ateşoğlu ve Bayraktar (2011), Oğuz (2018)

the figures of the travel agencies, price, natural beauties, historical fabric, and suggestions by friends. The survey conducted by Vetitnev et al. (2013) into the 3 major vacations spots in Russia helped to collect data regarding the factors considered in selecting domestic vacation spots. It was established that, from the tourists' viewpoints, the most important reasons for dissatisfaction with the vacation spots were related to health services, accessibility, and shopping issues. Furthermore, factors such as the selection of the accommodation spot, the method of meeting travel expenses, and the purpose of travel were presented as the ones that mostly affect the satisfaction of local tourists.

Another major study into the selection of the vacation spot was the study conducted by Albayrak and Özkul (2013), which assesses the views of the Y generation tourists. In this study in which, face-to-face surveys were conducted with 384 people, the most important factors in the destination image were identified, and their effect on the vacation spot was researched. Therefore, the factors assessed for the selection of a vacation spot for the Y generation were price, easy transportation, coast, and coastal characteristics, accommodation services, luxurious vacation spot, travel becoming a fashion, weather conditions, presence of entertainment places, the security of the region and, sports opportunities. The data obtained through descriptive statistics, t-test, and ANOVA analysis prove that the Y generation tourists act with the will to discover new places and prefer qualified and luxurious vacation centers at reasonable prices. Therefore, the most important factors in selecting the vacation center were price and easy accessibility. In the research conducted by Memiş (2016) into the factors used by domestic tourists in preferring the vacation center, the data collected with the help of a survey were assessed by the factor analysis method, and seven main factors were defined. These factors may be listed as service quality and security, entertainment, location, shopping and health services, promotion and brand image, opportunities related to children, price-transportation opportunities, and natural heritage and satisfaction. The study analyzed the findings in respect of tourism marketing and attempted to demonstrate the relationship between the factors affecting domestic tourists who are consumers and the demographic characteristics of consumers. Ayaz and Apak (2017) surveyed 480 domestic tourists to determine the factors considered by domestic tourists for winter tourism centers. The two main factors determined by the factor analysis are travel motivation and travel satisfaction of domestic tourists. The explanatory sub-factors of the travel motivation factor are relaxation, self-improvement, peace, and socialization whereas the explanatory sub-factors of the travel satisfaction factor are ski track, shopping, security, accommodation, trainer, and transportation factors. A test was conducted to understand whether a relationship exists between travel motivation and travel satisfaction. Therefore, a significant relation was identified between the pre- and post-travel views. The travel motivations of the domestic tourists were utilized in the selection of the winter tourism centers, and it was concluded that these may contribute to the development of winter centers. In their research on the perception of the destination image of Safranbolu, Uğur et al. (2018) delved into the factors that played a determining role on the decision of tourists to revisit a region for vacation purposes. The study conducted into 277 domestic tourists used the statistical assessment into service quality, problems related to environment and infrastructure, factors like historical fabric, since these constitute the perception of the destination image. A positive relation was detected between service quality and the revisiting of the region by tourists. It was concluded that the sub-factors of destination image, such as price, shopping opportunities, transport opportunities, accommodation opportunities, recognition of the tourism center, and natural beauties had a positive effect on the decision of the tourists to revisit the region.

3 Methodology

As one of the most important parts of the study, this part presents detailed information regarding the two assessment models of DEMATEL and AHP which are selected as the most appropriate methods out of the MCDM techniques involved.

3.1 DEMATEL (The Decision-Making Trial and Evaluation Laboratory)

DEMATEL, which was introduced and developed by Fontella and Gabus in 1972, 1974, and 1976, came to be used as a new method for the multi-criteria decision-making techniques. DEMATEL helps to divide the factors into groups to better understand the cause-effect relation, supports the analysis of complicated systemic problems, and also contributes to the planning in issues that have a hierarchical problem structure. However, Yang and Tzeng (2011: 1418) argue that the level that the criteria affect each other and the analysis of the coincidental relations between these criteria are the greatest contributions suggested for the multi-criteria decision-making techniques. These are used in the solution of various problems ranging from social security problems (Yang et al., 2018), supplier selection problems (Liu et al., 2017; Abdel-Basset et al., 2018), green supply chain (Wu & Chang, 2015; Gandhi et al., 2016), selection of the managerial strategies

(Wu, 2008), brand marketing (Wang & Tzeng, 2012; Lada et al., 2020), identification of critical success factors for lean manufacture (Attia et al., 2018), technology selection (Nabeeh, 2020), emergency management (Zhou et al., 2017), critical operations hazard method in petrol tanks petrol (Akyüz & Çelik, 2015). Yazdi et al. (2020: 5) explained the advantages of DEMATEL in three conditions. The first of these is the successful reflection of mutual effects as both direct and indirect effects. This will lead to a better comprehension of the reasons and effects of complicated problems. The second advantageous condition is the use of an effective relation map by DEMATEL. This map will display a broad view of the effects between the factors and will help them to understand their effects. The last advantage is that it presents results that may be used both for the listing and defining of the critical factors. DEMATEL method is composed of 6 basic steps (Tzeng et al., 2007: 1032):

Step 1: Calculate Group Decision Matrix
The purpose in the first step whereby a comparative scale is used to determine the relationship between the relevant factors is to develop a direct relation matrix with the responses given by the expert decision-makers who attend the survey to collect data. There are five levels in the comparison scale: 0 (ineffective), 1 (lowly effective), 2 (moderately effective), 3 (highly effective), 4 (extremely effective). But using these levels, the intensity and guidance of the effectiveness of the factors are determined. Each x_{ij} element used in the X matrix developed to show the comparisons, display the effectiveness of the i criterion on the j criterion. Direct relation matrices with a quantity of k are developed as a result of the assessments of the decision-makers.

$$X = \begin{bmatrix} 0 & x_{12} & \cdots & x_{1n} \\ x_{21} & 0 & \cdots & x_{2n} \\ \vdots & \vdots & \ddots & \vdots \\ x_{n1} & x_{n2} & \cdots & 0 \end{bmatrix}$$

If the number of decision-making experts is more than one, the arithmetic average of the entire assessments is considered and the group decision matrix A is obtained by using the below equation shown as 1 (Yang & Tzeng, 2011: 1418).

$$a_{ij} = \frac{1}{z} \sum_{k=1}^{z} x_{ij}^{k}$$ (Equation 1)

$$A = \begin{bmatrix} 0 & a_{12} & \cdots & a_{1n} \\ a_{21} & 0 & \cdots & a_{2n} \\ \vdots & \vdots & \ddots & \vdots \\ a_{n1} & a_{n2} & \cdots & 0 \end{bmatrix}$$

x_{ij}^k: decision-maker k shows the level where i factor affects the j factor, the n value in matrix A gives the number of the compared factor. Z shows the number of experts that attended the survey.

Step 2: Determination of the Normalized Direct and Total Relation Matrix
A matrix is normalized to reduce the number of the indefinitive decisions taken by the decision-makers in comparing the assessed criteria and the direct relation matrix (D) normalized with the use of Equation 2 will be as follows:

$$D = \frac{A}{max\left(max_{1 \leq i \leq n} \sum_{j=1}^{n} a_{ij}, max_{1 \leq j \leq n} \sum_{j=1}^{n} a_{ij}\right)} \quad \text{(Equation 2)}$$

Step 3: Determination of the Direct and Indirect Relation Matrix
Following the formation of the normalized direct relation matrix (D), (T) total relation matrix is created showing the direct and indirect relationship between the assessment criteria with the use of the n x n unit matrix (I) and the equation below.

$$T = D^1 + D^2 + D^3 + \ldots, \lim_{m \to \infty} (D)^m = \sum_{i=1}^{\infty} D^i = D(I - D)^{-1} \quad \text{(Equation 3)}$$

Step 4: Derive of Cause and Effect Factor Groups and Calculation of the Net Effect Degrees
n x 1 dimension R_i matrix is obtained by considering the sum of the i line in the T total relation matrix calculated in the previous step, and 1 x n dimension C_i matrix is obtained by using the below-cited equations and considering the sum of line i.

$$T = \left[t_{ij}\right]_{n \times n}, \quad i, j = 1, 2, \ldots, n$$

$$R_i = \left[\sum_{j=1}^{n} t_{ij}\right] = [t_i]_{n \times 1}, \quad i = 1, 2, \ldots, n \quad \text{(Equation 4)}$$

$$C_i = \left[\sum_{i=1}^{n} t_{ij}\right] = [t_j]_{n\times 1}, j = 1, 2, \ldots, n \qquad \text{(Equation 5)}$$

R_i vector shows the direct and indirect effect of the i factor on the other factors; while the C_i vector shows the levels of the effect of the other factors on the i factor. An equal value should be obtained to draw the graph that shows the causality relation. For that reason, the R-C and R + C values are calculated and the effect level and relation level of each factor on the other will be calculated. The fact that $R_i - C_i$ gains a positive value shows that factor i is causer that affects the other factors; the negative value of $R_i - C_i$ shows that it is the net affected. (Tamura et al., 2005: 140). If we attempt to draw an effect relation diagram, then the values that remain under the threshold in the T total relation matrix are shown in the coordinate as $(R_i + C_i, R_i - C_i)$.

Step 5: Determination of the Threshold Values and the Formation of the Effect-Relation Diagram

This is an optional step and by using the casuality values, the weighted values relating to the factors are calculated with the help of Equations 6 and 7 (Dalalah, 2011: 8387).

$$w_i = \sqrt{[(R_i + C_i)]^2 + [(R_i - C_i)]^2} \qquad \text{(Equation 6)}$$

$$W_i = \frac{w_i}{\sum_{j=1}^{n} w_i} \; i = 1, 2, \ldots, n \qquad \text{(Equation 7)}$$

3.2 AHP (Analytical Hierarchy Process)

One of the most frequently used multi-criteria decision-making techniques, the AHP was first offered by Myers and Alpert in 1968 and was further developed by Thomas L. Saaty with his studies in 1977 and 1980, and the mathematical steps of the technique were thus created. Used in almost all fields, the AHP is composed of 3 main sections according to Xia and Wu (2007: 496) development of a hierarchical structure, development of the pairwise comparison matrix, and the calculation of the weight. As mentioned by Saaty (2008: 85), the decision problem should fit the purpose and the main criteria should be determined for the creation of a hierarchical structure by giving the details about these three main steps. This stage will be followed by the criteria at the middle level of the hierarchy and the alternatives at the lowest level. The prerequisite of the pairwise comparisons conducted after the hierarchy is the filling of the

Tab. 19-2: Saaty's Scale of Relative Importances

Intensity of Importance	Definition	Explanation
1	Equal Importance	Two activities contribute equally to the objective.
3	Moderate Importance	Experience and judgement slightly favour one activity over another.
5	Strong Importance	Experience and judgement strongly favour one activity over another.
7	Very Strong Importance	An activity is favored very strongly over another; its dominance demonstrated in practice.
9	Absolute Importance	The evidence favoring one activity over another is of the highest possible order of affirmation.
2,4,6,8	Intermediate Values	When compromise is needed.

Source: (Saaty, 1977: 246)

survey by the relevant people. Otherwise, the expected pairwise comparisons will be inconsistent. If the decision is taken not by a single expert but by a group, then the judgments relating to these people should be individually analyzed and should be gathered by using the geometrical average (Saaty, 1989: 63). This will help to clarify the judgment and bring it to the same dimension.

The scale of importance developed by Saaty (1977) should be known before explaining the steps of the AHP method. These scale criteria given in Tab. 19-2 are of utmost importance because this helps to obtain the relative measurement values precisely and turn them into a matrix.

The following steps will come once the hierarchical structure of the AHP is formed with the survey data filled in under the existing scale:

Step 1: Construction of Pairwise Comparison Matrix
The matrix should be designed as an nxn dimension and since the value of the corresponding factor is 1, the diagonal elements should be written as 1. Below is the A pairwise comparison matrix formed as a result of the assessments of each decision-maker:

Selection of Winter Tourism Centers 383

$$A = \begin{bmatrix} 1 & a_{12} & \cdots & a_{1n} \\ a_{21} & 1 & \cdots & a_{2n} \\ \vdots & \vdots & \ddots & \vdots \\ a_{n1} & a_{n2} & \cdots & 1 \end{bmatrix}_{nxn}$$

While forming the comparison matrix A, the comparison is conducted for those values that remain over the diagonal line. The components below the diagonal line should be calculated by using Equation 8.

$$a_{ji} = \frac{1}{a_{ij}} \qquad \text{(Equation 8)}$$

Step 2: Development of the Priority Vector
Following the development of the pairwise comparison matrix A, the highest eigenvalue, and the eigenvector corresponding to this eigenvalue should be calculated and normalized to calculate the importance of every factor. There are four different methods in the calculation of the priority vector. Ertuğrul and Karakaşoğlu (2008: 32) suggest that the best way will be as: elements n in each line of the pairwise comparison matrix will be multiplied with each other, the root n will be found and the values that are obtained this way will be normalized. To determine the degrees of importance of these factors, comparison matrix A should be left, and by using equation 9, column vector B with a dimension of nxn will be formed.

$$b_{ij} = \frac{a_{ij}}{\sum_{i=1}^{n} a_{ij}} \qquad \text{(Equation 9)}$$

$$B_i = \begin{bmatrix} b_{1i} \\ b_{2i} \\ \cdot \\ \cdot \\ b_{ni} \end{bmatrix}$$

The number of column vectors B is equal to the number of vectors. If n column vector B is brought together under a matrix format, the matrix C below will be formed showing the percentage of the importance of factors.

$$C = \begin{bmatrix} b_{11} & b_{12} & \cdots & b_{1n} \\ b_{21} & b_{22} & \cdots & b_{2n} \\ \vdots & \vdots & \ddots & \vdots \\ b_{n1} & b_{n2} & \cdots & b_{11} \end{bmatrix}_{nxn}$$

The arithmetic average of the line values that form matrix C and the priority vector W is obtained by using the formula shown in Equation 10.

$$w_i = \frac{\sum_{j=1}^{n} b_{ij}}{n} \qquad \text{(Equation 10)}$$

$$W = \begin{bmatrix} w_1 \\ w_2 \\ . \\ . \\ w_n \end{bmatrix}$$

Priority vector W is the column vector that shows the percentile importance degrees that display the priorities of the entire factors involved in the assessment.

Step 3. Calculation of the Consistency Ratio
Consistency ratio (CR) is a rate that allows testing the consistency of the comparisons displayed by the priority vector, which was calculated at the previous step. CR is calculated by comparing the number of factors assessed, and the λ (Basic Value) coefficient. Matrix D is obtained by the multiplication of the comparison matrix A and the priority vector matrices W for the calculation of λ. Then, the basic value E of each vector is obtained from the column vector of the matrix D and the comparison elements of the column vector W. The arithmetic average relating to these values gives the λ value as shown in Equation 11.

$$D = \begin{bmatrix} a_{11} & a_{12} & \cdots & a_{1n} \\ a_{21} & a_{22} & \cdots & a_{2n} \\ \vdots & \vdots & \ddots & \vdots \\ a_{n1} & a_{n2} & \cdots & a_{nn} \end{bmatrix} \times \begin{bmatrix} w_1 \\ w_2 \\ . \\ . \\ w_n \end{bmatrix}$$

$$E_i = \frac{d_i}{w_i} \quad i = 1, 2, \ldots. n \quad \lambda = \frac{\sum_{i=1}^{n} E_i}{n} \qquad \text{(Equation 11)}$$

Tab. 19-3: Random Indicator Values

N	1	2	3	4	5	6	7	8	9	10	11	12
RI	0	0	0,58	0,90	1,12	1,24	1,41	1,45	1,49	1,51	1,48	1,56

Source: (Saaty, 1980: 54)

Following the calculation of λ, the calculation of the Consistency Indicator (CI) is completed. We need this rate for the calculation of the consistency rate. This indicator also known as the Random Indicator (RI) is the average value of the pairwise comparison matrices formed randomly based on the indicator factor number.

$$I = \frac{\lambda_{maks} - n}{n - 1} \qquad \text{(Equation 12)}$$

CI is divided by the Random Indicator (RI) value, and the CR consistency rate is obtained (Saaty, 1980: 21). RI indicated in Tab. 19-3 shows the average value of the pairwise comparison matrices produced randomly based on the number n.

$$R = \frac{CI}{RI} \qquad \text{(Equation 13)}$$

Saaty (1982: 82) underscored that the comparisons made by the decision-making mechanism will be consistent if the calculated consistency rate is under 10 % and is equal to 10 % in his study. Also, Saaty stressed that there would be an inconsistency or an error if this rate is not under 10 %. He added that the assessments should be revised and a new procedure should be followed if there is an error.

4 Assessment of the Relative Importance of the Selection Factors of Winter Tourism Centers

Tab. 19-1 provides the factors that have to be considered for the selection of the winter tourism centers obtained as a result of the literature review and discussions held with specialists. The goal is to assess the hierarchy based on factors with the use of DEMATEL and AHP methods and deliver suggestions on the prioritized factors to the managers of winter tourism. The five main factors will be the price (C_1), accessibility (C_2), comfort (C_3), destination image (C_4), and alternative tourism resources (C_5). The first example of the surveys is

Tab. 19-4: Direct Relation Matrix for the Main Factors

Main Factors	C_1	C_2	C_3	C_4	C_5
C_1	0	2,6	3	2,6	1,6
C_2	1,9	0	1,8	1,9	1,4
C_3	3,2	1,4	0	2	1,6
C_4	2,7	2,3	2,1	0	1,9
C_5	2	1,1	1,3	1,8	0

Source: Author

given in Appendix 1 and Appendix 2, where these 5 main factors and 22 sub-factors were used.

Considering the profiles of the 10 expert decision-makers who accommodated at least 3 times at different winter tourism centers of the world, we see that they hold undergraduate and graduate degrees; that they work as academicians, engineers, senior executives, and sales experts; and that these people are interested in skiing as they regularly go on vacation during the wintertime. The data obtained as a result of the surveys based on multi-criteria decision-making techniques and the results obtained by the use of DEMATEL and AHP methods are given below. Tab. 19-4 gives the direct relation matrix of the decision-making experts which were merged with the arithmetic average and used in the DEMATEL method.

The direct relation matrix considers specialists as a group. Tab. 19-5 shows the total relation matrix normalized by the use of Equation 2 and Equation 3.

Since the total direct relation matrix is used for the calculation of the casuality values, R+C value calculated with the data in Tab. 19-5 or the degree of importance of the factors and the R-C value or the effect level of each factor were shown in Tab. 19-6. These values will help to interpret the findings obtained as affected and affecting groups.

If we would interpret the factors one by one, we can say that having a high R+C value and positive R-C value, the price factor has a high effect over the factors of accessibility, comfort, destination image, and alternative tourism resources. The values prove that the price factor is a factor where the decision-makers have to focus on to obtain the desired results. Following the price factors, it is striking to see the effect of the destination image. The values in Tab. 19-6 illustrate that the destination image has a medium effect over the other factors. With these values, it is possible to say that the comfort factor is effective over

Tab. 19-5: Total Direct Relation Matrix for the Main Factors

Main Factors	C_1	C_2	C_3	C_4	C_5
C_1	1,184	1,163	1,278	1,245	0,973
C_2	1,056	0,721	0,942	0,945	0,755
C_3	1,292	0,970	0,922	1,083	0,873
C_4	1,316	1,079	1,146	0,964	0,937
C_5	0,977	0,757	0,832	0,865	0,571

Source: Author

Tab. 19-6: Cause and Effect Groups for the Main Factors

Main Factors	R	C	$R + C$	$R - C$
C_1	5,843	5,825	11,668	0,018
C_2	4,418	4,691	9,109	-0,272
C_3	5,140	5,120	10,260	0,020
C_4	5,443	5,101	10,544	0,341
C_5	4,003	4,109	8,112	-0,107

Source: Author

the accessibility and alternative tourism factors. If we evaluate these affected factors, we may say that alternative tourism is a relatively independent factor. Out of the main factors, alternative tourism factor was the factor that had the least effect over the selection of the winter tourism center, as indicated by the assessment conducted by decision-makers. Considering the accessibility factor, it is obvious that this factor should be indirectly developed.

The price factor ranks the first position for the order of importance of the main factors as shown in Tab. 19-7 with the use of the DEMATEL method. The ordering of the main factors this way is consistent with the assessment of affecting and affected factors. Therefore, the price factor is the first factor that attracts the attention of the visitors in selecting winter tourism centers. So, it is a factor that should be developed and paid utmost attention to tourism managers. The affecting factors of price, comfort, and destination image are those factors that need special focus. On the other hand, accessibility and alternative tourism resources are classified as the affected factors. Therefore, since the

Tab. 19-7: Weights and Ranking Result of Main Factors Calculated with DEMATEL

Main Factors	$R + C$	$R - C$	Weights of Factors (w)	Ranking of Importance
C_1	11,668	0,018	0,235	1
C_2	9,109	-0,272	0,183	4
C_3	10,260	0,020	0,206	3
C_4	10,544	0,341	0,212	2
C_5	8,112	-0,107	0,163	5

Source: Author

accessibility factor will be positively affected by the affecting pricing, differentiation of comfort, and destination image, the alternative tourism resources will not be much affected by these changes. It is possible to interpret this in various ways by analyzing the elements explaining the main factor. A detailed assessment is shown in Tab. 19-8 regarding the price sector.

According to the information given in Tab. 19-8, the accommodation fee was considered as the factor with the highest degree of effect, and this factor was also suggested as a factor to which the decision-making mechanism should pay the utmost attention. Since the factors whose R-C value is positive are considered as the affecting factors, another important factor other than the accommodation fee was set as the factor of transportation fee. Therefore, the effect of the transportation factor is expected to be higher compared to other factors. The money paid to the facility during the vacation has a great effect on training and the hired equipment. Therefore, the managers of the tourism centers should indirectly revise the fees of training and hired equipment. Once the mechanical facility fees are compared with the accommodation and transport fees, it is seen that this will not be a factor that may affect and change the decision of the decision-making mechanism. A general ordering of the factors shows that the importance levels of the factors are close to one another.

The factor of distance to the airport has a higher degree of affecting other factors considering the sub-factors which account for the main factor of accessibility. Furthermore, the distance to the accommodation center is highly affected by other factors. Considering the entire factors, the factor of distance ranks first instead of accommodation. It is followed by the factors of distance to the ski track, distance to the city center, and distance to the airport as shown

Tab. 19-8: Cause and Effect Groups of the Sub-Factors of Price and Its Weights

Sub-Factors of Price	$R + C$	$R - C$	Weights of Factors (w)	Ranking of Importance
C_{11}	13,431	0,760	0,256	3
C_{12}	11,805	0,607	0,225	4
C_{13}	13,468	-1,220	0,258	2
C_{14}	13,675	-0,147	0,261	1

Source: Author

in Tab. 19-9. These important values which are close to one another prove that accessibility is the most important issue here.

On comfort, the assessment indicates that the most important factor is whether security measures were taken at the ski track and the presence of medical teams. This is followed by other factors as to whether the mechanical facilities are at an adequate level, whether alternative tracks are available, whether services are rendered for the training of skiing, whether the accommodation facilities are adequate and whether the daily accommodation facilities are adequate. Considering the level of sensitivity, the importance degrees of all these factors are close to one another, but the adequacy of the mechanical facilities, the ski track security measures, the presence of medical teams, and the presence of the daily accommodation facilities are considered as the affecting group. This group includes the major factors that are required by the decision-making mechanism to reach the expected results. Any improvement or change in the factors of the focused issue as the adequacy of the accommodation facilities, presence of alternative ski tracks and ski training, etc. will be affected. The factor of ski training will be the least affected, as shown by the calculated R+C and R-C values in Tab. 19-10. This leads to the conclusion that the presence of ski training is an advantage at the winter tourism centers, and leads to no disadvantages.

One of the leading exploratory sub-factors of the destination image is the personal views of the persons who previously had a vacation in the center. This is followed by the recognition of the center and the environment-friendly approach of the center. As shown in Tab. 19-11, focusing on an environmental-friendly center was described as a major focal point as shown by group calculations. If this idea is developed, it is evident that it will also affect other factors.

Tab. 19-9: Cause and Effect Groups of the Sub-Factors of Accessibility and Its Weights

Sub-Factors of Accessibility	$R + C$	$R - C$	Weights of Factors (w)	Ranking of Importance
C_{21}	6,046	1,412	0,221	4
C_{22}	6,438	-0,011	0,230	3
C_{23}	8,004	-0,550	0,286	1
C_{24}	7,322	-0,852	0,263	2

Source: Author

As the winter tourism center was selected as the main factor of alternative tourism, it was determined as the factor with the least importance among the factors that may be effective as shown in Tab. 19-12. However, the most important factor out of the possible effective factors will be the presence of entertainment tourism centers. It is not surprising that the presence of entertainment activities and tourism locations is identified as important during the summer and winter vacations. As nature tourism is attributed to importance, the presence of the cultural tourism locations and thermal tourism was determined as the factors with the least effect. The idea that entertainment tourism locations will be further developed if cultural and natural tourism locations are rendered more effective correspond with the group values of the affecting or the affected.

Following the assessment of the main and sub-factors through the use of the DEMATEL method, the survey regarding the AHP method mentioned in Appendix-2 was conducted on the same specialized group, and the views of the specialists inside the group were merged with the use of the geometrical method, as the most frequently used theoretical method, and pairwise comparison matrices were formed for both the main factor and the sub-factors. Tab. 19-13 indicates the pairwise comparison matrix as well as the precedence values and order of importance regarding the main factors.

The concept used as precedence values in the AHP method represents the factor weights. The consistency rate calculated by the use of Equations 11, 12, and 13, detailed in the methodology section, will be needed to interpret the results of the comparison conducted by the use of Saaty's 1–9 scale (1980). In case the consistency rate is less than or equal to 0,1, the values in the table are considered as consistent and reliable, and these values may thus be interpreted. As indicated in Tab. 19-13, the replies of the specialist decision-makers relating to the mutual assessment of the main factors as the selection factors of the winter tourism centers are considered consistent. As the price factor is determined as the most important factor as a result of the assessment of the consistent

Tab. 19-10: Cause and Effect Groups of the Sub-Factors of Comfort and Its Weights

Sub-Factors of Comfort	$R + C$	$R - C$	Weights of Factors (w)	Ranking of Importance
C_{31}	9,799	-0,038	0,162	4
C_{32}	10,493	0,083	0,173	2
C_{33}	10,985	0,786	0,182	1
C_{34}	9,553	0,223	0,158	6
C_{35}	10,101	-0,311	0,167	3
C_{36}	9,667	-0,744	0,160	5

Source: Author

matrices, this factor is followed by other factors like comfort, the destination image, accessibility, and alternative tourism resources.

Tab. 19-14 including the importance levels of the main and sub-factors and the entire consistent assessments is presented below, instead of forming individual tables for the sub-factors of the entire main factors as shown in Tab. 19-13.

In ordering the sub-factors of the main factor of the price that draws importance, the accommodation fee comes first; and it is followed by the fee paid to the mechanical facilities, transport fee, and training and hired equipment fee. Any improvement in the accommodation fees will play an important role in the tourists planning to visit the winter tourism centers. For this goal, the opening of several centers should be supported, the fee should be improved, and therefore more tourists should be attracted. Since winter tourism always creates the perception of an expensive vacation, it has different importance levels on the main factor of price and its sub-factors. Considering the winter tourism for the factor of accessibility, the most important sub-factor was identified as the distance to the ski track. On the other hand, the distance to the accommodation facility was determined as the second most important factor. The distance to the ski track and the distance to the accommodation center constitute 80 % of the weights of the entire factors and prove the findings that they attract the utmost importance. Ensuring the security measures of the ski track and the presence of the medical teams are amongst the most important sub-factors of the comfort factors. The fact that the issue of security comes first to mind when comfort is concerned proves that this is a consistent assessment. The sufficiency of the mechanical facilities and the characteristics relating to the accommodation facilities has precedence to the other factors. The assessment into the

Tab. 19-11: Cause and Effect Groups of the Sub-Factors of Destination Image and Its Weights

Sub-Factors of Destination Image	$R + C$	$R - C$	Weights of Factors (w)	Ranking of Importance
C_{41}	14,796	-0,742	0,337	2
C_{42}	13,402	1,088	0,306	3
C_{43}	15,642	-0,346	0,356	1

Source: Author

destination image prioritizes the views of the people who went on a vacation during the winter time as an indication of the level of customer satisfaction. This shows that the services rendered during winter tourism bear importance. Though the presence of alternative tourism resources ranks the last position among the main factors it proved to be more important than the presence of entertainment tourism locations, cultural tourism locations, nature tourism locations, and thermal tourism locations. It was noted that if there would be alternative tourism in winter tourism centers, this would be focusing on entertainment and would be more important than visiting and spending time in natural places, cultural and touristic locations. Given this fact, the notions of skiing and entertainment come to mind when winter tourism center is concerned.

Conclusion

One of the main objectives of the study is the global-scale growth of the winter tourism industry in recent years. The need for the development of winter tourism centers in Turkey in the last decade and the targets set and projects created by the Ministry of Culture and Tourism should also be pointed out. A detailed literature review was thus conducted into the selection of winter tourism centers, and selection factors were determined to help the development and preferability of these centers. The reason for selecting Turkey (holding 16 winter tourism centers, 9 of which are active and 7 of which are partially active) as the country of implementation in this study was that the winter tourism centers in Turkey do not have the capacity to compete with the mostly-preferred winter tourism centers in the world. The goal is to ensure that the tourism centers work with a full capacity of visitors to raise competitiveness. Therefore, the assessment of the visitors regarding the selection factor of the tourism centers bears utmost importance. Hence, the results of the assessment

Tab. 19-12: Cause and Effect Groups of the Sub-Factors of Alternative Tourism and Its Weights

Sub-Factors of Alternative Tourism	$R + C$	$R - C$	Weight f Factors (w)	Ranking of Importance
C_{51}	16,301	0,649	0,234	3
C_{52}	18,533	-0,438	0,266	1
C_{53}	18,124	0,290	0,260	2
C_{54}	16,816	-0,501	0,241	4

Source: Author

Tab. 19-13: Weights and Ranking Result of Main Factors Calculated with AHP

Main Factors	C_1	C_2	C_3	C_4	C_5	Weights of Factors (w)	Ranking of Importance
C_1	1,564	1,000	1,403	1,484	2,141	0,2869	1
C_2	1,000	0,639	1,142	0,740	1,249	0,1824	4
C_3	0,876	0,713	1,000	1,115	1,915	0,2084	2
C_4	1,352	0,674	0,664	1,000	1,353	0,1897	3
C_5	0,801	0,467	0,522	0,739	1,000	0,1327	5
Consistency Ratio:							$0,0 \leq 0,1$

Source: Author

that give an idea to the managers of tourism centers are aimed at understanding which factors are more important as observed by visitors.

The five main factors for the selection of winter tourism centers, which are identified as price, accessibility, comfort, destination image, and alternative tourism resources, were assessed with the DEMATEL and AHP methods, one of the most important weighting techniques of the MCDM methods. This was followed by comparing results with one another. AHP uses a special scale where the factors are compared and assessed with pairwise comparison scale (Saaty, 1977), and this scale was preferred in this study as it comes into mind when relative prioritization is concerned. DEMATEL is distinguished from other weighting techniques since it is a technique that prioritizes the relations of the factors and the importance of the effects that they have on one another. The technique is preferred since it provides guidance to the affecting and affected factor groups and the decision-making mechanism and it is compared with the weighting method results like AHP since the assessments into the factors are enriched with supporting results. Totally 22 explanatory sub-factors of the 5 main factor groups were subjected to an assessment under the relevant main factor, and decision matrices were developed for each main and sub-factor under the assessment of the specialist decision-makers. Tab. 19-15 indicates the comparative results of the main factor were obtained through the DEMATEL and AHP methods.

Compared to the results obtained by the two methods, the price factor ranks the first position in the selection of the winter tourism center. This is followed by comfort, destination image, accessibility, and alternative tourism resources. The order of precedence of the results obtained by DEMATEL and supported by the AHP technique ensures that the study obtains consistent and strong results. DEMATEL suggests the interaction levels of the main and sub-factors as affecting and affected levels and presents results that may be more precisely interpreted. The findings in the explanatory sub-factors prove that any changes into the factors of price, comfort, and destination image that the visitors pay utmost importance to in the selection of winter tourism centers as an affecting category, will put the centers into a more preferable position. Therefore, the factor related to the accessibility of the winter tourism center as a main affecting factor is at a level that may be improved. It is considered that the presence of alternative tourism resources will lead to a framework that is less affected or unaffected by the decided policies. Since the inter-factor order regarding the DEMATEL and AHP methods were identified the same, Tab. 19-16 gives a joint precedence value and order.

The government may provide incentives to raise the diversity of tourism centers and increase the capability of attracting more visitors, to ensure that tourism center managers may make adjustments to the prices and open new tourism centers over the location that covers the vacation area. This will also help prices to be adjusted in a manner that may be advantageous for visitors. It was further established that any improvements to the accommodation fee will also affect the money paid for training and equipment. Managers of winter tourism centers should also ponder over the ways of not distinguishing between the accommodation fees paid for the visitors who do not receive daily services and the fee paid for training and equipment. Since the mechanical facilities are in joint use, the improvements in the mechanical facility fees are not affected by any changes in the other fee units.

Visitors also attribute importance to the dimensions of accessibility to winter tourism centers. It was proven that having to use alternative routes to

Tab. 19-14: Weights of Main Factors and Sub-Factors

Main Factors	Weights of Factors	Sub-Factors	Weights of Factors
C_1	0,2869	C_{11}	0,3498
		C_{12}	0,1908
			0,3195
		C_{13}	0,1399
		C_{14}	
	0,1824	C_{21}	0,0681
			0,1057
C_2		C_{22}	0,2929
			0,5334
		C_{23}	
		C_{24}	
		C_{31}	0,1733
		C_{32}	0,1762
			0,3376
C_3	0,2084	C_{33}	0,1109
		C_{34}	0,1245
			0,0775
		C_{35}	
		C_{36}	
C_4	0,1897	C_{41}	0,3079
		C_{42}	0,2303
		C_{43}	0,4619
		C_{51}	0,2216
C_5	0,1327	C_{52}	0,3539
			0,2508
			0,1738
		C_{53}	
		C_{54}	

Source: Author

reach a tourism center or to travel across cities to reach the destination did not have any major effects over the selection of the tourism center. As indicated in Tab. 19-16, the distance to the ski track was determined as the most important factor. This is the reason why skiing tourism comes to mind when winter tourism is concerned. The proximity of the ski track and accommodation to the winter center indicates that visitors prefer the center in the vicinity to spend

Tab. 19-15: Comparison of the Weights of Main Factors

Main Factors	DEMATEL Weights of Factors	DEMATEL Ranking of Importance	AHP Weights of Factors	AHP Ranking of Importance
C_1	0,235	1	0,2869	1
C_2	0,183	4	0,1824	4
C_3	0,206	2	0,2084	2
C_4	0,212	3	0,1897	3
C_5	0,163	5	0,1327	5

Source: Author

the short vacation efficiently. Therefore, designing the planned winter centers at a location closer to the ski track will help to increase the number of visitors.

The diversity in the services provided by winter tourism centers to their visitors under the main title of comfort aims at ensuring that visitors feel to be on vacation in a reliable environment. Therefore, according to the views of visitors, the most important factor out of the 6 explanatory sub-factors of the main factor of comfort is taking safety measures for the ski track and the availability of medical trams at the center. Comfort has gained significance particularly regarding the security concerns on the ski track since the relative weights of the factors of the adequacy of mechanical facilities, adequacy of the accommodation facilities, presence of alternative ski tracks, adequacy daily accommodation facilities and the availability of skiing training, are close to one another. When the sub-factors of the main factor of the image of destination that includes the impressions of the people who stayed in these centers, the impression obtained through ads and sensitivity towards the environment are concerned, the most important factor for the visitors is the observations of previous customers of the center. Recognition and factors related to the environment follow each other in the order of importance. These factors are part of the destination image, but it was supported by these results that it is not possible to prevent the effectiveness of the experiences of visitors. Entertainment tourism and nature tourism are the two important factors in the selection of winter tourism centers depending on the assessment results related to the explanatory sub-factors of the factor of alternative tourism resources which are classified at the lowermost level in the ordering of the main factors. Such factors as the presence of cultural tourism centers and thermal tourism centers on the location of

Tab. 19-16: Average Weights of the Sub-Factors and Final Ranking

Sub-Factors	DEMATEL Weights of Factors	AHP Weights of Factors	Average Weights of Factors	Final Ranking of Importance
C_{11}	0,256	0,349	0,303	1
	0,225	0,191	0,208	3
C_{12}	0,258	0,319	0,289	2
C_{13}	0,261	0,139	0,200	4
C_{14}				
C_{21}	0,221	0,068	0,145	4
	0,230	0,106	0,168	3
C_{22}	0,286	0,293	0,289	2
C_{23}	0,263	0,533	0,398	1
C_{24}				
C_{31}	0,162	0,173	0,167	3
C_{32}	0,173	0,176	0,175	2
	0,182	0,338	0,260	1
C_{33}	0,158	0,111	0,134	5
C_{34}	0,167	0,125	0,146	4
	0,160	0,078	0,119	6
C_{35}				
C_{36}				
C_{41}	0,337	0,308	0,323	2
C_{42}	0,306	0,230	0,268	3
C_{43}	0,356	0,462	0,409	1
C_{51}	0,234	0,227	0,228	3
	0,266	0,354	0,310	1
C_{52}	0,260	0,251	0,255	2
C_{53}	0,241	0,174	0,207	4
C_{54}				

Source: Author

the winter tourism center were not determined as factors that may be regarded as those visitors selecting the winter tourism centers. Considering this fact, a vacation comes into the mind where skiing may be done safely and visitors may spend time in nature by skiing and other entertainment activities.

This study presented the main and sub-factors assessed for the selection of winter tourism centers, the concepts that the managers should attribute importance to, and the ideas suggested for these concepts. In future research, a service performance assessment may be done into the active winter tourism facilities under certain factor groups. Based on the active centers, suggestions may be delivered by this present study for the new centers under integrity that may be created by bringing together the factors that were relatively considered as important.

Appendices

Appendix 1 Sample Survey Question in DEMATEL Method

FACTORS	is ineffective	is lowly effective	is moderately effective.	is highly effective.	is extremely effective

The Effect of Price
on accessibility
on comfort
on destination image
on alternative tourism opportunities
The effect of accessibility
on price
on comfort
on destination image
on alternative tourism opportunities
The effect of comfort
on price
on accessibility
on destination image
on alternative tourism opportunities
The effect of destination image
on price
on accessibility
on comfort
on alternative tourism opportunities
The effect of alternative tourism opportunities

FACTORS	is ineffective	is lowly effective	is moderately effective.	is highly effective.	is extremely effective
on price					
on accessibility					
on comfort					
on destination image					

Appendix 2 Sample Survey Question in AHP Method

FACTOR 1	Absolute Importance (9)	(8)	Very Strong Importance (7)	(6)	Strong Importance (5)	(4)	Moderate Importance (3)	(2)	Equal Importance (1)	(2)	Moderate Importance (3)	(4)	Strong Importance (5)	(6)	Very Strong Importance (7)	(8)	Absolute Importance (9)	FACTOR 2
Price																		Accessibility
Price																		Comfort
Price																		Destination Image
Price																		Alternative Tourism Opportunities
Accessibility																		Comfort
Accessibility																		Destination Image

	Alternative Tourism Opportunities	Destination Image	Alternative Tourism Opportunities	Alternative Tourism Opportunities
Accessibility				
Comfort				
Comfort				
Destination Image				

Bibliography

Abdel-Basset, M., Manogaran, G., Gamal, A., and Smarandache, F. (2018), "A Hybrid Approach of Neutrosophic Sets and DEMATEL Method for Developing Supplier Selection Criteria", *Design Automation for Embedded Systems*, 22, pp. 257–278.

Adıgüzel, O., Çetintürk, İ., and Er, O. (2009), "Konaklama İşletmelerine Olan Müşteri Tercihinin Analitik Hiyerarşi Prosesi Yöntemi İle Belirlenmesi", *Süleyman Demirel Üniversitesi Vizyoner Dergisi*, 1, pp. 17–35.

Akyüz, E. and Çelik, E. (2015), "A Fuzzy DEMATEL Method to Evaluate Critical Operational Hazards During Gas Freeing Process in Crude Oil Tankers", *Journal of Loss Prevention in the Process Industries*, 38, pp. 243–253.

Albayrak, A. and Özkul, E. (2013), "Y Kuşağı Turistlerin Destinasyon İmaj Algıları Üzerine Bir Araştırma", *International Periodical For The Languages, Literature and History of Turkish or Turkic*, 8 (6), pp. 15–31.

Altaş, N. T., Çavuş, A., and Zaman, N. (2015), "Türkiye'nin Kış Turizmi Koridorunda Yeni Bir Kış Turizm Merkezi: Konaklı", *Marmara Coğrafya Dergisi*, 31, pp. 345–365.

Arslan-Muhacir, E. S., and Yaman, Y. K. (2016), Turizmin Çeşitlendirilmesi Stratejisi Bağlamında Artvin'de Kış Turizmi Odaklı Rekreasyon Faaliyetlerinin Değerlendirilmesi, *IWCS International Winter Cities Symposium*, Erzurum, pp. 381–388.

Atay, L. and Akyurt, H. (2009), "Uludağ Destinasyonun İmajına Yönelik Ziyaretçi Algı ve Tutumlarını Belirlemeye Yönelik Bir Araştırma", *Seyahat ve Otel İşletmeciliği Dergisi*, 6 (3), pp. 67–76.

Ateşoğlu, İ. and Bayraktar, S. (2011), "Ağızdan Ağıza Pazarlamanın Turistlerin Destinasyon Seçimindeki Etkisi", *ZKÜ Sosyal Bilimler Dergisi*, 7 (14), pp. 95–108.

Attia, E. A., Seleem, S., and El-Assal, A. (2018), "Identification of Critical Success Factors for Lean Manufacturing Using Fuzzy Dematel Method", *Journal of Engineering and Applied Science*, 62 (4), pp. 141–163.

El-Awady, A., Seleem, S., and El-Assal, A. (2018), "Identification of Critical Success Factors for Lean Manufacturing Using Fuzzy Dematel Method", *Journal of Engineering and Applied Science*, 62 (4), pp. 141–163.

Ayaz, N. and Apak, Ö. C. (2017), "Kış Turizmine Katılan Yerli Ziyaretçilerin Seyahat Davranışları: Erciyes Kayak Merkezi Örneği", *Erciyes Üniversitesi İktisadi ve İdari Bilimler Fakültesi Dergisi*, 49, pp. 27–43.

Baldemir, E. and Akyurt Kurnaz, H. (2013), "İlçelerin Turizm Potansiyellerinin Analitik Hiyerarşi Yöntemi İle Sıralanması: Muğla Örneği", *Sosyal ve Beşeri Bilimler Araştırmaları Dergisi*, 30, pp. 51–67.

Chou, T.-Y., Hsu, C.-L. and Chen, M.-C. (2008), "A Fuzzy Multi-Criteria Decision Model for International Tourist Hotels Location Selection", *International Journal of Hospitality Management*, 27, pp. 293–301.

Çakmak, F. and Yılmaz, Ö. (2018), "Turizmin İktisadi Sürdürülebilirliği Açısından Kış Turizmi", *Hitit Üniversitesi Sosyal Bilimler Enstitüsü Dergisi*, 11 (1), pp. 267–286.

Dalalah, D., Hajaneh, M., and Batieha, F. (2011), "A Fuzzy Multi-criteria Decision Making Model for Supplier Selection", *Expert Systems with Applications*, 38, pp. 8384–8391.

Doğan, N. Ö. and Gencan, S. (2013), "Seyahat Acentası Yöneticilerinin Bakış Açısıyla En Uygun Otel Seçimi: Bir Analitik Hiyerarşi Prosesi (AHP) Uygulaması", *Erciyes Üniversitesi İİBF Dergisi*, 41, pp. 69–88.

Erbaş, E. (2016), "Competitive Determinance–Performance Analysis: An Illustration on Turkish Winter Tourism Destinations", *Tourism Analysis*, 21, pp. 93–106.

Ertuğrul, İ. and Karakaşoğlu, N. (2008), "Comparison of Fuzzy AHP and Fuzzy TOPSIS Methods for Facility Location Selection", *The International Journal of Advanced Manufacturing Technology*, 39, pp. 783–795.

Erzincanlı, T. and Aksakal, E. (2019), Kayak Merkezi Seçimi Problemine Entropy ve Irp Yöntemleri ile Bütünleşik Bir Yaklaşım: Türkiye Uygulaması, *1. Uluslararası Kış Turizmi Kongresi*, 19–21 December, pp. 20–27.

Eşitti, B. (2018), "Sürdürülebilir Kış Sporları Turizmi: Sarıkamış'ın Potansiyelinin SWOT Analizi ile Değerlendirilmesi", *Turizm Akademik Dergisi*, 5 (1), pp. 205–220.

Evren, S., Zeybek, H. İ., and Taş, S. (2019), Destinasyon Paydaşlarının Bakış Açısıyla Zigana Kış Turizmi Merkezi'nde Mevcut Durum ve Gelecek İçin Öneriler, *1. Uluslararası Kış Turizmi Kongresi*, 19–21 December, pp. 78–86.

Fontela, E. and Gabus, A. (1974), *DEMATEL, Innovative Methods. Report No. 2 Structural Analysis of the World Problematique* (pp. 67–69), Geneva, Switzerland: Battelle Geneva Research Institute.

Fontela, E. and Gabus, A. (1976), *The DEMATEL Observer: Battelle Institute* (pp. 56–61), Geneva, Switzerland: Geneva Research Center.

Forman, E. H. (1990), "Random Indices for Incomplete Pairwise Comparison Matrices", *European Journal of Operational Research*, 48, pp. 153–155.

Gabus, A. and Fontela, E. (1972), *World Problems an Invitation to Further Thought within the Framework of DEMATEL*, Geneva, Switzerland: Battelle Geneva Research Centre.

Gandhi, S., Kumar Mangla, S., Kumar, P., and Kumar, D. (2016), "A Combined Approach Using AHP and DEMATEL for Evaluating Success Factors in Implementation of Green Supply Chain Management in Indian Manufacturing Industries", *International Journal of Logistics Research and Applications*, 19 (6), pp. 537–561.

Göktuğ, T. and Arpa, N. (2015), "Korunan Alanlar Yönetimi Bağlamında Kayak Merkezilerinin Fiziksel ve Sosyal Taşıma Kapasitelerinin Analizi: Ilgaz Dağı Milli Parkı, Ilgaz Kış Sporları Turizm Merkezi", *Kastamonu Üniversitesi Orman Fakültesi Dergisi*, 15 (1), pp. 104–119.

Hsu, T.-K., Tsai, Y.-F., and Wu, H.-H. (2009), "The Preference Analysis for Tourist Choice of Destination: A Case Study of Taiwan", *Tourism Management*, 30, pp. 288–297.

İbragimov, M. (2001), *Ürün Çeşitlendirmesi Açısından Kış Turizmi ve Almatı Çimbulak – Medeu Örneği* (Unpublished Master's Dissertation), Dokuz Eylül Üniversitesi, İzmir.

İbret, Ü. (2006), "Ilgaz Dağlarında Kış Turizmi", *Türk Coğrafya Dergisi*, 44, pp. 61–78.

İlban, M. O. and Kaşlı, M. (2011), Kış Turizmi. In N. Hacıoğlu and C. Avcıkurt (Eds.), *Turistik Ürün Çeşitlendirmesi* (pp. 319–342), Ankara: Nobel Basım ve Dağıtım Yayınları.

Kadıoğlu, Y. (2017), "Samsun'da Yöresel Öneme Sahip Bir Kış Turizm Merkezi: Akdağ (Ladik)", *Doğu Coğrafya Dergisi*, 22 (38), pp. 161–172.

Karaman, A. and Gül, M. (2016), "Alternatif Turizm Kapsamında Kahramanmaraş İli Yedikuyular Bölgesi Dağ ve Kış Sporları Turizmi İçin Öneriler", *Atatürk Üniversitesi Sosyal Bilimler Enstitüsü Dergisi*, 20 (2), pp. 583–599.

Koca, H., Zaman, S., and Coşkun, O. (2011), "Erzurum'un Spor-Kamp Turizmi Potansiyeli", *Doğu Coğrafya Dergisi*, 12 (18), pp. 205–224.

Korgavuş, B. (2017), "Erzurum Palandöken'in Kış Turizmi Açısından Değerlendirilmesi", *Ata Planlama ve Tasarım Dergisi*, 1 (1), pp. 13–23.

Ministry of Culture and Tourism (2007), "Türkiye Turizm Stratejisi 2023 Eylem Planı", Ankara.

Ministry of Culture and Tourism (2020), "Turizm İstatistikleri 1 Raporu", Ankara.

Ministry of Development (2018), "Onbirinci Kalkınma Planı (2019–2023) Özel İhtisas Komisyonu Raporu", Ankara.

Lada, S., William, J. J., and Azaze-Azizi, A. A. (2020), "Application of AHP and DEMATEL Procedure on Brand Experience", *International Journal of Business and Technology Management*, 2 (1), pp. 1–7.

Liu, T., Deng, Y., and Chan, F. (2017), "Evidential Supplier Selection Based on DEMATEL and Game Theory", *International Journal of Fuzzy Systems*, 20 (4), pp. 1321–1333.

Lo, H.-W., Liou, J. J. H., Huang, C.-N., Chuang, Y.-C., and Tzeng, G.-H. (2020), "A New Soft Computing Approach for Analyzing the Influential Relationships of Critical Infrastructures", *International Journal of Critical Infrastructure Protection*, 28, pp. 1–16.

Manap, G. (2006), "Analitik Hiyerarşi Yaklaşımı ile Turizm Merkezi Seçimi", *Ticaret ve Turizm Eğitim Fakültesi Dergisi*, 2, pp. 157–170.

Memiş, S. (2016), "Tatil Yeri Seçiminde Etkili Olan Faktörler: Yerli Turistler Üzerine Bir Araştırma", *Kesit Akademi Dergisi*, 4, pp. 212–226.

Myers, J. H. and Albert, M. I. (1968), "Determinant Buying Attitudes: Meaning and Measurement", *Journal of Marketing*, 32 (4), pp. 13–20.

Nabeeh, N. A. (2020), "A Hybrid Neutrosophic Approach of DEMATEL with AR-DEA in Technology Selection", *Neutrosophic Sets and Systems*, 31 (1), pp. 17–30.

Oğuz, A. (2018), "Kış Turizm Merkezi Seçimi İçin Kriter Ağırlıklarının Bulanık AHP Kullanılarak Belirlenmesi", *Uluslararası İktisadi ve İdari İncelemeler Dergisi*, (18. EYI Special Issue), pp. 789–802.

Özçoban, E. (2019), "Türkiye'de Kış Turizmi ve Erciyes Kayak Merkezi Üzerine Bir Değerlendirme", *Journal of Tourism and Gastronomy Studies*, 7 (3), pp. 1606–1625.

Price, M. F., Borowski, D., Macleod, C., Rudaz, G., and Debarbieux, B. (2011), *From Rio 1992 to 2012 and beyond: 20 Years of Sustainable Mountain Development What Have We Learnt and Where Should We Go?* The Alps, Regional Report, Swiss Federal Office for Spatial Development.

Saaty, T. L. (1977), "A Scaling Method for Priorities in Hierarchical Structures", *Journal of Mathematical Psychology*, 15 (3), pp. 234–281.

Saaty, T. L. (1980), *The Analytic Hierarchy Process*, New York: McGraw-Hill.

Saaty, T. L. (1982), *Decision Making for Leaders*, Bellmont, CA: Lifetime Learning Publications, Wadsworth.

Saaty, T. L. (1989), Group Decision Making and the AHP. In B.L. Golden., E.A. Wasil, P. T. Harker (Eds), *The Analytic Hierarchy Process*, Berlin, Heidelberg: Springer.

Saaty, T. L. (1990), "How to Make a Decision: The Analytic Hierarchy Process", *European Journal of Operational Research*, 48, pp. 9-26.

Saaty, T. L. (2008), "Decision Making with the Analytic Hierarchy Process", *International Journal of Services Sciences*, 1 (1), pp. 83-98.

Stojčetović, B., Nikolić, D., and Velinov, V. (2015), "Application of Swot-Ahp Method in Strategy Selection: Case of Ski Centre Brezovica", *Act a Oeconomica*, 4 (1), pp. 106-113.

Tamura, H. and Akazawa, K. (2005), "Stochastic DEMATEL for Structural Modeling of A Complex Problematique for Realizing Safe, Secure and Reliable Society", *Journal of Telecommunications and Information Technology*, 4, pp. 139-146.

Tzeng, Gwo-Hshiung, Chiang, Cheng-Hsin, and Li, Chung-Wei (2007), "Evaluating İntertwined Effects in E-Learning Programs: A Novel Hybrid MCDM Model Based on Factor Analysis and DEMATEL", *Expert Systems with Applications*, 32 (4), pp. 1028-1044.

Uğur, İ., Gökkaya, S., and Acar, A. (2018), "Yerli Turistlerin Destinasyon İmajına İlişkin Tekrar Ziyaret Etme Niyetleri: Safranbolu Üzerine Bir Çalışma", *Journal of Saffron Culture and Tourism Researches*, 1 (1), pp. 29-40.

Vanat, L. (2020), "2020 International Report on Snow & Mountain Tourism". https://www.vanat.ch/RM-world-report-2020.pdf, (03.06.2020).

Vetitnev, A., Romanova, G., Matushenko, N., and Kvetenadze, E. (2013), "Factors Affecting Domestic Tourists' Destination Satisfaction: The Case of Russia Resorts", *World Applied Sciences Journal*, 22 (8), pp. 1162-1173.

Yang, J. L. and Tzeng, G.-H. (2011), "An Integrated MCDM Technique Combined with DEMATEL for A Novel Cluster-Weighted with ANP Method", *Expert Systems with Applications Journal*, 38 (3), pp. 1417-1424.

Yang, J., Han, J. and Zhang, X. (2018), Information System Security Risk Assessment Based on IDAV Multi-Criteria Decision Model, *12th IEEE International Conference on Anti-counterfeiting, Security, and Identification* (ASID), Xiamen, China, 121-127, Business School of Jilin University, Changchun, China.

Yazdi, M., Khan, F., Abbasi, R., and Rusil, R. (2020), "Improved DEMATEL Methodology for Effective Safety Management Decision-Making", *Safety Science*, 127, pp. 1-17.

Yıldız, S. and Kaya, F. (2016), "Yerel Paydaşların Bakış Açısıyla Bingöl İlinin Kış Turizmi Açısından Değerlendirilmesi", *Journal of Recreation and Tourism Research*, 3 (2), pp. 50-59.

Wang, Y. (2011), Destination Marketing and Management: Scope, Definiton and Structures. In Y. Wang and A. Pizam (Eds.), *Destination Marketing and Management: Theories and Applications* (pp. 1–20), Preston: CABI.

Wang, Y.-L. and Tzeng, G.-H. (2012), "Brand Marketing for Creating Brand Value Based on A MCDM Model Combining DEMATEL with ANP and VIKOR Methods", *Expert Systems with Applications*, 39, pp. 5600–5615.

Williams, A. M. and Shaw, G. (Eds.). (1988), *Tourism and Economic Development, Western European Expehences*, London: Pinter Publishers Limited.

Wu, H. H. and Chang, S. Y. (2015), "A Case Study of Using DEMATEL Method to Identify Critical Factors in Green Supply Chain Management", *Applied Mathematics and Computation*, 256, pp. 394–403.

Wu, W.-W. (2008), "Choosing Knowledge Management Strategies by Using A Combined ANP and DEMATEL Approach", *Expert Systems with Applications*, 35, pp. 828–835.

Xia, W. and Wu, Z. (2007), "Suppliers Selection with Multi-Criteria in Volume Discount Environment", *Omega*, 35 (5), pp. 494–504.

Zhou, X., Shi, Y., Deng, X., and Deng, Y. (2017), "D-DEMATEL: A New Method to İdentify Critical Success Factors in Emergency Management", *Safety Science*, 91, pp. 93–104.

Yasin KELEŞ, Seden DOĞAN and Mutlu KAYA

Graduation Metaphors of Undergraduate Students Studying Tourism: A Study at Ondokuz Mayıs University

1 Introduction

In the world, business life has begun to differentiate rapidly over time with globalization, speed of technological development, and changes in social life. In the early 1900s, business life, which was based on physical strength and technical knowledge, in other words, where industrial power was the determinant, became more complex and diverse, especially in the 1960s. Over time, the positive impacts of international trade and tourism on the country's economies have forced both businesses and employees to follow the diary and an understanding that spreads learning in all areas of their life. However, education at the university level based on newly developing professions has started to become widespread.

While the requirement for university graduation was less sought in jobs that require technical knowledge, the effects of the population, production, and consumption capacity, the need for employees studying at the university level has increased with the different reflections of the mentioned factors and the professionalization of the jobs. However, the trained workforce has become a vital force in the fields of production. As Porter (1990) puts it, human resources are amongthe most important competitive factors of an industry or nation. Accordingly, it is possible to evaluate that the educated workforce is a critical strategic force in production. In the tourism sector that has been developing and proliferating since the 1960s, the need for a qualified workforce continued to increase with the increasing employment capacity each year.

The number of students at the graduate level training in tourism since the 1980s rapid development of tourism in Turkey also began to increase gradually. Considering the historical development of tourism education in Turkey, since the republic began developing the first year of formal education, which organizes formal education in tourism, it has seen training with the 1960s. The change of the Trade Higher Teacher Training School as "Trade and Tourism Higher Teacher Training School" since 1965 forms the basis of tourism education at the undergraduate level. Considering the date range of organizing formal education in tourism, it is striking that it is parallel to the transition to the planned period and the tourism movements in the world. Since this period, many steps have been taken in tourism education; since 2009, it has been organized as the Faculty of Tourism at the undergraduate level.

As it is generally accepted, getting an undergraduate education in a particular field means also deciding to do the related profession for an individual. The university graduation period is considered a multifaceted transition process in social, economic, and psychological terms (Overton-Healy, 2010). It is possible for students studying in the field of tourism, as in many other fields, to have various social, psychological, and economic concerns and expectations after graduation. The correct determination of these concerns and expectations and developing plans are essential for the employment and training of tourism. Accordingly, the students who received tourism education at the undergraduate level in the research were evaluated through the metaphors.

2 Problems Related to Working Life in Tourism and Anxiety Sources of Students Receiving Tourism Education

The tourism industry offers 24-hours service and is considered challenging in terms of working conditions. In the previous studies (Kokko & Guerrier, 1994; Hjalager & Andersen, 2001; Jolliffe & Farnsworth, 2003; Szivas et al., 2003; Lindsay & McQuaid, 2004; Richardson, 2008; Jiang & Tribe, 2009; Richardson, 2010; Keles, 2018a;) the reasons such as the low wages of the tourism industry, long/uncertain working hours, sexual discrimination, limited task characteristics/low-skilled jobs seasonality, high employee turnover rate, insufficient career opportunities, and the necessity to work under intense stress may be expressed as the negativities experienced by those employed in the tourism industry. It is also stated in some studies (Guerrier, 1999; Boella, 2000) that working in tourism is considered low prestigious in many societies. These disadvantages are that even those who receive tourism education stay away from working in the tourism industry (Pavesic & Brymer, 1990; Ağaoğlu, 1991; Baron & Maxwell, 1993; Altman & Brothers, 1995; Kızılırmak, 2000; Kuşluvan & Kuşluvan, 2000; Kozak & Kızılırmak, 2001; King et al., 2003; Tuyluoglu, 2003; Keleş, 2018a).

The tourism industry's challenges and the low image of respectability for jobs in the tourism industry can also affect students' career orientation who choose the university exam. Increasing the number and quota of institutions that give undergraduate level tourism education caused the students, whose academic achievement is in average of Turkey and have different targets and/

or no targets, prefer to study at the tourism faculties. This situation can be explained by students' success levelswho have settled in tourism faculties (Keles, 2018b). When the exam success levels of students who have settled in tourism faculties since 2017 are examined, it is possible to evaluate that students, especially at the tourism management department, prefer the tourism faculties only to settle in an undergraduate program. In addition to the ongoing problems in the tourism sector and tourism education, this situation further deepens the problems and necessitates questioning the students' aims and future perspectives.University preferences are an essential turning point in terms of individuals' career preferences. While Kozak (2001) states that many variables such as personality, intelligence, self-confidence, value, beliefs, and attitudes are useful in the career choices of individuals, the determining factor in the students who prefer tourism faculties is the point; It is noteworthy that most of the students prefer tourism faculties because they cannot find another option (Keles, 2018a). However, as Fleet (1991) stated, the self-recognition in the career selection process, the profession or career they want to choose and as a result needs and expectations, and career needs. In this case, the reflection of career choices made without considering these criteria into the working life after graduation becomes uncertain.

It should be emphasized that students may have some university graduation concerns regardless of what field they study. Regardless of tourism-related professions or anxiety about employment in this field, it can be said that students in each field may feel intense anxiety when they realize that they will leave the university experience, which can be considered relatively free, and that they will move to a period in which they should take more responsibility as an adult (Hunter et al., 2012). Lane (2013) states that graduating from university and moving to another life stage is a significant developmental challenge and often stressful transition period for graduate students. Again, Yazedjian et al. (2010) state that senior students are frightened by being out of the university environment they are accustomed to, becoming obliged to obscure, having to make career plans, and the thoughts of being economically independent / not supported. As will be noted, the biggest concern of students who will graduate from university is uncertainty (Moritimer et al., 2002). When the uncertainty in question is combined with employment issues in the tourism sector, students' anxiety about graduation may increase.

Arnett (2015) states that in the period that students call adulthood, which also includes their university years, people have a more variable and less structured lifestyle. They have had a relatively free life period because they are exposed to fewer responsibilities and demands. As can be seen, in these years between the most free period of students and their responsibilities, there can be a sharp transition between the aim of entering the school and the life built after graduation. The idea of graduation or concept becomes an important concept for students in terms of transition and occupational, economic, psychological, and social concerns. Accordingly, the research aims to determine the point of view of the students who receive tourism education to the concept of graduation. Accordingly, the research focuses on the meaning of the students who received tourism education at the undergraduate level on the concept of graduation; in other words, what the graduation means for themselves and/or how it can have results. Therefore, students' metaphors about graduation have been evaluated as the most effective way to explain what the concept of graduation means to them.

Lakoff and Johnson (2005) define the concept of metaphor in Turkish (TDK, 2019) as "understanding and experiencing something from the perspective of something else." Metaphors, which are used as a cognitive tool to understand the real world and to explain how people see life, environment, events, and objects using different metaphors, show how people perceive the world in their daily lives (Göçen, 2019). Metaphors are used to make the meaning of the thought intended to be more effective, bring vitality to the meaning, andexpress the thought through another concept (Korkut & Keskin, 2016). Metaphor is a robust mental mapping and modeling mechanism for individuals to understand and construct their worlds (Karataş et al., 2017). Metaphors save reality from boring for both listening and speaking and offer them a new way of thinking with concepts that the target audience can understand and believe (Limon & Durnalı, 2018).

It is provided through metaphors to enrich the expression and express the thoughts by using other concepts to describe a concept (Sezgin et al., 2017: 601). Although there are academic researches that use metaphor analysis in different fields of social sciences, especially in the field of education (Kulakoğlu Dilek et al., 2016; Taş et al., 2016; Şahin et al., 2018; Erol & Düşmezkalender, 2019; Ertaş, 2019; Oğuzbalaban, 2019; Belhassen et al., 2020) and research for students, studying tourism is limited (Alyakut & Küçükkürürler, 2018; Doğan & Keleş, 2018; Köroğlu et al., 2018a, 2018b). In this direction, the metaphors developed by students who undertake tourism education at the undergraduate level will clearly understand the feelings and thoughts about the graduation concept.

3 Methodology

This study aims to determine the thoughts of students who continue their tourism education at the undergraduate level regarding graduation. "Phenomenology," one of the qualitative research patterns, was used to determine the students' metaphorical thinking status for the concept of "graduation." Phenomenology (phenomenology) pattern focuses on cases that are aware but do not have an in-depth and detailed understanding. In many studies, it is stated that metaphors can be a qualitative data collection tool and can be used to obtain robust and rich findings (Korkut & Keskin, 2016; Sezgin et al., 2017; Doğan & Keleş, 2018).

The study group of the research consists of students studying in the Tourism Guidance and Tourism Management departments of Ondokuz Mayıs University Faculty of Tourism. The application of the research was carried out during the lesson hours of one of the researchers. First of all, the students were informed about the metaphor technique. Then students were asked to complete the sentence, "Graduation is like . . . . because . . . .." Students were asked to provide a rationale for the metaphors they developed. These metaphors created by students constitute the primary data source of the research.

In the analysis of the data obtained in the study, the path specified in different studies in the literature was followed (Doğan & Keleş, 2018; Köroğlu et al., 2018b; Şahin et al., 2018). For this, first of all:

1- Metaphors were determined, sorted, and coded.
2- Sample metaphors were compiled from metaphors, and primary metaphors were determined.
3- The primary metaphors were grouped as positive, negative, and uncertainty among them.

According to these stages, a list of metaphors produced by students was made first, and then it was checked whether they formed meaningful metaphors and sentences. Forms where no metaphor is used, irrelevant connections are established, and metaphors with unexplained reason are marked for exclusion. Each metaphor was then broken into pieces and reviewed one by one in terms of similarities or common features with other metaphors. Thus, every metaphor that students wrote was analyzed in terms of the relationship between the subject and the metaphor's source. After that, each metaphor was associated with a particular theme and according to the perspectives on the concept of "graduation" (Akgül et al., 2017; Baş & Kıvılcım, 2019; Beyoğlu & Ergin, 2019; Ergün & Kıyıcı, 2019; Saraç, 2019) different categories have been created.

4 Findings

Information about the students participating in the research is given in Tab. 20-1.

Tab. 20-1: Distribution of the Participants according to Their Demographic Characteristics

	Group	F	%
Gender	Female	22	35,5
	Male	40	64,5
Class	First	11	17,7
	Second	17	27,4
	Third	14	22,6
	Fourth	20	32,3
Department	Tourism Management	36	58,1
	Tourism Guidance	26	41,9
View to the concept of graduation	Positive	35	56,5
	Negative	19	30,6
	Containing uncertainty	8	12,9

As shown in Tab. 20-1, 40 of 62 students participating in the study are male, and 22 are female. 20 of the students are fourth-grade students. 36 of the students study in Tourism Management and 26 of them in Tourism Guidance. When the students' metaphors regarding the concept of graduation were examined, it was evaluated that more than half (56.5 %) of them were positive, and 30.6 % of them were negative metaphors. The rate of metaphors containing uncertainty was 12.9 %. Information about the metaphors developed by students is presented in Tab. 20-2, and evaluations regarding the grouping of metaphors are presented in Tab. 20-3.

According to Tab. 20-2, 62 students developed 52 metaphors in total. The most repeated metaphor is freedom (n= 6). Students' metaphorsare grouped in three different categories, namely positive, negative and uncertain metaphors, and are given in Tab. 20-3.

As can be seen in Tab. 20-3, 34 of 62 students developed metaphors with a positive meaning to the concept of graduation, while 19 developed metaphors with a negative meaning. The metaphor developed by nine students contains neither positive nor negative meaning, but it shows uncertainty.

Some examples of students' expressions are given below:

"Graduation is like a clean sheet because we will start new beginnings, new responsibilities, and business life." Male, Tourism Guidance 2nd year student
"Graduation is like the last step of the ladder because the real-life struggle then begins." Male, Tourism Management 4th Year student

Tab. 20-2: Metaphors Developed for the Concept of Graduation

Item No	Metaphor Name	Number of Students Developing Metaphor		Item No	Metaphor Name	Number of Students Developing Metaphor	
		N	%			N	%
1	Quittance	1	1,6	27	Tricycle to be two-wheeled	1	1,6
2	Success	2	3,2	28	Reborn	1	1,6
3	Baby's first step	1	1,6	29	Balloon	1	1,6
4	Baby walking	1	1,6	30	Glass ceiling	1	1,6
5	Baby	1	1,6	31	Gasoline finished car	1	1,6
6	Knowledgeable people	1	1,6	32	Reborn	2	3,2
7	Doctor	1	1,6	33	Fenerbahce	1	1,6
8	Awareness	1	1,6	34	Storm	1	1,6
9	Entrepreneurship	1	1,6	35	Dark	1	1,6
10	Sun	1	1,6	36	Unemployment	2	3,2
11	Alleviation	1	1,6	37	Worry	2	3,2
12	Weekend	1	1,6	38	Fear	1	1,6
13	Coming to dream	1	1,6	39	Cold water	1	1,6
14	The beginning of dreams	1	1,6	40	Responsibility*	1	1,6
15	Starting life	1	1,6	41	Fish out of water	2	
16	Approaching targets	1	1,6	42	A new life	1	1,6
17	Cargo package	1	1,6	43	Load	1	1,6
18	Liberation	1	1,6	44	Starting business life	1	1,6
19	The last step of the ladder	1	1,6	45	Wedding	1	1,6
20	Happiness	1	1,6	46	The beginning of life	1	1,6
21	Freedom	6	9,7	47	The last step of the ladder*	1	1,6
22	Welfare	1	1,6	48	Eat orange	1	1,6
23	Exclusiveness	1	1,6	49	Oracle	1	1,6
24	Level up	1	1,6	50	New beginning	1	1,6
25	Responsibility	1	1,6	51	A new step	1	1,6
26	Clean sheet	1	1,6	52	Newborn baby	1	1,6

* Metaphors of the same name with different meanings are given separately.

Tab. 20-3: Classification of Graduation Metaphors

	Category	Developed Metaphors	Number of Metaphors	f	%
1	Positive Metaphors	Laundering, Success, Baby's first step, baby's walking, baby, Knowledgeable human, doctor, awareness, entrepreneurship, sun, easing, weekend, dreaming, beginning of dreams, starting life, approaching goals, cargo package, salvation, ladder the last step, happiness, freedom, prosperity, exclusivity, leveling up, responsibility, clean sheet, tricycle being two wheels, reborn	28	34	54,8
2	Negative Metaphors	Balloon, gasoline-finished car, glass roof, Fenerbahce, Storm, darkness, unemployment, anxiety, fear, cold water, responsibility, fish out of water, new life, reborn, a burden	15	19	30,6
3	Uncertain metaphors	Starting business life, wedding, life beginning, last step of stairs, eating an orange, torpedo (explosive), a new beginning, a new step, a newborn baby	9	9	14,5
TOTAL			52	62	100

"Graduation is like a newborn baby, because nine months of pregnancy and four years of reading are equivalent, and happiness is similar." Male, Tourism Management 4th Year student

"Graduation is like freedom because we will start real life." Women, Tourism Guidance 2nd year student

In addition to the above positive or non-negative metaphors, some negative metaphors are presented below:

"Graduation is like fear because there is unemployment; we can be unemployed for a long time." Woman, Tourism Management 2nd Year student

"Graduation is like a storm, because job opportunities will be difficult, there will be family pressure and a storm will break." Woman, Tourism Management 4th Year student

"Graduation is like unemployment because most people are unemployed, and unemployment psychology is very bad in the country and my neighborhood." Male, Tourism Management 3rd year student

"Graduation is like unemployment because we will be unemployed." Male, Tourism Guidance 1st year student

General evaluations about the metaphors developed by students are evaluated in the conclusion and discussion section.

Conclusion and Discussion

In this research, when students' metaphors are evaluated, significant (30.6 %) negative metaphors stand out besides positive metaphors. Positive metaphors generally point to new beginnings and completion of studentship, while a remarkable result specific to tourism has not emerged. It is worrisome that negative metaphors turn to unemployment. In previous studies (Pavesic & Brymer, 1990; Baron & Maxwell, 1993; Altman & Brothers, 1995; Kozak & Kızılırmak, 2001; King et al., 2003; Tuyluoglu, 2003; Keleş, 2018a) and participation in employment within the industry negative evaluations about the desire are in line with the negative results in this research. With this research, a significant portion of the students were concerned about unemployment. Similarly, according to the results of metaphor analysis carried out to determine the future and career perceptions of university students in other fields (Korkut & Kesin, 2016; Topgül, 2017), metaphors indicating uncertainty and difficulties after graduation emerged. Although concerns about life in general, it should not be considered ordinary considering the specific problems of the tourism field.

Students' concerns about graduation require studies on university-sector cooperation, especially by considering the academic units. University-sector cooperation studies should come to the fore with internships that are suitable for their purpose. Here, it is necessary to underline the expression of internships suitable for its purpose. Internships are among the most influential models of applied education in tourism education (Yiu & Law, 2012). Through internships, students can discover career options, develop knowledge learned in the traditional classroom environment, and establish the link between theoretical knowledge and practice (Kim & Jeong, 2018). In this direction, students who discover career opportunities through internship activities can be minimized and have the opportunity to see the opportunities for employment in tourism and the aspects appropriate for their personalities closely. On the contrary, negative perspectives that may occur in internships may lead students to develop themselves in different fields, and create an anxiety-reducing effect.

University-sector cooperation is not just a case for universities to solve. It is considered as an imperative to conclude the studies, especially on occupational tourism law. In addition to this, it is essential to develop a policy considering the quotas of tourism faculties, the types of points, and the profile of the students who are settled in the context of YOK-OSYM-Tourism Faculties (Keleş, 2018b) will contribute to minimizing the anxiety that may arise at the end of their education long in advance.

Since qualitative research design was used in this study, comparisons were not made according to the students' demographic characteristics and individual qualitiesIn future research, the differences in departments in tourism faculties can be revealed by making using quantitative data. On the other hand, opinions and thoughts on graduation and post-graduation can be analyzed comprehensively by conducting in-depth interviews with students.

The data obtained in this research is limited to the students of Ondokuz Mayıs University Tourism Faculty. More general and detailed results can be produced by using the mixed method or by expanding the research universe with a quantitative method. At the same time, what the concept of graduation means for students can be examined compared to students studying in other faculties.

Bibliography

Ağaoğlu, O. K. (1991), *Türkiye'de Turizm Eğitimi ve Etkenliği*. Milli Prodüktivite Merkezi Yayınları, 439, Ankara.

Akgül, B. M., Kaya, S., Ayyıldız, T., and Karaküçük, S. (2017), Rekreasyon Uzman Adaylarının Mesleklerine İlişkin Algıları: Bir Metaforik Çalışma, *4. Disiplinler Arası Turizm Araştırmaları Kongresi*, 9–12 Kasım, Kusadasi, Turkiye.

Altman, L. A. and Brothers, L. R. (1995), "Career Longevity of Hospitality Graduates", *FIU Hospitality Review*, 13 (2), pp. 77–83.

Alyakut, Ö. and Küçükkömürler, S. (2018), "Gastronomi Eğitimi Alan Üniversite Öğrencilerinin Mesleklerine Yönelik Metafor Algılarının Değerlendirilmesi", *OPUS Uluslararası Toplum Araştırmaları Dergisi*, 9 (6), pp. 823–852.

Arnett, J. J. (2015), *Emerging Adulthood: The Winding Road from the Late Teens through the Twenties*, New York: Oxford University Press.

Baron, P. and Maxwell, G. (1993), "Hospitality Management Students' Image of the Hospitality Industry", *International Journal of Contemporary Hospitality Management*, 5 (5), pp. 5–8.

Baş, G. and Kıvılcım, Z. S. (2019), "Türkiye'de Öğrencilerin Merkezi Sistem Sınavları ile İlgili Algıları: Bir Metafor Analizi Çalışması", *Eğitimde Nitel Araştırmalar Dergisi*, 7 (2), pp. 639–667.

Belhassen, Y., Caton, K., and Vahaba, C. (2020), "Boot Camps, Bugs, and Dreams: Metaphor Analysis of Internship Experiences in the Hospitality Industry", *Journal of Hospitality, Leisure, Sport & Tourism Education*, https://doi.org/10.1016/j.jhlste.2019.100228.

Beyoğlu, A. and Ergin, D. Y. (2019), "Öğretmen Adaylarının Şiddet Kavramına İlişkin Algılarının Metafor Analii Yoluyla İncelenmesi", *Trakya Üniversitesi Sosyal Bilimler Dergisi*, 21 (1), pp. 283-294.

Boella, M. J. (2000), *Human Resource Management in the Hospitality Industry* (7th Ed.), Cheltenham: Stanley Thornes.

Doğan, S. and Keleş, Y. (2018), Turizm Rehberliği Öğrencilerinin "Turist Rehberi"ne İlişkin Algılarının Metafor Analizi ile İncelenmesi, *The Second International Congress on Future of Tourism: Innovation, Entrepreneurship and Sustainability*, 27-29 Eylül, Mersin, Türkiye.

Ergün, A. and Kıyıcı, G. (2019), "Fen Bilgisi Öğretmen Adaylarının STEM Eğitimine İlişkin Metaforik Algıları", *Kastamonu Eğitim Dergisi*, 27 (6), pp. 2513-2527.

Erol, G. and Düşmezkalender, E. (2019), Analysis of the Concepts of Tourism and Tourist by Metaphors, *4th International Tourism Congress*, 16-19 Ekim, Eskişehir, Türkiye.

Ertaş, Ç. (2019), "Şırnak'ta Yaşayan Yerel Halkın Turizme Dönük Metaforik Algısı", *Elektronik Sosyal Bilimler Dergisi*, 18 (70), pp. 720-733.

Fleet, V. (1991), *Contemporary Management*, Boston, MA: Houghton Mifflin Company.

Göçen, G. (2019), "Türkçeyi Yabancı Dil Olarak Öğrenenlerin "Türkçenin Dil Bilgisi"ne Yönelik Metaforik Algısı", *Başkent University Journal of Education*, 6 (1), pp. 28-45.

Guerrier, Y. (1999), *Organizational Behavior in Hotels and Restaurants: An International Perspective*, Chichester: John Wiley.

Hjalager, A. and Andersen, S. (2001), "Tourism Employment: Contingent Work or Professional Career?" *Employee Relations*, 23 (2), pp. 115-129.

Hunter, M. S., Keup, J. R., Kinzie, J., and Maietta, H. (Eds.). (2012), *The Senior Year: Culminating Experiences and Transitions*. Columbia, SC: University of South Carolina, National Resource Center for the First-Year Experience and Students in Transition.

Jiang, B. and Tribe, J. (2009), 'Tourism Jobs-Short Lived Professions': Student Attitudes towards Tourism Careers in China", *Journal of Hospitality, Leisure, Sport & Tourism Education*, 8 (1), pp. 4-19.

Jolliffe, L. and Farnsworth, R. (2003), "Seasonality in Tourism Employment: Human Resource Challenges", *International Journal of Contemporary Hospitality*, 15 (6), pp. 312-316.

Karataş, K., Ardıç, T., and Oral, B. (2017), "Öğretmenlik Mesleğinin Yeterlikleri ve Geleceği: Metaforik Bir Analiz", *Turkish Studies (International Periodical for the Languages, Literature and History of Turkish or Turkic)*, 12 (33), pp. 291-312.

Keleş, Y. (2018a), "Neden Turizm Eğitimi? Lisans Düzeyinde Turizm Öğrencilerine Yönelik Bir Araştırma", *Journal of Tourism and Gastronomy Studies*, 7 (2), pp. 229-236.

Keleş, Y. (2018b), "Turizm Fakültelerine Yerleşen Öğrencilerin Üniversite Sınavlarındaki Başarı Durumunun Değerlendirilmesi", *Journal of Recreation and Tourism Research*, 5 (4), pp. 11-20.

Kızılırmak, İ. (2000), "Yüksekokulların Turizm ve Otelcilik Programlarının Turizm Sektörünün Beklentileri Doğrultusunda Değerlendirilmesi", *Milli Eğitim Dergisi*, 147, pp. 54-60.

Kim, H. J. and Jeong, M. (2018), "Research on Hospitality and Tourism Education: Now and Future", *Tourism Management Perspectives*, 25, pp. 119-122.

King, B., McKercher, B., and Waryszak, R. (2003), "A Comparative Study of Hospitality and Tourism Graduates in Australia and Hong Kong", *International Journal of Tourism Research*, 5 (6), pp. 409-420.

Kokko, J. and Guerrier, Y. (1994), "Overeducation, Underemployment and Job Satisfaction: A Study of Finnish Hotel Receptionists", *International Journal of Hospitality Management*, 13 (4), pp. 375-386.

Korkut, A. and Keskin, İ. (2016), "Üniversite Öğrencilerinin Kariyer Algıları: Metaforik Bir Analiz Çalışması", *Mustafa Kemal Üniversitesi Sosyal Bilimler Enstitüsü Dergisi*, 13 (33), pp. 194-211.

Kozak, M. A. (2001), *Konaklama İşletmelerinde Kariyer Planlaması* Eskişehir: Anadolu Üniversitesi, Eskişehir Meslek Yüksekokulu Yayınları.

Kozak M. and Kızılırmak, İ. (2001), "Türkiye'de Meslek Yüksekokulu Turizm Otelcilik Programı Öğrencilerinin Turizm Sektörüne Yönelik Tutumlarının Demografik Değişkenlere Göre Değişimi: Anadolu, Akdeniz ve Karadeniz Teknik Üniversitesi Öğrencileri Üzerine Bir Uygulama", *Anatolia Turizm Araştırmaları Dergisi*, 12 (2), pp. 9-16.

Köroğlu, Ö., Ulusoy Yıldırım, H., and Avcıkurt, C. (2018), "Kültürel Miras Kavramına İlişkin Algıların Metafor Analizi Yoluyla İncelenmesi", *Turizm Akademik Dergisi*, 5 (1), pp. 98-113.

Köroğlu, Ö., Manav, S., and Karaca, K. Ç. (2018), "Turizm Rehberliği Öğrencilerinin "Türk Mutfağı" Kavramına İlişkin Algılarının Metaforlar Yöntemi ile Belirlenmesi", *Sosyal ve Beşeri Bilimleri Dergisi*, 10 (2), pp. 114-129.

Kulakoğlu Dilek, N., Dilek, S. E., and Gümüş, M. (2016), "Otel Çalışanlarının Turizm ve Barış İlişkisine Yönelik Metaforik Algıları", *Batman Üniversitesi Yaşam Bilimleri Dergisi*, 6 (2/1), pp. 1-15.

Kuşluvan, S. and Kuşluvan, Z. (2000), "Perceptions and Attitudes of Undergraduate Tourism Students towards Working in the Tourism Industry in Turkey", *Tourism Management*, 21 (3), pp. 251-269.

Lakoff, G. and Johnson, M. (2005), *Metaforlar: Hayat, Anlamve Dil* (Çev: G. Y. Demir), İstanbul: Paradigma.

Lane, J. A. (2013b), "Group Counseling for Students Transitioning out of Post Secondary Education: A Narrative Approach", *Group work: An Interdisciplinary Journal for Working with Groups*, 23 (1), pp. 34-55.

Limon, İ. and Durnalı, M. (2018), "Doktora Öğrencilerinin Doktora Eğitimi ve Öğretim Üyelerine Yönelik Metaforik Algıları", *Sakarya University Journal of Education*, 8 (1), pp. 26-40.

Lindsay, C. and McQuaid, R. W. (2004), "Avoiding the 'McJobs': Unemployed Job Seekers and Attitudes to Service Work", *Work, Employment and Society*, 18 (2), pp. 297-319.

Mortimer, J. T., Zimmer-Gembeck, M. J., Holmes, M., and Shanahan, M. J. (2002), "The Process of Occupational Decision Making: Patterns during the Transition to Adulthood", *Journal of Vocational Behavior*, 61 (3), pp. 439-465.

Oğuzbalaban, G. (2019), "Karadeniz Ereğli'de Yaşayan Yerel Halkın "Turizm" Kavramına İlişkin Algılarının Metaforlar Yardımıyla Analizi", *Journal of Tourism and Gastronomy Studies*, 7 (4), pp. 2566-2581.

Overton-Healy, J. (2010), *First-Generation College Seniors: A Phenomenological Exploration of the Transitional Experience of the Final College Year* (Unpublished Doctorate Thesis), Indiana University of Pennsylvania.

Pavesic, D. V. and Brymer, R. A. (1990), "Job Satisfaction: What's Happening to the Young Managers?", *The Cornell Hotel and Restaurant Administration Quarterly*, 30 (4), pp. 90-96.

Porter, M. E. (1990), *The Competitive Advantage of Nations*, London and Basingstoke: The MacMillan Press.

Richardson, S. (2010), "Tourism and Hospitality Students' Perceptions of a Career in the Industry: A Comparison of Domestic (Australian) Students and International Students Studying in Australia", *Journal of Hospitality and Tourism Management*, 17, pp. 1-11.

Richardson, S. A. (2008), "Undergraduate Tourism and Hospitality Students' Attitudes towards a Career in the Industry: A Preliminary Investigation", *Journal of Teaching in Travel and Tourism*, 8 (1), pp. 23-46.

Saraç, H. (2019), "Ortaokul 8. Sınıf Öğrencilerinin Akıllı Tahta ve Cep Telefonu Hakkında Görüşleri: Metafor Analizi Çalışması", *Pamukkale Üniversitesi Eğitim Fakültesi Dergisi*, 45, pp. 22-37.

Sezgin, F., Koşar, D., Koşar, S., and Er, E. (2017), "Öğretmenlerin Öğrenciye Yönelik Metaforlarının Belirlenmesine İlişkin Nitel Bir Araştırma", *Hacettepe Üniversitesi Eğitim Fakültesi Dergisi*, 32 (3), pp. 600-611.

Szivas, E., Riley, M., and Airey, D. (2003), "Labor Mobility into Tourism: Attraction and Satisfaction", *Annals of Tourism Research*, 30 (1), pp. 64-76.

Şahin, S., Tezcan, A. E., and Bekçi, M. (2018), "Yerli Turistlerin Türkiye, İstanbul, Turizm ve Turist Rehberi ile İlgili Metaforları", *Turizm Akademik Dergisi*, 5 (1), pp. 251-264.

Taş, M., Düz, İ., and Ünlü, E. (2016), "Ortaöğretim Öğrencilerinin "Alternatif Turizm" Kavramına İlişkin Algılarının Metaforlar Yardımıyla Analizi", *Eğitim ve Öğretim Araştırmaları Dergisi*, 5 (Özel Sayı), 352-360.

Topgül, S. (2017), Çalışma Ekonomisi ve Endüstrsi İlişkileri Öğrencilerinin Gelecek, İstihdam ve Mesleğe İlişkin Metaforik Algıları. *Çalışma İlişkileri Dergisi*, 8 (1), pp. 100-117.

Türk Dil Kurumu Sözlükleri (2019), Metafor. Retrieved from https://sozluk.gov.tr/.

Tüylüoğlu, T. (2003), *Türkiye'de Turizm Eğitiminin Niteliği* (Yayımlanmamış Yüksek Lisans Tezi), Ankara Üniversitesi Sosyal Bilimler Enstitüsü, Ankara.

Yazedjian, A., Kielaszek, B., and Toews, M. (2010), "Students' Perceptions Regarding Their Impending Transition out of College", *Journal of the First Year Experience & Students in Transition*, 22 (2), pp. 33-48.

Yiu, M. and Law, R. (2012), A Review of Hospitality Internship: Different Perspectives of Students, Employers, and Educators", *Journal of Teaching in Travel and Tourism*, 12 (4), pp. 377-402.

YÖK Atlas. (2019), "Yükseköğretim Program Atlası". Retrieved from https://yokatlas.yok.gov.tr/.

Eda Rukiye DÖNBAK

Tour Guiding Profession Perceptions of the Students Receiving Tourism Guidance Education

1 Introduction

The tourism industry, which is largely labor intensive (Kim & Park, 2013; Grobelna, 2017), is one of the most developed industries in the world today (Bahçelerli & Sucuoğlu, 2015). It is also seen as an economic development tool by many countries (Kuo et al., 2018). The number of tourists and visitors, as well as the diversity and richness of its natural, cultural and historical wealth, indicates that the tourism sector in our country is important for the country's economy. The quality of the services of the employees who produce services by direct contact with tourists depends on the fact that university education is in the field of tourism and this education must be suitable for technical and professional competence (Ünlüönen, 2004; Ehtiyar & Üngüren, 2008; Aymankuy & Tetik, 2013). In addition, service quality is possible with high commitment staff and their motivation (Kuşluvan & Kuşluvan, 2000; Grobelna, 2017). It is considered that the university education that tourist guides receive is important for the country's tourism. Today, it is still a sensitive issue that the highly motivated employees, who have good education in the field mentioned above, who have professional competence and who are committed to their job, can be brought to the hospitality and tourism sector (Lucas & Johnson, 2003; Richardson & Butler, 2012; Wan et al., 2014; Grobelna, 2017).

The first tourist guide profile in our country is the guides who are only capable of foreign language translation (Ahipaşaoğlu, 2006). As a result of the regulations made by the Ministry of Tourism and Promotion in 1965, the profile of tourist guide who received undergraduate and associate degree tourist guidance training became possible after 1995 (Hacıoğlu, 2008; Tolga et al., 2015). Tourist guides have an important place in the tourism industry as they can directly affect the perceptions and experiences of tourists (Leclerc & Martin, 2004; Huang et al., 2010; Kuo et al., 2018)..

The aim of the study is to evaluate the perceptions of the students who do not have experience in the sector and preferred their university education in the field of tourism guidance and to interpret the results to contribute to their education. For this purpose, research questions will be created by searching similar studies in the literature, and the stages of obtaining data, analyzing and summarizing the findings will be applied with the qualitative research method.

2 Conceptual Framework

In the literature review, together with studies investigating the attitudes of tourism students towards the sector (Kuşluvan & Kuşluvan, 2000; Aksu & Köksal, 2005; Roney & Öztin, 2007; Kozak, 2009; Avcı, 2011; Richardson & Thomas, 2012; Wang, 2014) motivational sources (Wong & Ladkin, 2008; Grobelna, 2017) and the perception of working conditions in the sector (Bahçelerli & Sucuoğlu, 2015) were evaluated. Other studies that focus on students' perceptions and attitudes towards the tourist guiding profession (Tolga et al., 2015; Şahin & Acun, 2016; Aloudat, 2017). were also examined.

Avcı (2011) determined that the students studying in the tourism guidance department have the intention to work in the sector with the most important determinants of human relations, business characteristics, business environment, innovation and leadership. In her study, the perception of the business value of students with a bachelor's degree in tourism was evaluated. In the literature, it is considered that the query items related to the nature of the job, such as independence and interestingness, are used for students who are trained in other fields of tourism and tourism guidance (Bahçelerli & Sucuoğlu, 2015; Tolga et al., 2015; Aloudat, 2017).

Tolga et al. (2015) reported the main factors affecting student attitudes at the undergraduate as the nature of work and working conditions, social status, wages and additional income, managers, private life, cooperation, person-industry compliance and agency managers. The expressions with the highest value in the sub-dimensions are:

- The person is in the industry compliance dimension: *"My personality is suitable to work as a guide"* and *"I can use my knowledge and skills in guidance."*
- In terms of wages and additional income: *"Guidance fee is insufficient to lead an economically comfortable life"* (reverse code).
- In the nature of work and working conditions: *"Working hours are very long in guidance."*
- In private life (reverse code): *"It is difficult to maintain family life by working as a guide"* and *"my private life is important to me."*

- Social status dimension (reverse code): *"Working as a guide requires compromising moral values"* and *"Working as a guide does not get much respect in society."*
- In terms of agency managers: *"Agency managers empower their guides to do their jobs better"* (Tolga et al., 2015).

The statements with the highest item average in the studies of Şahin and Acun (2016) on the attitudes of students studying in the tourism guidance section toward the profession:

- *"I like to converse with guiding people".*
- *"I think that guidance will give me opportunities to create and create something".*
- *"The thought of introducing people to the region they visit makes me happy".*
- *"The working conditions of guidance are attractive to me".*
- *"I believe that guidance will give me dignity in the society".*
- *"It seems that I have a special talent for guidance".*

Aloudat (2017), in his research, evaluated the perceptions of students who received undergraduate education in the tourist guidance department before they started working in the sector, about their tourist guidance career. In his study, Aloudat (2017) created expressions under the dimensions of the nature of work, social status, roles under tourist guidance, wage issues and the degree of participation of the students who participated in the survey were expressed as percentages and averages.

- *"I find it interesting tourist guidance"* and *"I think my profession would be free and independent"* about the nature of work.
- *"I am proud of my family for doing this profession in the future"* and *"I am proud of talking to my friends and relatives about this profession"* regarding social status.
- *"I am aware that I will have important roles when I do this profession"* and *"I will be the cultural ambassador of my country at the same time while guiding,"* professional roles.
- *"Commissions and gratuities are an important source of income"* are statements with the highest percentage of wages.

2.1 Research Questions

In the literature, as a result of the evaluation of the studies focusing on the attitudes and perceptions of the tourism guidance students toward the tourist guidance profession, obtained frame of items are summarized below:

- Thoughts and comments about what is inherent in the guidance profession,
- What are the perspectives on guidance profession in society?,
- The wages of the profession,
- Personal characteristics suitable for the profession,
- Professional roles of guides,
- Assessment of the experiences of meeting the guidance professors was provided.

Each of the research questions addressed to the students was written on the first line in the tables below.

3 Method

In this study, a qualitative research method was applied to evaluate the perceptions of students studying in the field of tourism guidance. Qualitative research is a type of approach to investigate and understand the meaning that individuals or groups attribute to social issues (Creswell, 2014). Yıldırım and Şimşek (2018: 41) define the qualitative research method as, "uses qualitative data collection methods such as observation, interview and document analysis, followed by a qualitative process for the realization of the perceptions and events in a realistic and holistic way."

3.1 Research Pattern

The data were analyzed by the phenomenology pattern. It was preferred for investigating the perceptions of the students studying tourism guidance toward their profession in depth and thematically.

3.2 Research Group

The research group consists of two male students and eight female students who were in their first academic year in the 2018–2019. Unlike quantitative research, in qualitative research methods, the number of samples can be determined as judicial, non-random. The interviewees were determined using the non-random sampling method. The students who included in this research were in the first term of their school.

3.3 Data Collection

All of the students in the research group participated in the study voluntarily. The main qualitative data collection types are basic and supportive data collection methods; data were obtained through in-depth phenomenology interview (Özdemir, 2010), which is included in the basic data collection method. In the phenomenology meeting, the researcher makes sense of how individuals attribute meaning to the facts (Greasley & Ashworth, 2007; Özdemir, 2010); for this purpose, a quiet and spacious environment where students can express themselves comfortably and a timetable suitable for the lesson times was created. After the purpose of the study was explained to the participants, it was informed to the participants that the audio recording would be done and that the records would be used in the study without disclosing their names. In the interviews held with semi-structured pre-prepared questions in the acquisition of the data, the side and sub-questions were used to open the views of the participants, if the answer to the question was expressed in another question, it was not directed as a question again (Türnüklü, 2000). Six questions were prepared by evaluating the literature about tourism and tourism guidance education.

3.4 Data Analysis

The data were analyzed by content analysis. Interviews were recorded while recording the necessary points. After the sound recordings were written in the computer environment, the first step in induction-type content analysis and the process of creating themes were analyzed. Yıldırım and Şimşek (2018: 242) state that the main goal in content analysis is *to reach the concepts and relationships that can explain the collected data*. The data were arranged on the basis of questions, and the coding of the answers given by the participants to each answer was made manually with the notes on the edges of the papers. Codes were created by repeating and interpreting the responses at certain intervals of days instead of the pre-formed code list and concepts. In order to ensure control of the data underlying the themes that emerged in this way and to ensure that all of the themes can explain the data obtained in the research in a meaningful way, a comparison of code and themes was made and an 80 % agreement was achieved in case studies. In the findings part of the study, it was arranged in tables to correspond to the codes and themes by using labels such as S1, S2, (...) and their expressions were presented under the tables (Tab. 21-1 to 21-6).

4 Results

S1: I think guidance is different from a profession. It is necessary to love. Because it requires both enthusiasm. It cannot be done when a person does not like it, especially if he/she does not like guidance, it is reflected in the interaction with people; this pushes us back, and we cannot do the profession positively. The positive shape also has an impact on people's character. A person can develop himself/herself, develop differently with different people and cultures.

S2: I think the person who will be a guide should pay attention to his appearance because he socializes with different people. In this case, the appearance must be beautiful in order to express himself/herself. Image can be said anything. Guidance on general cultural knowledge will add a lot. The guides are really traveling with a lot of information, and they are in contact with a lot of societies and a lot of people. So I think it will add something more in the positive direction, not in the negative direction.

S3: Everything is inherent in this profession, in terms of being able to be talented with each other, to be able to communicate with a new person, to introduce them to a new place. What attracts me is that learning more than one language in this school will be advantageous for me.

S4: Working opportunities are very wide. We can do the job anytime, anywhere, with a broader perspective. We can do all over the country; a tiring profession is a profession that the willing person who loves can do in a reliable way.

S5: I evaluated negatively when I chose this profession. I saw the good aspects of the profession by evaluating the working areas with working hours. My family was opposed when I preferred. I convinced them. You are constantly traveling, and there is no working hours. I chose this profession in order to explore myself. I was curious about the profession, to see new information, to see new places, to develop myself in different cultures. This profession brings some contributions to the person.

S6: There are many interesting places in the profession, as far as I have heard, you are visiting many places and meeting many people. Who lived in a historical region, what is their life like in this region? According to him/her, his perspective on history also changes. This is reflected in your life. It is interesting not only to read history, but also to live. The profession of a guide can even change his own life.

S7: The thing that attracts my attention the most is traveling a lot. It caught my attention. It is very proud to represent the region and my country. I would like to talk to foreign people, get to know them and know their lifestyle. I love to relate to them. That's why I like it.

Tab. 21-1: Nature of the Profession

Sub Themes	Question 1: Can You Talk about the Nature of the Guidance Profession?				
S1	Do it fondly and sincerely, do it eagerly		Self-improvement		
S2	Communication, appearance, personal image	Ability to move freely, socialize	Expressing yourself, being knowledgeable		
S3	Understand of others, communicate with them		Being talented	Promote places	New contacts, traveling
S4	Do it eagerly, do it fondly and sincerely	Being flexible, being independent, Different work schedule			Different profession, around the country
S5		Different work schedule	New knowledges, new places		Traveling, new places, different cultures, personal achievements
S6			Knowing history, living history, being knowledgeable	Promotion	Changing life, meeting, engaging,
S7	Communication, lifestyle			Represent	Traveling, meeting, foreign people
S8		Independence, individuality, self-study	Being knowledgeable		Wander
S9	Communication, loving, problem solving, honesty, loyalty to work			Country image, revisiting	Entertainment
S10	Communication, listening, broadness, comfort, dedication to work		Knowing yourself, adding value to yourself		
Themes	*Communication ability and Empathy*	*Being Independent*	*Being Creative*	*Promotion*	*Being fun and engaging*
Category	*The Nature of Guidance Profession*				

S8: Singles can do the guidance profession more comfortably than family members. Because they are constantly traveling. They have less responsibility than married people. Time needs to be spent in guidance. I also think it is an independent profession. The guide does his/her own job on his/her own. He/she does it with his/her own information.

S9: As a guide candidate, a guide who loves his/her job should be fun and language development should be active among people. Having the ability to talk to people, loyal to their job, honestly love their job, making them feel comfortable in the tourists who come to their own country from outside, and entertain them. Having the ability to produce solutions in the face of a problem. Their image becomes more effective in terms of their own country. The basis of guidance is that he loves his job.

S10: The person who will guide should be identified with her/his profession. Our communication needs to be good because people are our business. We communicate, and I connect this profession completely to communication. Communication with foreigners is the nature of the profession. It adds a lot to the person doing the profession. You can see different places, you know yourself in terms of self-development. We are different even in the same province, whether it is cultural food, tradition and custom.

S1: "Of course, money has a certain value. In my opinion, enjoying one's own profession is more important than the money she earns. Earnings in this profession are protected by a legal framework. Besides, if he develops himself/herself, he/she can make big money in a short time; personal development is important.

S2: "I think it is enough, even the guides who have grown up in a good way can earn many times more. In this case, I don't think they get too low. This can happen in a season, like working for a month and not for months. If we evaluate it according to its conditions, there may be such a contrast, but I don't think.

S3: I haven't researched the money they earned. They can provide a certain level of income, if not very good, but not too bad.

S4: They make money with the trips they make, but travel agencies have a lot of money in terms of profit. I think that if he is in the foreground, he will have beautiful reflections. They have seasonal earnings. Or it happens in four seasons. Of course, even if we come to the fore with their own savings, they gain.

Tab. 21-2: Financial Gain from the Profession

Sub Themes	Question 2: What Do You Think about the Income Status of the Guides?		
S1	Satisfaction, enjoyment	Law, minimum amount,	Development, performance, increase
S2	Seasonal	Enough	Experience, More earnings
S3		Certain level, average income	Self-cultivation, income growth
S4	Four season Winning by traveling, Travel agencies, support	Average income	Accumulation, Increase in income,
S5		Sufficient amount, family livelihood, being married	
S6	Not easy	They win well	Three different languages, income increase
S7	Making concessions, not civil servants, overwork, independent work, across the country	Good income level	
S8	Independent work	Can make a living	Amount may vary
S9	Trust		Amount may vary, travel agencies, collaboration, tourist satisfaction
S10		Makes a living	Experience, wage growth, not stable, experience, performance
Themes	*Working Conditions and Wages*	*Subsistence Level and Wages*	*Professional Performance and Earnings*
Category	*Financial Gain from the Profession*		

S5: I think it's enough. Even if we are married, it will be enough.
S6: I did not research economically, I researched how to be a tourist guides. I think that such an important profession is well paid because not an easy profession, not everyone can do it. Not everyone can speak three languages.
S7: I don't know very well. But they compromise more than their lives. They are not doing eight o'clock in the morning and five in the morning like the officers in everyone's minds. They do not have a room or a desk in every profession. So I think they get a good salary, they get a salary they are satisfied. They cannot be tied somewhere. They are promoting the country and the region. They introduce the four sides of the country.
S8: They can do their job without an agency. We can introduce tourist places to tourists for a certain fee. Their earnings could support a family. If they work independently, their amount of gain may vary.
S9: They could earn their money at the end of a tour from travel agency. The satisfaction of the tourists is very important; it affects both the agencies and the guide, because they work in a common way; it is also important for the next generations.
S10: I have no idea. They can continue their livelihood with the amount they earn. Even teachers get along very hard with what they earn. As far as I know, it has no fixed price. A teacher earns the same salary, whether good or not, in the profession. But this does not apply to the tourist guide. It is related to the location with the firm it works with and his/her performance.

S1: I think eighty percent has little idea of tourist guidance. I think they will respect him/her when they receive guidance service. As tourist guidance appeals to a specific audience, guides go to a higher level in people's eyes because their level of knowledge is high. I think he/she is considered a knowledgeable person in society.
S2: I think it's not where it deserves much now. Being a civil servant is more important in our country. I do not know, and I always see the positive aspects since I want to be a guide. Even though the job is tiring, my opinion does not change.
S3: In my opinion, guidance does not seem to be a profession from the perspective of the society. This profession perceived in a simplicity that everyone can achieve. It is considered a profession in my environment. In general, it is seen as a disrespect when looking at the society.

Tab. 21-3: Social Status

Sub Themes	Question 3: Can We Learn Their Thoughts on How the Guidance Profession Is Evaluated in the Community?		
S1	A respectable, knowledgeable, important person	Being a witness, no idea	
S2	Officer, reputation, comfort	One-way thinking, generalization	Where it does not deserve / will deserve
S3	Not respectable, anyone can do it, the majority		
S4	It is not respected to see, to see, not to be a profession, to be a translator, to tiring	The profession is not recognized	Reputable, future, process
S5	Irregular, unemployed, job opportunities are limited		
S6	Knowledgeable, cultured, teacher, leader, all-knowing, able to establish a dialogue		
S7	Respected, knowledge intensive communication, excessive relationship, conservative segment		
S8	Important place, businesses, business challenges, communication, overcoming		Country, nation, representation, impression
S9	Knowledgeable, needed, experienced, experienced		Advanced times, respect, tourism industry, development
S10	No value		
Themes	*Prestige*	*Public Awareness*	*Guidance in the Future*
Category	*Social Status*		

S4: I think it is not accepted as a profession. They are known as not just doing business but just traveling. As a result of this, the profession is not respected. They think you will be an interpreter. Of course, they think so because they do not have information about this profession. I believe that it will gradually become a respected profession in the society.

S5: They said there is no order in this profession. Those who receive tourism education are thought to be unemployed. I think it should be investigated.

S6: When I said I study in this field, 2–3 people probably talked to me about this subject. They were people who had traveled with a tour guide before; they said the guide was very knowledgeable, multicultural. A tourist guide said it was a teacher rather than introducing a place. I am proud of saying this profession to my family.

S7: It is perceived as very sincere with people who are not in a suitable environment. Some people say that it is a very good profession. People who are respected in terms of knowledge.

S8: A tourist guide communicates with many people in the community in terms of his work. This is a constant relationship with many people, whether it is a business, institution, organization, individual or tradesman, and this is one of the challenges of his business, and this relationship does not progress at a fixed level; in some cases, it may fall in reverse with people. It should not be forgotten that the guide is a human after all. A good guide can overcome all difficulties. Even if it is a hard job to manage all the places, the tour group (tourists) and the agency it works with, whether it is a public, it will have a great place and importance in the society. The place of the guide in the society is known to the world as a big and important result with the impression that it represents, promotes and leaves our country and nation.

S9: Today, conversations with audio recordings are preferred in the historical places visited, but on the other hand, I think someone who comes to our country for the first time wants to have someone with experience. Because it doesn't just happen with the knowledge of sound, where it doesn't know. It is a good thing to have a guide at the place where they will stay, where they will go and where to go.

S10: I guess it is not very important. A teacher is more respected in the community. It is mostly valid in the departments focused on foreign languages.

S10: I think families need to be proud. Because a certain effort was spent in it. There is a long way to go in this profession. In fact, we make more effort than ordinary professions. Because we have to go to new places and dedicate ourselves to this profession. It is not something that will happen immediately, both experience and time are necessary.

Tab. 21-4: Roles of Tourist Guides

Sub Themes	Question 4: What Are the Roles of Tourist Guides in Tourism and Social Areas?		
S1		Promotion, influence	History, research, forgotten traditions, preservation of culture, social memory, local culture, tourist, bridge, guiding
S2	Communication, problem solving, attitude, style, smiling		
S3	The way of expression is remarkable,	Islam, Turkish society, image, impression, history, foreign tourist, Turkish culture	
S4		Promotion, own opinions, accumulation	
S5	Package tour, trust	Promotion	
S6		Presentation, perspective, prejudice, image, Islam, fanaticism, Turkish people, aggressive	History, past life, region
S7		American, tourist, promotion, Islam, attitude, behavior	
S8		Promotion, behavior, attitude	
S9	Tour group, welcome, appearance, influence		
S10	Opposite view, discussion, conciliation		
Themes	*Mediation*	*Promotion*	*Representation*
Category	*Roles of Tourist Guides*		

S1: The tourist guide conducts research into the past. It develops itself in this field; it prevents people from forgetting their own culture and future when they introduce tradition or anything to the people by researching the past. It can also guide people. It can also lead a good way in society.

S2: Since we have to be very strong in communication, our main job is to communicate with people, and we should be careful about the style, when there is something, both sides should be smiling without breaking their hearts or humiliating their problems, and they should be able to maintain the distance from people, and should not enter unnecessary intimacy when they meet someone immediately.

S3: I think most guides have such a mission. I think that it has a mission to improve our country, to introduce our country to others from a historical point of view. If there is more accurate information and expression that will attract the attention of that person, it will have more positive effects. We can introduce Muslims to the world better. We can introduce Turkish culture better. As prejudices are formed for Muslims, there is a bad impression for Turks as well.

S4: Yes, it has a feature that distinguishes it from other professions. His/her own accumulation is on his/her way. He/she organizes excursions based on his/her own opinions. It has a good mission when introducing its own country.

S5: You are attending a tour, people around you trust you. We need to know how to introduce our country against foreign people. He/she can promote his own religion. It would be better for us to know our culture better.

S6: The primary mission of the Turkish guide is to promote the country well and to break the prejudices of foreign tourists. Every word can change the perspective of the tourists that may arise from the mouth against Turkey directory that has the effect of which is not a tourist guide on the country. We can change the perspective of Islam. Turkish people have the idea that they are angry. We can tell that it is not actually the case in the West. This judgment changes with the impression of tourist guides.

S7: Yes. If an American tourist comes to our country, the human aspect of that tourist guide recognizes our country with his attitude toward Islam. That's why it's important.

S8: Every family is proud of the profession their child does. I think we are introducing an important occupational country with different people. We introduce us to other cultures. Each guide represents his own country. Guides are known just as they introduce our country to the tourist.

Tab. 21-5: Individual Features

Sub Themes	Question 5: Do You Think You Have Personal Qualities Suitable for the Guidance Profession?			
S1	Curiosity			
S2		Children and the elderly,	Language learning, diction, communication	
S3	Traveling, meeting	Helpful, love traveling, making people happy	Language learning	Family addiction,
S4	Traveling, enjoy meeting with different people			Stay away from the family
S5	Loving the history, researching, traveling		Language learning	
S6	Traveling the world, transferring information, curious, exploring		Language learning, communication	
S7	Traveling, seeing, *wondering*	Welcoming, helpful	Language learning	
S8	Traveling, different cultures, different people			Individual study
S9			Contact	Confident
S10		To be fair	Language learning	
Themes	*Curiosity and Discovery*	*Helpfulness*	*Learning a foreign language*	*Being Independent*
Category	*Individual Characteristics Associated with Profession*			

S9: It can be by age difference. It can be an example for the incoming tourist organization, making it nice to welcome the tourist group, as well as expressing facial expressions with their particular gestures and facial expressions affect the other person. Using your tongue would be negative. There is a fight with the other person and he is unhappy when he does not explain himself/herself. Using an effective communication with the confidence that language power will give to the other person.

S10: It should be completely neutral, and if it has an opposite view, it should be created in a common point rather than discussion. It is difficult to change people's minds, so going over it can backfire. So it should be respected.

S1: Yes, I think it has personal characteristics suitable for this profession because the enthusiasm, excitement and curiosity I have in this profession increase my interest in this profession even more.

S2: I think it is appropriate, I have diction properly, I take care to be well cared, I am good with the elderly and children, and I love learning languages.

S3: I made a connection with the guidance profession that I wanted to visit places all the time. I also wanted to learn a language, so I connected. I am very dependent on the family. I am trying to beat with this. I like to meet new people all the time. So I help people. I help according to my mood at the moment. Making people happy also makes me happy.

S4: I love to see different places and meet different people. I have different experiences and I am very happy. I will be able to work away from my family, my situation of traveling at all times and my situation of promotion. My family has always been respectful about my decisions. My choice of a different profession also gave them a different feeling. Being a little tiring has created negative impressions. Usually they wanted deskwork but I preferred this.

S5: I like research from the past. When a place is mentioned in a conversation, I wonder what kind of place history has lived in the past and how it has come to present. I have an interest in language learning. That's why I miss this job.

S6: My personal trait is a curious person. My desire to discover new places is very... I love to convey information to a person. As far as I can see, visiting foreign languages and discovering places are also in the characteristics of the guidance profession. That's why tourism guidance came very close to me. It was very similar to my personal traits. I will be promoting not only from abroad but also to people in the country. It enables us to empathize with different people by meeting with different people.

Tab. 21-6: Tourist Guide Recognition Experience

Sub Themes	Question 6: Have You Had the Experience of Meeting Tour Guides or Participating in Tours?		
S1	No		
S2	No		
S3	Yes	Lecture, listening, impressive	
S4	No		
S5	No		
S6	Yes	To choose	
S7	Yes		
S8	Yes	To choose	
S9	No		
S10	No		Touristic place
Themes	*Profession Preference*		*Place of Residence*
Category	*Tourist Guide Recognition Experience*		

S7: Yes. I don't like history and geography subjects very much. But I love learning foreign languages. I love to see and wonder. I'm welcoming. I love helping people.

S8: I like to travel more. When I chose this profession, it was interesting to get to know different cultures and different people, so I chose to do my own job alone.

S9: The most basic feature of a guide is self-confidence and confidence in front of you and doing its job lovingly. It is a good communication with the other person and I have established a relationship with this feature.

S10: Yes, there are some conditions in guidance; for example, I am just, I do not tolerate injustice, I think I will act justly, I have an enthusiasm for learning languages.

S1 Unfortunately it did not happen but I would like it.

S2 No it didn't. I can say about myself about this subject. My friends came from Italy, we were in a crowded place, I was trying to explain myself, so I had no such experience.

S3 Once, a group of Korean tourists was touring. He was promoting a religious space. I wondered when I watched it. He was telling very well, but someone other than the group of tourists was angry at asking questions. This was wrong for me. Korean group but happy. It was very compatible with those who wanted to take pictures.

S4 It hasn't happened yet, but I talked to foreign people in museums.
S5 No, I haven't met. I had met someone who read the profession and had learned about the profession.
S6 No, I did not participate in the tours, but I met several tourist guides. A friend's uncle was a tourist guide. He was taking Turkish people abroad and introducing them abroad. He said he loved his job. It also reflects the interest of another culture in Turks. Meeting him/her also influenced his decision to become a tourist guide.
S7 No I did not agree. But I got to know the tourist guide, the room is a little bit conservative, so it has left its profession. He realizes that he personally has a lot of positive aspects, but he has come out because he does not like the environment in terms of his business area.
S8 I did not participate in the tour, but I met the tourist guide and it was also effective to choose the profession. I met while touring a Russian group in Antalya. It was nice to know the language. I liked that they had a holiday and earned money. Nice profession.
S9 No it didn't. I hope this year will happen the most.
S10 No, I did not come across much about where I live.

Conclusion and Discussion

Students' perceptions on the nature of the profession were evaluated in the first question as the students were asked "what are their opinions about the nature of the guidance profession." The thematic evaluation gave themes of "empathy and communication ability," "being independent," "being creative," "promotion" and "fun and engaging." The empathy ability of the tourist guides is evaluated in the literature in the sense that a tourist guide develops coping mechanism in unexpected or extraordinary events during the tour. And tries to care for and satisfy the psychological needs of group members (Huang et al., 2010; Kuo et al., 2018). In the analyzed part of the literature, the sub-dimension of empathy ability associated with the nature of the profession was not found. Unlike other studies in the literature, empathy ability is among the findings under the nature of the profession in this study. Kuo et al. (2018) stated that the empathy ability of the guide is aimed at satisfying the psychological need of the tourists and the communication ability is aimed at providing an effective interpersonal relationship. In the literature, the independence of the guideline in this study has been associated with its knowledge and work and the difference of working hours, in a different sense than independence (Sanchez & Fraser, 1993;

O'Connor & Shewchuk, 1995; Kuşluvan & Eren, 2011), which has been handled within the framework of service tendency and skills. In the study of Şahin and Acun (2016), most of the students who received guidance training evaluated positively "I think that guidance will give me opportunities to create and create something." Among the findings of the studies in the literature regarding the perceptions and attitudes of the students who receive tourism guidance education toward the guidance profession, being fun and interesting within the nature of the profession, similar to our study (Aloudat, 2017). Aloudat (2017) emphasizes that students develop a sense of attraction and interest about the nature of the profession in relation to meeting foreigners creating an opportunity to learn new things. The statements of the students interviewed in our study also support Aloudat's (2017) findings of attractiveness and interestingness of the nature of the profession. In the study of Şahin and Acun (2016), the attractiveness of the profession is within the scope of the working conditions, while Tolga et al. (2015) consisted of sub-dimensions such as the nature of the work and the dimension of working conditions, the very long working hours, and the stressful and tiring work environment. In the study of Aksu and Köksal (2005), most of the students who received tourism education stated that the profession is interesting.

When the statements of the students who participated in the interview about the income status of the guides were evaluated, the themes of "working conditions and wages," "subsistence level and wages" and "professional performance and earnings" were obtained. Sub-themes such as seasonal, overwork and four seasons are associated with the working conditions and wages of the guides. Aloudat (2017) and Tolga et al. (2015) reported different findings in terms of the relationship between wages and working conditions in the wage-related findings. Sub-themes related to subsistence level and wage theme were as good earnings, good income, law, livelihood, sufficient amount. Most of the students interviewed had positive evaluations about the earnings and livelihood level of the guides. Kuşluvan and Kuşluvan (2003), on the other hand, reported negative perceptions of tourism students about salaries paid to employees in the tourism sector. We have observed the opposite perception of this result in our study. In addition, this finding was in line with the findings of Aloudat's (2017) study. The sub-themes of development, self-cultivation, knowledge of three different languages and increase in wages were the themes of professional performance and amount of earnings. It can be said that the students establish a parallelism between the amount of earning of the guidance profession and their professional performance.

When the question "how the guidance profession is interpreted in the society" is evaluated, themes of respectability, public awareness and guidance in the future were obtained. The theme of reputation, respect, value and dignity have been reached. The sub-themes such as idea formation, witnessing, generalization and unrecognition were reached in the society. In the study of Aksu and Köksal (2005), it was reported that the students receiving tourism education agree with the statement that being a tourism worker was seen as having a profession in the society in a little more than half. On the other hand, Tolga et al. (2015), reported the positive evaluation of the students about social status dimension. In the study of Şahin and Acun (2016), it was stated that the students agree with the statement "I believe that guidance will give me respect in the society." Considering guidance as a profession in the society has been evaluated positively in Aloudat (2017) study in a similar way to other studies.

The fourth question is about the roles of guide. Considering the answers to that question mediation, promotion and representation themes were obtained regarding the interpretation of the roles of the guides. A mediation theme has been created from the sub-themes of package tour, trust, influence, opposite view, style and expression. Introductory theme was obtained from Islam, Turkish people, behavior, impression and foreign tourist sub-themes. The theme of representation has been reached from the sub-themes of history, bridge, history, culture and conservation. In the study of Aloudat (2017), students have participated positively in their role in contributing to the development of their country's tourism, promotion ambassador, being a mediator between local people and tourists and affecting tourist behaviors for environmental protection. However, none of the students who took part in this study stated any expression regarding the role of protecting the environment.

When the question of what personal characteristics appropriate for the guidance profession is evaluated, the themes of curiosity and discovery, benevolence, foreign language learning and independence were reached. Traveling, meeting, researching, seeing, different people and different cultures sub-themes are the theme of curiosity and discovery. Children, old people, making people happy and sub-themes have created the theme of helpfulness. Self-confidence, self-study, being away from the family and family addiction sub-themes have created themes of independence. This theme does not appear to be included in other studies. In the study of Şahin and Acun (2016), most of the students who study tourist guidance stated that the guidance profession is suitable for their lifestyles, personalities and abilities. In this study, students also expressed their suitability for foreign language learning abilities. The theme of benevolence obtained in this research is similar to Tolga et al. (2015); the person who was formed with sub-dimensions such as "it makes me happy to serve people" was similar to the positive assessment of the industry adaptation dimension by students studying tourism guidance.

Considering the experience of meeting with tour guides or participating in tours, the choice of profession and place of residence were reached. While the narrative, impressive, preference sub-themes and the choice of profession were reached, whereas the sub-theme of the place of residence was reached from the sub-contact of the tourist place. It is seen that half of the students who participated in the research did not have the experience of meeting a guide or participating in their tours before. It is observed that those who have such an experience share their experiences in choosing the profession. In addition, the students who saw the lack of this experience as a deficiency explained this deficiency as the place where they live is not a touristic place. As a conclusion, we suggest that the future studies on this issue should investigate the professional perceptions of those with and without industry experience.

Bibliography

Ahipaşaoğlu, H. S. (2006), *Guidance in Tourism* (2nd Ed.), Ankara: Gazi Bookstore.

Aksu, A. A. and Köksal, C. D. (2005), "Perceptions and Attitudes of Tourism Students in Turkey", *International Journal of Contemporary Hospitality Management*, 17 (5), pp. 436–447.

Aloudat, A. S. (2017), "Undergraduate Students' Perceptions of a Tour Guiding Career", *Scandinavian Journal of Hospitality and Tourism*, 7 (4), pp. 333–344.

Avcı, N. (2011), "Business Values of Undergraduate Students Receiving Tourism Education: Example of Çeşme School of Tourism and Hotel Management", *Anatolia: Journal of Tourism Research*, 22 (1), pp. 7–18.

Aymankuy, Y., Tetik, N., Girgin, K. G., and Aymankuy, Ş. (2013), "Students 'and Academicians' Views on Internship Practice in Undergraduate Tourism Education (Application in BTIOYO)", *International Journal of Human Sciences*, 10 (1), pp. 101–128.

Bahçelerli, N. M. and Sucuoğlu, E. (2015), "Undergraduate Tourism Students' Opinions Regarding the Work Conditions in the Tourism Industry", *Procedia Economics and Finance*, 26, pp. 1130–1135.

Creswell, J. W. (2014), *Research Design, Qualitative, Quantitative and Mixed Methods Approaches* (4th Ed.), Washington, DC: Sage.

Ehtiyar, R. and Üngüren, E. (2008), "A Research on the Determination of the Relationship Between the Hopelessness and Anxiety Levels of Students Receiving Tourism Education and Their Attitudes towards Education", *International Journal of Social Research*, 1 (4), pp. 159–181.

Greasley, K. and Ashworth, P. (2007), "The Phenomenology of "Approach to Studying": The University Student's Studies within the Lifework", *British Educational Research Journal*, 32, pp. 819–843.

Grobelna, A. (2017), "The Perception of Job-related Motivators when Choosing a Career in the Tourism and Hospitality Industry – A Comparative Study between Polish and Spanish Students", *International Journal of Management and Economics*, 53 (2), pp. 84–106.

Hacıoğlu, N., Kaşlı, M., Şahin, S., and Tetik, N. (2008), *Tourism Education in Turkey*, Ankara: Detay Publishing.

Huang, S., Hsu, C. H. C., and Chan, A. (2010), "Tour Guide Performance and Tourist Satisfaction: A Study of the Package Tours in Shanghai", *Journal of Hospitality and Tourism Research*, 34 (1), pp. 3–33.

Kim, H. and Park, E. J. (2013), "The Role of Social Experience in Undergraduates' Career Perceptions through Internships", *Journal of Hospitality, Leisure, Sport and Tourism Education*, 12, pp. 70–78.

Kozak, M. A. (2009), "A Situation Analysis on Academic Tourism Education", *Journal of Muğla University Institute of Social Sciences*, 22, pp. 1–20.

Kuo, N.-T., Cheng, Y.-S., Chang, K.-C., and Chuang, L.-Y. (2018), "The Asymmetric Effect of Tour Guide Service Quality on Tourist Satisfaction", *Journal of Quality Assurance in Hospitality and Tourism*, 19 (4), pp. 521–542.

Kusluvan, S. and Kusluvan, Z. (2000), "Perceptions and Attitudes of Undergraduate Tourism Students towards Working in the Tourism Industry in Turkey", *Tourism Management*, 21 (3), pp. 251–269.

Kuşluvan, S. and Eren, D. (2011), "Susceptibility and Measurement of Workers as Personality Traits: A Literature Review", *Anatolia: Journal of Tourism Research*, 22 (2), pp. 139–153.

Kuşluvan, S. and Kuşluvan, Z. (2003), "Perceptions and Attitudes of Undergraduate Tourism Students Towards Working in The Tourism Industry in Turkey", *Tourism Management*, 21, pp. 251–269.

Leclerc, D. and Martin, J. N. (2004), "Tour Guide Communication Competence: French, German and American Tourists' Perceptions", *International Journal of Intercultural Relations*, 28, pp. 181-200.

Lucas, R. and Johnson, K. (2003), Managing Students as a Flexible Labor Resource in Hospitality and Tourism in Centraland Eastern Europe and the U. K. In S. Kusluvan (Ed.), *Managing Employee Attitudes and Behaviors in the Tourism and Hospitality Industry* (pp. 153-170), New York: Nova Publishers.

O'Connor, S. J. and Shewchuk, R. M. (Winter 1995). "Service Quality Revisited: Striving for a New Orientation", *Hospital & Health Services Administration*, 4 (4), pp. 535-552.

Özdemir, M. (2010), "Qualitative Data Analysis: A Study on the Problematic of Methodology in Social Sciences", *Eskişehir Osmangazi University Journal of Social Sciences*, 11 (1), pp. 323-343.

Richardson, S. and Butler, G. (2012), "Attitudes of Malaysian Tourism and Hospitality Students' towards a Career in the Industry", *Asia Pacific Journal of Tourism Research*, 17 (3), pp. 262-276.

Richardson, S. and Thomas, N. J. (2012), "Utilizing Generation Y: United States Hospitality and Tourism Students' Perceptions of Careers in the Industry", *Journal of Hospitality and Tourism Management*, 19, pp. 1-13.

Roney, A. S. and Öztin, P. (2007), "Career Perceptions of Undergraduate Tourism Students: A Case Study in Turkey", *Journal of Hospitality, Leisure, Sport and Tourism Education*, 6 (1), pp. 4-17.

Sanchez, J. and Fraser, S. (1993). *Development and Validation of the Corporate Social Style Inventory: A Measure of Customer Service Skills (Report Number 93-108)*. Cambridge, MA: Marketing Science Institute.

Şahin, S. and Acun, A. (2016) "Attitudes of Tourism Guidance Students towards Profession", *Gaziantep University Journal of Social Sciences*, 15 (2), pp. 563-580.

Tolga, Ö., Korkmaz, H., and Atay, L. (2015), "Attitudes of Undergraduate Tourist Guidance Students", *A Journal of Research, Travel and Hotel Management for Professional*, 12 (2), pp. 26-41.

Türnüklü, A. (2000), "A Qualitative Research Method to Be Effectively Used in Scientific Research in Education: Interview", *Educational Administration: Theory and Practice*, 6 (24), pp. 543-559.

Ünlüönen, K. (2004), "Comparison of Tourism Management Teaching Programs in Terms of Student Expectations and Perceptions (1998-1999 and 2003-2004 Education Years)", *Gazi University Journal of Commerce and Tourism Education Faculty*, 1, pp. 108-130.

Wan, Y. K. P., Wong, I. A., and Kong, W. H. (2014), "Student Career Prospect and Industry Commitment: The Roles of İndustry Attitude, Perceived Social Status, and Salary Expectations", *Tourism Management*, 40, pp. 1–14.

Wang, S. (2014), "Tourism Career Perceptions and Implications, A Case Study of Chinese Undergraduate Students", *Tourism Today*, Fall, pp. 55–67.

Wong, S. C. and Ladkin, A. (2008), "Exploring the Relationship between Employee Creativity and Job Related Motivators in the Hong Kong Hotel Industry", *International Journal of Hospitality Management*, 27 (3), pp. 426–437.

Yıldırım, A. and Şimşek, H. (2018), *Qualitative Research Methods in Social Sciences* (11th Ed.), Ankara: Seçkin Publishing.

List of Figures

Fig. 2-1:	Learning Pyramid-a	53
Fig. 2-2:	Learning Pyramid	53
Fig. 2-3:	Learning Area Process – Improving Economic and Social Dimensions in the Tourist Sector	56
Fig. 2-4:	Provisional Learning Area Sites Used to Develop the Handbook	59
Fig. 4-1:	The Relationship between Characteristics of Service and Culture	86
Fig. 6-1:	Kid-friendly Rooms and Theatrical Shows at TLOL	120
Fig. 6-2:	Customer Profile of TLOL	122
Fig. 6-3:	The Revenue Streams of the Theme Park in the Four Different Accounts	123
Fig. 6-4:	Linking Growth Strategy to Business Model Renewal – The Case of Rixos Tourism Group	124
Fig. 9-1:	Number of Foreign Visitors in 2019	183
Fig. 9-2:	Tourism Revenue in Total Exports	185
Fig. 10-1:	The Factors Affecting the Destination Preferences of Tourists	205
Fig. 11-1:	Travel Motivation	222
Fig. 11-2:	Effective Factors on Formation of Gastronomy Image	224
Fig. 13-1:	Regional Distribution of Second Homes by Years. Source: General Directorate of Population and Citizenship Affairs of Turkey.	258
Fig. 13-2:	Second Home Patterns according to Geographical Regions of Turkey in 2013.	260
Photo 13-1:	Development of Second Homes in Göller Highland of Kozan District.	261
Scheme 18-1:	Design View of Data Preprocessing	353

List of Tables

Tab. 2-1: Comparison of Traditional and Learning Organizations 45
Tab. 5-1: Power Distance Level and Working Life 97
Tab. 5-2: Colectivism/Individualism and Working Life 97
Tab. 5-3: Masculinity/Feminity and Working Life 98
Tab. 5-4: Uncertainty Avoidance Level-Short/Long Term Orientation and Working Life 100
Tab. 5-5: Culture of Hotel Employees 103
Tab. 6-1: Business Portfolio of TLOL 119
Tab. 7-1: Findings by Explanatory Factor Analysis and Reliability Level of Emotional Labor Scale 141
Tab. 7-2: Descriptive Statistics for Emotional Labor Scale 143
Tab. 7-3: Emotional Labor Behaviors of Participants and Analysis of Variance for the Comparison of Age and Tourism Education Level Variables 145
Tab. 7-4: Analysis of Variance for Comparison of Emotional Labor Behaviors of Participants and Income Level and Type of employment in the business Variables 148
Tab. 9-1: Tourism Receipts per Person 186
Tab. 9-2: Tourism Revenue in GNP (%) 188
Tab. 9-3: Number of Insured Persons by SGK in Tourism and Hospitality 188
Tab. 10-1: Destination Types and Markets 203
Tab. 16-1: The Main Effects of ICT on Tourism Events 316
Tab. 17-1: Results on Demographic Characteristics of the Respondents 339
Tab. 17-2: Respondents by Centers of Attraction 339
Tab. 17-3: Results on Travel and Accommodation Experiences of the Respondents 340
Tab. 17-4: Scoring Limits of the Questionnaire Questions 341
Tab. 17-5: Extent of Agreement of Respondents with Brand Loyalty 343
Tab. 18-1: Data Set Sample of Hotel Comments 352
Tab. 18-2: Word Frequencies Rating Obtained from Room Comments Related to the 4- and 5-Star Hotels 354
Tab. 18-3: Word Frequencies Related to Food and Drink Obtained from 4- and 5-Star Hotels' Comments 355
Tab. 18-4: Comment Frequencies for Adjectives for 4- and 5-Star Hotels. 356

Tab. 18-5: Word Frequencies Related Stuff That Is Obtained from 4- and 5-Star Hotel Comments in Izmir. 358
Tab. 18-6: Word Frequencies Related to Transportation Obtained from 4- and 5-Star Hotels Comments in Izmir 359
Tab. 18-7: Word Frequencies in Service Comments Obtained from 4- and 5-Star Hotel Comments in Izmir 360
Tab. 18-8: Word Frequencies of Actions in Hotel Comments for 4- and 5-Star Hotels in Izmir 361
Tab. 18-9: Word Frequencies about Price Comments of 4- and 5-Star Hotels in Izmir 361
Tab. 19-1: The Main and Sub-Factors of Selection of Winter Tourism Centers 376
Tab. 19-2: Saaty's Scale of Relative Importances 382
Tab. 19-3: Random Indicator Values 385
Tab. 19-4: Direct Relation Matrix for the Main Factors 386
Tab. 19-5: Total Direct Relation Matrix for the Main Factors 387
Tab. 19-6: Cause and Effect Groups for the Main Factors 387
Tab. 19-7: Weights and Ranking Result of Main Factors Calculated with DEMATEL 388
Tab. 19-8: Cause and Effect Groups of the Sub-Factors of Price and Its Weights 389
Tab. 19-9: Cause and Effect Groups of the Sub-Factors of Accessibility and Its Weights 390
Tab. 19-10: Cause and Effect Groups of the Sub-Factors of Comfort and Its Weights 391
Tab. 19-11: Cause and Effect Groups of the Sub-Factors of Destination Image and Its Weights 392
Tab. 19-12: Cause and Effect Groups of the Sub-Factors of Alternative Tourism and Its Weights 393
Tab. 19-13: Weights and Ranking Result of Main Factors Calculated with AHP 393
Tab. 19-14: Weights of Main Factors and Sub-Factors 395
Tab. 19-15: Comparison of the Weights of Main Factors 396
Tab. 19-16: Average Weights of the Sub-Factors and Final Ranking 397
Tab. 20-1: Distribution of the Participants according to Their Demographic Characteristics 414
Tab. 20-2: Metaphors Developed for the Concept of Graduation 415
Tab. 20-3: Classification of Graduation Metaphors 416
Tab. 21-1: Nature of the Profession 429
Tab. 21-2: Financial Gain from the Profession 431

List of Tables

Tab. 21-3: Social Status 433
Tab. 21-4: Roles of Tourist Guides 435
Tab. 21-5: Individual Features 437
Tab. 21-6: Tourist Guide Recognition Experience 439

About Editors

Elbeyi PELİT is Prof. Dr. at Faculty of Tourism, Afyon Kocatepe University, Afyonkarahisar (Turkey)
Hasan Hüseyin SOYBALI is Prof. Dr. at Faculty of Tourism, Afyon Kocatepe University, Afyonkarahisar (Turkey)
Ali AVAN is Asst. Prof. at Faculty of Tourism, Afyon Kocatepe University, Afyonkarahisar (Turkey)

www.ingramcontent.com/pod-product-compliance
Ingram Content Group UK Ltd.
Pitfield, Milton Keynes, MK11 3LW, UK
UKHW022121230426
12048UKWH00011BA/644